# FLUID MACHINERY
## (HYDRAULIC MACHINES)

# FLUID MACHINERY

## (HYDRAULIC MACHINES)

A Text Book for B.E. I B.Tech. students of all Indian Universities and Institutions.
Also useful for Engineering Services, Civil Services Examinations conducted by UPSC , GATE and other Competitive Examinations


**DR. SADHU SINGH**

*Formerly*, **Professor & Head**, Mechanical Engineering Department
& **Dean**, Faculty of Engineering & Technology
Govind Ballabh Pant University of Agriculture & Technology
Pantnagar - 263 145 ( Uttarakhand )
*Formerly*, **Director** (Colleges)
Punjab Technical University, Jalandhar

**KHANNA BOOK PUBLISHING CO. (P) LTD.**

Publisher of Engineering and Computer Books

4C/4344, Ansari Road, Darya Ganj, New Delhi-110002
**Phone** : 011-23244447-48 **Mobile:** +91-9910909320
**E-mail** : contact@khannabooks.com
**Website** : www.khannabooks.com

Dear Readers,

To prevent the piracy, this book is secured with HIGH SECURITY HOLOGRAM on the front title cover. In case you don't find the hologram on the front cover title, please write us to at contact@khannabooks.com or whatsapp us at +91-99109 09320 and avail special gift voucher for yourself.

Specimen of Hologram on front Cover title:

Moreover, there is a SPECIAL DISCOUNT COUPON for you with EVERY HOLOGRAM.

How to avail this SPECIAL DISCOUNT:

Step 1: Scratch the hologram
Step 2: Under the scratch area, your "coupon code" is available
Step 3: Logon to www.khannabooks.com
Step 4: Use your "coupon code" in the shopping cart and get your copy at a special discount
Step 5: Enjoy your reading!

**Price: ₹ 475.00**

**ISBN:** 978-93-82609-28-5

**Fluid Machinery** *by*
Sadhu Singh

**Edition: 2025**

To view complete list of books, Please scan the QR Code:

*Published by:*
**Khanna Book Publishing Co. (P) Ltd.**
*CIN: U22110DL1998PTC095547*

Visit us at: **www.khannabooks.com**
Write us at: **contact@khannabooks.com**

*Printed in Bharat.*

# Preface

Fluid Machinery is a Text Book for B.E. / B.Tech. students of all Indian Universities and Institutions. The book contains fifteen chapters comprising of Basic Concepts of Fluid Machinery; Whirling of fluids; Hydro Power Generation Fundamentals; Unit and Specific Quantities Pelton; Turbine; Francis Turbine; Kaplan Turbine; Draft Tube and Cavitation; Governing, Selection and Performance of Hydraulic Turbines; Reciprocating Pumps; Centrifugal Pump; Fluid Devices and Systems; Fluid Machinery Laboratory Practice; Conventional Questions From IES, Multi - Choice Questions From IES.

The book contains a large number of solved and unsolved problems. The special features of the book are : Summary, Review Questions, Multi-choice Questions and end of chapter Numerical Problems.

The help provided by Mr. Mukul Seth and Puneet Khanna during the preparation of the manuscript and otherwise is highly appreciated and acknowledged.

My grand children, Kanupreet Kaur and Amitoej Singh have been a source of great inspiration during the preparation of the manuscript. They kept my mood cheerful, which helped in improving the text.

It is hoped that the book in its present form shall be quite useful to the students and teachers. Further suggestions for the improvement of the book are invited by the publishers and the author.

*Dr. Sadhu Singh*

# Contents

# CHAPTER 1 Basic Concepts of Fluid Machinery

## 1.1 Introduction

Fluid machinery – Fluid machinery is a general term used for all devices / machines handling fluids. This covers machinery used for power generation, transporting fluids, lifting devices, and control devices.

Turbo machines – These are devices in which energy is transferred either to, or from, a continuously flowing fluid by the dynamic action of moving blades on the runner. The word turbo or turbines is of Latin origin and implies that which spins or whirls around. Hydraulic turbines and pumps, generally known as rotodynamic machines, are called turbo machines.

### Types of fluid machinery

Fluid machinery consist of the following:

1. Turbo machines – Hydraulic turbines: Pelton, Francis, Kaplan, propeller, etc.

   Pumps: centrifugal, submersible, etc.
2. Reciprocating machines – Reciprocating pumps, hand pumps. These are positive displacement type.
3. Water lifting devices – Jet pump, air – lift pump, hydraulic ram, etc.
4. Hydraulic controls and systems – Gear pump, vane pump, volute pump, various appliances and accessories.

### Types of turbomachines

The turbo machines may be classified as:

1. Open and enclosed machines:
   (a) Open turbo machines are those which influence an indefinite quantity of fluid, e.g, propellers, windmills and unshrouded fans. Such machines come generally under the category of aerodynamics.
   (b) Enclosed turbo machines are those in which a finite quantity of fluid passes through a casing in unit time.
2. Absorption and production of power:
   (a) Those which absorb power to increase the fluid pressure or head, e.g., pumps, ducted fans and compressors

(b) Those which reduce power by expanding fluid to a lower pressure or head, e.g. hydraulic, steam and gas turbines.

3. Type of fluid handled:
   (a) Those which handle water, e.g., pumps and hydraulic turbines.
   (b) Those which handle steam, e.g., steam turbines.
   (c) Those which handle gas or air, e.g, ducted fans, compressors and gas turbines

## 1.2 Newton's Second Law of Motion

Newton's second law of motion states that the rate of change of momentum is directly proportional to the applied force and takes place in the direction of the force. For rectilinear motion the linear momentum is the product of mass m of the fluid and its linear velocity v with which it is moving. If the velocity changes by an amount $dv$ during an interval of time dt, then

Change in momentum = $m \times dv$

Rate of change of momentum = $m \times \dfrac{dv}{dt}$

According to the Newton's second law of motion,

Dynamic force applied = rate of change of momentum

$$F = m \times \frac{dv}{dt} \qquad \text{...(1.1)}$$

Eq. (1.1) is known as the linear – momentum equation

If the velocity of the fluid changes by equal amounts in equal intervals of time, then $\dfrac{dv}{dt}$ is called the linear uniform acceleration, '$a$'.

$$\therefore \qquad F = m \times a$$

In the S.I. system of units, m is measured in kg, '$a$' in m/s$^2$ and $F$ in Newtons (N).

## 1.3 Impulse – Momentum Principle

Eq. (1.1) can be written as:

$$F \times dt = m \times dv \qquad \text{...(1.2)}$$

The L.H.S. of Eq. (1.2) is the product of force and time increment during which it acts. This is known as impulse of applied force. The R.H.S. is resulting change in momentum. Eq. (1.2) is known as the impulse – momentum equation. This leads to the impulse – momentum principle, which can be stated as:

The impulse of a dynamic force is equal to the resulting change in momentum of the body.

## 1.4 Applications of Linear Momentum Equation

The applications of linear – momentum equation are in the following areas:

1. To determine the force exerted by the flowing fluid on the boundaries of flow passage due to change of momentum
2. To determine the flow characteristics when there is some loss of unknown quantity of energy in the system such as sudden enlargement of a pipe cross – section, hydraulic pump in an open channel, etc.

## 1.5 Definitions

*Hydraulic Machine* – any device or machine handling liquids.

*Vortex* – a mass of fluid in rotation about a fixed axis. In a free vortex the fluid mass rotates without any external impressed contact force whereas in a forced vortex the fluid mass is made to rotate by means of some external agency.

*Gross head* – is the difference in elevation between the head race level and the tail race level, both being measured simultaneously.

*Net (or effective) head* – is obtained by subtracting from gross head all the losses in carrying water from the head race to the entrance of the turbine.

*Impulse turbine* – in which whole pressure energy of water is transformed into kinetic energy. It was invented by LesterAllen Pelton and patented in 1889. It is a tangential flow turbine whose specific speed ranges from 10 to 35. It is used for high head and low discharge.

*Reaction turbine* – in which part of the head acting on the turbine is transformed into kinetic energy and the rest remains as pressure head. Francis and Kaplan turbines are of this type. Francis is a mixed flow and Kaplan an axial flow turbine. Francis turbine is used for medium heads and medium quantities of water and the specific speed varies from 60 to 300. Kaplan turbine is used for low heads and large quantities of flow and its specific speed varies from 300 to 1000.

*Specific speed* – of a turbine is the speed of a geometrically similar turbine that would develop unit brake power ($P_t$) under a unit head

$$N_s = \frac{N\sqrt{P_t}}{H^{5/4}}$$

Similarly tho specific speed of a centrifugal pump is the speed of a geometrically similar pump when delivering one m$^3$/s against a head of one metre.

$$N_s = \frac{N\sqrt{Q}}{H^{3/4}}$$

*Draft tube* – is a conduit which connects the runner exit of a reaction turbine to the tail race.

*Runaway speed* – of turbine is the maximum speed attained by the runner under maximum head and full gate opening, when the external load is disconnected from the system and the governor ceases to function.

Penstock – is a pipe of large diameter which carries water from the reservoir to the turbine.

*Surge tank* – is a storage reservoir fitted at some opening made on a long penstock to receive the rejected flow when the penstock is suddenly closed by a valve fitted at its steep end.

*Fore bay* – is a storage reservoir at the beginning of a penstock made by enlarging the canal to store the rejected water when the load on the turbine is reduced.

*Pump* – is a mechanical device to increase the pressure energy of a liquid.

*Priming* – is a process of filling the impeller and suction pipe with water in order to remove any air, gas or vapour present form the water ways of the pump.

*Hydraulic ram* – is a contrivance to raise a part of large amount of water available at some height, to a greater height.

*Hydraulic system* – is a circuit in which force and power are transmitted through a fluid.

*Hydraulic motor* – is a device to convert hydraulic energy to mechanical energy.

Static head- is the sum of the suction and delivery heads.

*Manometric head* – is the head measured across the pump inlet and outlet flanges. It is the difference between the readings of pressure and vacuum gauges plus the vertical distance between the pressure toppings. It includes all losses against which the pump has to work except the kinetic head.

*Total, gross or effective head* – is the sum of the manometric head and the kinetic head.

## 1.6 Impact of Jets

A fluid jet is a stream of fluid issuing from a nozzle with a high velocity. The jet possesses high kinetic energy. When the jet impinges on a plate or vane, it exerts force on it. This forms the basis for an impulse turbine.

### 1.6.1 Jet Impinging on a Stationary Flat Plate

***(a) plate held normal to the jet***

Consider a fluid jet impinging on a flat plate held stationary and perpendicular to the jet, as shown in Fig.1.1

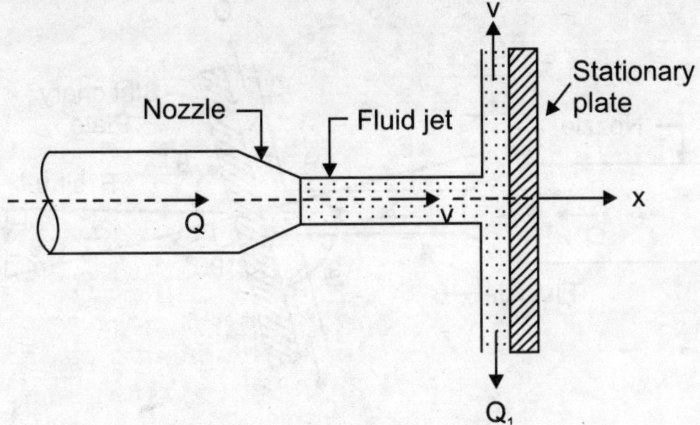


**Fig. 1.1 Fluid jet impinging on a flat stationary plate.**

let $v$ = velocity of jet, m/s

$\rho$ = density of fluid, kg/m$^3$

$A$ = area of cross – section of jet, m$^2$

$= \frac{\pi}{4}d^2$

$d$ = diameter of jet, m

Then mass of fluid striking the plate per second,

$$m = \rho A v = vQ \text{ kg/s}$$

Where $Q$ = quantity of fluid falling on the plate per second, m$^3$/s

After striking the stationary plate, the fluid jet deflects 90°. By applying Newton's second law of motion, we have

Rate of charge of momentum = impressed force

or $$m(v - 0) = F_x$$

or $$F_x = \rho A v^2 = \rho Q v \text{ N}$$

$F_x$ is called the hydrodynamic thrust. Since the plate is stationary, the work done by the fluid jet on the plate is zero.

The jet fluid gets divided into two equal parts after striking the plate.

$$Q_1 = \frac{Q}{2}$$

***(b) plate held inclined to the jet***

Consider a fluid jet striking a stationary flat plate held inclined at an angle $\theta$ to the jet, as shown in Fig. 1.2.

Applying the impulse – momentum principle,

Force exerted by jet normal to the plate,

$$F_n = \rho A v \times v \sin\theta = \rho A v^2 \sin\theta$$

Component of normal force along the direction of jet,

$$F_x = F_n \sin\theta = \rho A v^2 \sin\theta \times \sin\theta = \rho A v^2 \sin^2\theta$$

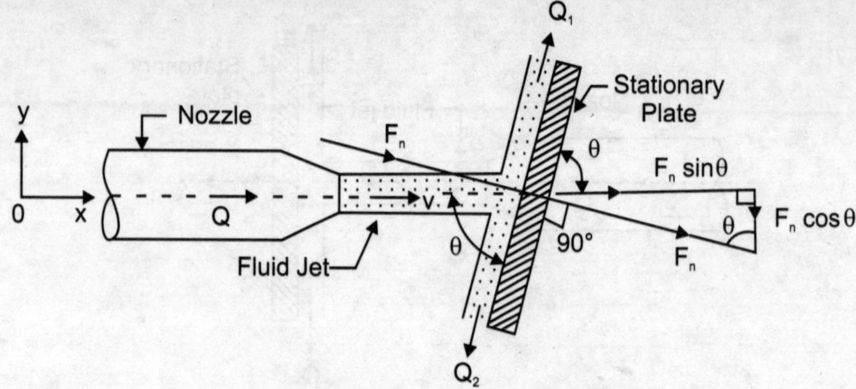


**Fig. 1.2 Fluid jet striking an inclined stationary flat plate.**

Component of normal force perpendicular to the direction of jet, in this case also, the work done is zero, as the plate is stationary.

$$F_y = F_n \cos\theta$$

$$= \rho A v^2 \sin\theta \cos\theta$$

The fluid jet gets divided into two parts $Q_1$ and $Q_2$.

Assuming frictional resistance of plate to be negligible the resultant force in the direction tangential to the plate is zero. Applying the impulse – momentum equation in the direction tangential to the plate, we have

$$(\rho Q_1 v - \rho Q_2 v) - \rho Q v \cos\theta = 0$$

Or $$Q_1 - Q_2 - Q\cos\theta = 0$$

Also $$Q_1 + Q_2 = Q$$

$\therefore$ $$Q_1 = \frac{Q}{2}(1+\cos\theta)$$

$$Q_2 = \frac{Q}{2}(1-\cos\theta)$$

Ratio of discharges, $$\frac{Q_1}{Q_2} = \frac{1+\cos\theta}{1-\cos\theta}$$

### 1.6.2 Jet Impinging on Stationary Curved Plate

#### *(a) Jet striking at centre of plate*

Consider a fluid jet striking at the centre of a curved plate, as shown in Fig.1.3. The jet after striking the plate comes out with the same velocity in the tangential direction of the curved plate. The velocity at the plate outlet can be resolved into the following two components:

$$v_x = -v\cos\theta$$

$$v_y = v\sin\theta$$

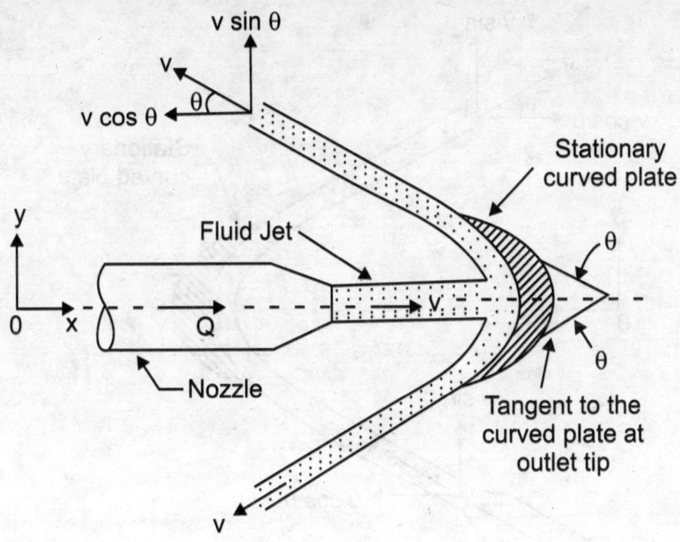


**Fig. 1.3 Fluid jet impinging on stationary curved plate.**

Applying impulse – momentum equation, we have

$$F_x = \rho A v\,(v_{1x} - v_{2x})$$

Now $\quad v_{1x} = v$ and $v_{2x} = v_x = -\,v \cos\theta$

$\therefore \quad F_x = \rho A v\,(v + v\cos\theta) = \rho A v^2\,(1 + \cos)$

Where $v_{1y} = 0,\ v_{2y} = v_y = v \sin\theta$

$$F_y = \rho A v\,(0 - v\sin\theta) = -\,\rho A v^2 \sin\theta$$

Angle of deflection of jet = 180° – θ

Resultant force on the curved plate,

$$F = \sqrt{F_x^2 + F_y^2}$$

$$= \rho A v^2 \sqrt{(1+\cos\theta)^2 + (-\sin\theta)^2}$$

$$= \rho A v^2 \left[1 + \cos^2\theta + 2\cos\theta + \sin^2\theta\right]^{1/2}$$

$$= \rho A v^2\,[2 + 2\cos\theta]^{1/2}$$

$$= \sqrt{2}\rho A v^2\,(1+\cos\theta)^{1/2}$$

$$\left[\because \cos\theta = 2\cos^2\frac{\theta}{2} - 1\right]$$

$$= 2\rho A v^2 \cos\left(\frac{\theta}{2}\right)$$

***(b) Jet striking at one end of plate tangentially***

Consider the fluid jet striking the curved plate tangentially at one end, as shown in Fig.1.4. Applying impulse – momentum equation, the forces in x- and y-directions on the curved plate are:

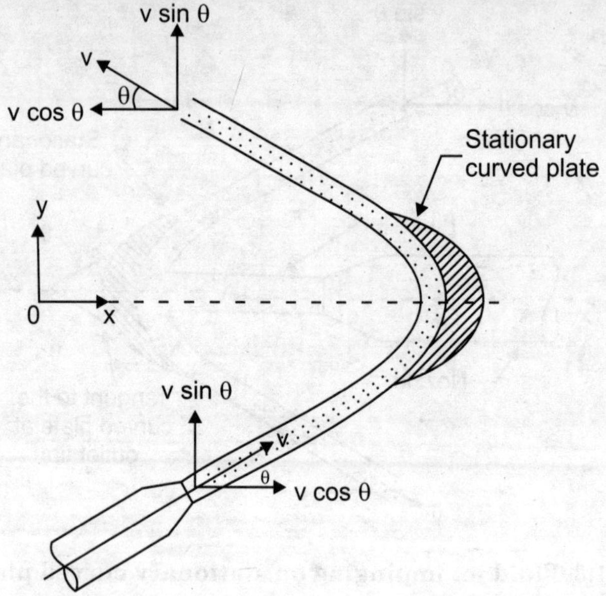


**Fig. 1.4 Fluid jet striking stationary curved plate tangentially.**

$$F_x = \rho Av\left[v\cos\theta - (-v\cos\theta)\right] = 2\rho Av^2\cos\theta$$

$$F_y = \rho Av\left[v\sin\theta - v\sin\theta\right] = 0$$

***(c) Jet striking at one end of unsymmetrical plate tangentially***

Let θ, ϕ = angles made by the tangents at inlet tip and outlet tip respectively with x- axis.

Then
$$F_x = \rho Av\left[v\cos\theta - (-v\cos\phi)\right]$$
$$= \rho Av^2\left[\cos\theta + \cos\phi\right]$$
$$F_y = \rho Av\left[v\sin\theta - v\sin\phi\right]$$
$$= \rho Av^2\left[\sin\theta - \sin\phi\right]$$

**Example 1.1** A jet of water strikes with a velocity of 30 m/s a flat plate inclined at 30° with the axis of the jet. The cross – sectional area of the jet is 20 cm². Calculate: (a) the force exerted by the jet on the plate, (b) the components of force in the direction normal to the jet, and (c) the ratio in which the discharge gets divided after striking the plate.

**Solution.**

Given: $v = 30$ m/s, $\theta = 30°$, $A = 20$ cm²

Normal force on the plate,

$$F_n = \rho Av^2\sin\theta$$
$$= 1000 \times 20 \times 10^{-4} \times 30^2 \sin 30° = 900 \text{ N}$$

$$F_x = F_n \sin\theta = 900 \sin 30° = 450N$$

$$F_y = F_n \cos\theta = 900 \cos 30° = 779.4N$$

$$\frac{Q_1}{Q_2} = \frac{1+\cos\theta}{1-\cos\theta} = \frac{1+\cos 30°}{1-\cos 30°} = 13.928$$

**Example 1.2** A jet of water of 15 mm diameter and moving at 20 m/s strikes at the centre of a symmetrical vane. After striking, the jet gets deflected through 150° by the vane. Calculate the force exerted by the jet on the vane, if (a) the vane is smooth, and (b) the ratio of outlet velocity to inlet velocity is 0.85.

**Solution.**

Given: $d$ = 15 mm, $v$ = 20 m/s, $\theta$ = 180° – 150° = 30°

Theoretical force exerted by jet on the vane,

$$F_x = \rho A v^2 (1+\cos\theta)$$

$$= 1000 \times \frac{\pi}{4} \times (15 \times 10^{-3})^2 \times 20^2 (1+\cos 30°)$$

$$= 131.9 \text{ N}$$

$$F_y = 0$$

$$k = \frac{v_2}{v_1} = 0.85$$

$$v_2 = kv_1 = 0.85 \times 20 = 17 m/s$$

$$F_x = \rho A v_1^2 (1 + k\cos\theta)$$

$$= 1000 \times \frac{\pi}{4} (15 \times 10^{-3})^2 \times 20^2 (1+0.85 \cos 30°)$$

$$= 122.72 \text{ N}$$

**Example 1.3** A rectangular plate weighing 60 N is suspended vertically by a hinge on the top horizontal edge. The centre of gravity of the plate is 100 mm from the hinge. A horizontal jet of water 20 mm diameter, whose axis is 150 mm below the hinge impinges normally on the plate with a velocity of 5 m/s. Determine:

(a) The horizontal force applied at the centre of gravity to maintain the plate in its vertical position.

(b) The corresponding velocity of the jet, if the plate is deflected through 30°and the same force continues to act at the centre of gravity of the plate.

**Solution.**

Given: Weight of plate, $W$ = 60 N, $y_1$ = 100 mm, $y_2$ = 150 mm, $d$ = 20 mm, $v$ = 5 m/s

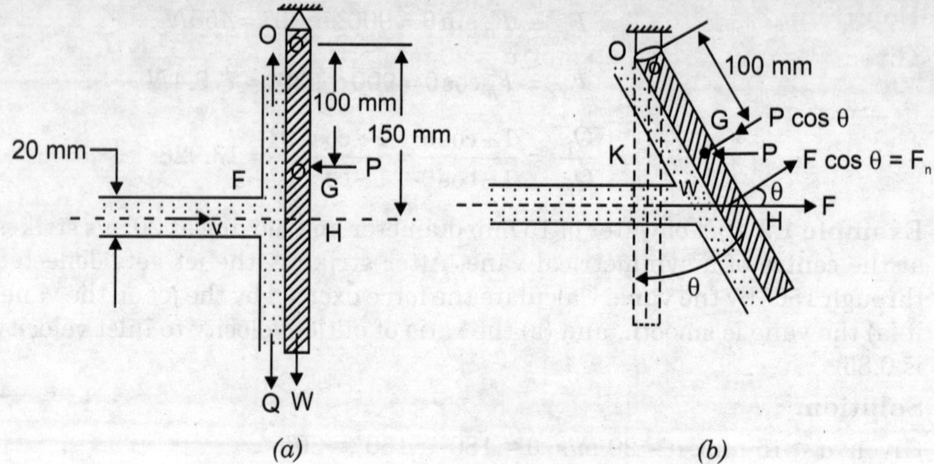


**Fig. 1.5 Fluid jet striking a hinged plate.**

Normal force exerted by the jet on the plate (Fig. 1.5a)

$$F = \rho A v^2 = 1000 \times \frac{\pi}{4}(0.020)^2 \times 5^2 = 7.85N$$

Let $P$ = force applied at the $CG$ of the late to keep it in vertical position Taking moments about hinge point $O$, we have

$$F \times 150 = P \times 100$$

$$P = 1.5 \times 7.85 = 11.77 \text{ N}$$

$$\theta = 30^o$$

The deflected position of plate is shown in Fig.1.5 (b)

$$GK = OG \cos\theta = 100 \cos 30° = 86.6 \text{ mm}$$

Taking moments about hinge $O$, we have

$$W \times GK + P\cos\theta \times 100 = F\cos\theta \times 150$$

$$60 \times 86.6 + 11.77 \times \cos 30° \times 100 = F \times \cos 30° \times 150$$

$$F = 26.8 \text{ N}$$

$$= \rho A v^2 = 1000 \times \frac{\pi}{4}(0.020)^2 \times v^2 = 0.314 v^2$$

$$v = 9.24 \text{ m/s}$$

**Example 1.4** A square plate weighing 150 N and of uniform thinness and 300 mm edge is hung so that it can swing freely about the upper horizontal edge. A horizontal jet of water 20 mm diameter and having a velocity of 15 m/s impinges on the plate. The centre line of the jet is 150 mm below the upper edge of the plate, When the plate is vertical the jet strikes the plate normally and at its centre.

(a) Find what force must be applied at the lower edge of the plate in order to keep the plate vertical.

(b) If the plate is allowed to swing freely, find the inclination to the vertical which the plate will assume under the action of the jet.

**Solution:**

Given: $W = 150$ N, $b = 300$ mm, $d = 20$ mm, $v = 15$ m/s, $h = 150$ mm.

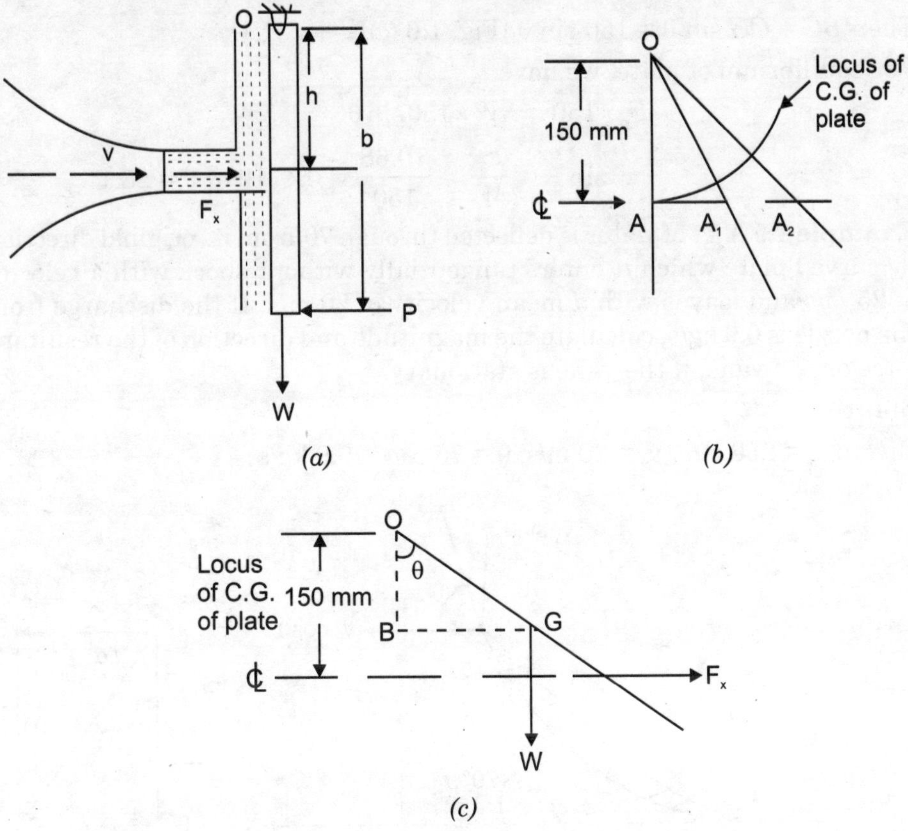


**Fig. 1.6 Fluid jet striking a swinging plate.**

(a) Dynamic force exerted by the fluid jet on the plate at its centre in $x$-direction,

$$F_x = \rho A v^2$$

$$= 1000 \times \frac{\pi}{4} \times \left(20 \times 10^{-3}\right)^2 \times (15)^2 = 70.68 \text{ N}$$

Let $P$ = force applied horizontally at the bottom edge of the plate [Fig.1.6 (a)] to keep the plate in vertical position.

Taking moments about the point of suspension $O$ of the plate, we have

$$F_x \times h = P \times b$$

$$P = \frac{F_x h}{b} = \frac{70.68 \times 150}{300} = 35.34 \text{ N}$$

(b) When the plate is allowed to swing freely, the vertical distance from the fulcrum $O$ to the centre line of water jet remains fixed at 150 mm

[Fig. 1.6 (b)], however the distance from $O$ to the point where the jet strikes the plate varies with the angle of inclination θ of the plate.

Let $O$ = actual angle of inclination.

Then $BG = OG \sin\theta = 150 \sin\theta$ [Fig. 1.6 (c)]

For equilibrium of plate, we have

$$F_x \times 150 = W \times 150 \sin\theta$$

$$\sin\theta = \frac{F_x}{W} = \frac{70.68}{150} = 0.4712 \text{ or } \theta = 28.1^\circ$$

**Example 1.5** A jet of water is deflected through 70°from its original direction by curved plate which it enters tangentially without shock with a velocity of 25 m/s and leaves with a mean velocity of 20 m/s. If the discharge from the nozzle is 0.9 kg/s, calculate the magnitude and direction of the resultant force on the vane, if the vane is stationary.

**Solution.**

Given: $v_1 = 256$ m/s, $v_2 = 20$ m/s, $\theta = 70^\circ$, $\dot{m} = 0.9$ kg/s

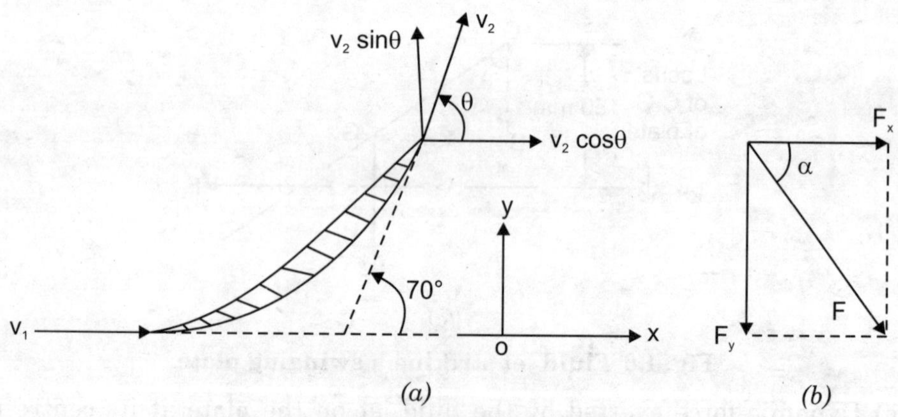


**Fig. 1.7 Water jet striking a curve plate.**

Force in the direction of jet, $F_x = \dot{m} \ (v_{1x} - v_{2x})$

$$= \dot{m} \ (v_1 - v_2 \cos 70^\circ)$$
$$= (25 - 20 \cos 70^\circ)$$
$$= 18.16 \text{ N}$$

Force normal to jet, $F_y = \dot{m}\left(v_{1y} - v_{2y}\right) = \left(0 - v_2 \sin\theta\right)$

$$= 0.9\ (0 - 20 \sin 70^\circ)$$
$$= -16.91 \text{ N}$$

Resultant force on curve vane, $F = \sqrt{F_x^2 + F_y^2}$

$$= \sqrt{(18.16)^2 + (-16.91)^2} = 24.81 \text{ N}$$

From Fig. 1.7 (b) we have

$$\tan\alpha = \frac{F_y}{F_x} = \frac{-16.91}{18.16} = -0.93117$$

$$\alpha = 42.96°$$

Since numerator is – *ve* and denominator is + *ve*, therefore, α lies in the 4th quadrant.

### 1.6.3 Jet Impinging on a Moving Flat Plate

(a) Flat plate normal to jet

Consider a flat plate moving with velocity $u$ to the right on which a fluid jet having velocity $\upsilon$ is impinging on it, as shown in Fig.1.8.

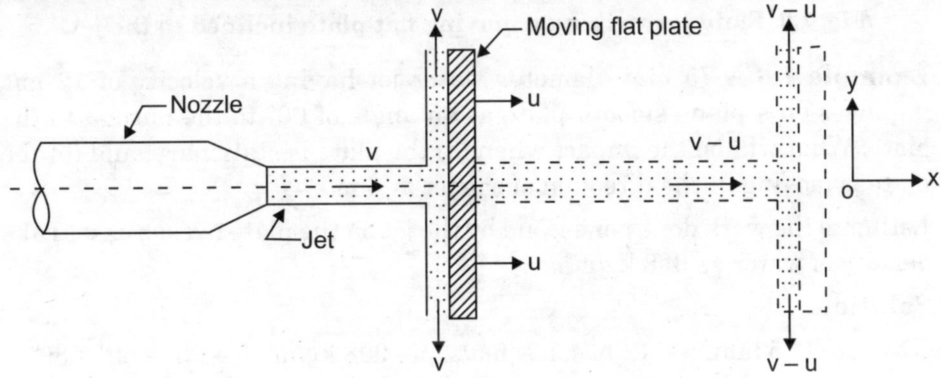


**Fig. 1.8 Fluid Jet striking a moving flat plate.**

Mass of fluid striking the plate per second,

$$\dot{m} = \rho A(v-u)$$

Where $A$ = area of cross – section of jet

Force exerted by the jet on the plate,

$$F_x = \dot{m} \times \text{change in velocity of jet}$$

$$= \dot{m} \times \left[(v-u)-0\right]$$

$$= \rho A(v-u)^2$$

Work done, $$W = F_x \times u$$

$$= \rho A(v-u)^2 u$$

(b) Flat plate inclined to the jet

When the flat plate is inclined to the jet at an angle θ (Fig.1.9) then the component of relative velocity of jet normal to the plate = $(v - u)\sin\theta$.

Mass of fluid striking the plate per second = $\rho A(v - u)$

Force exerted by the jet on the plate in the direction normal to the plate,

$$F = F_n \sin\theta = \rho A(v-u)^2 \sin^2\theta$$

$$\text{Work done} = F_x u = \rho A(v-u)^2 \sin^2\theta . u$$

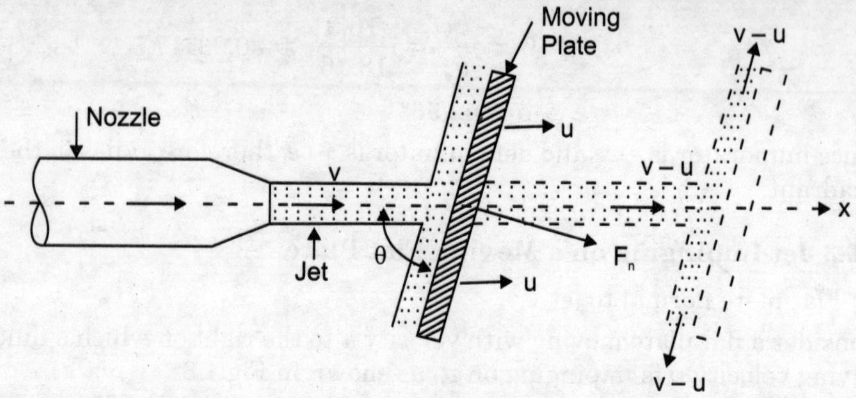


**Fig. 1.9 Fluid jet striking a moving flat plate inclined to the jet.**

**Example 1.6** A 75 mm diameter water jet having a velocity of 12 m/s impinges on a plane smooth plate at an angle of 60° to the normal to the plate. What will be the impact when (a) the plate is stationary, and (b) the plate is moving in the direction of the jet at 6 m/s?

Estimate the work done per second by the jet on the plate in each case. Take density of water as 998 kg/m³.

**Solution.**

Given: $d = 75$ mm, $v = 12$ m/s, $u = 6$m/s, $\rho = 998$ kg/m³, $\theta = 90° - 60° = 30°$

***(a) Force on inclined stationary flat plate***

$$F_x = \rho A v^2 \sin\theta$$

$$= 998 \times \frac{\pi}{4}(0.075)^2 \times 12^2 \times \sin 30°$$

$$= 317.45 \text{ N}$$

$$\text{Work done} = 0$$

***(b) Force on inclined moving plate***

$$F_x = \rho A(v-u)^2 \sin^2\theta$$

$$= 998 \times \frac{\pi}{4}(0.075)^2 \times (12-6)^2 \sin^2 30°$$

$$= 39.68 \text{ N}$$

$$\text{Work done} = F_x.u = 39.68 \times 6$$

$$= 238.08 \text{ N. m/s}$$

***(c) Series of plates mounted on a rotating wheel***

Fig. 1.10 shows a water jet impinging on a series of flat plates mounted on a rotating wheel. In such a system, each plate appears successively before the jet in the same position and always moving with a velocity $v$ in the direction of jet. Then whole flow from the jet is utilized by the plates.

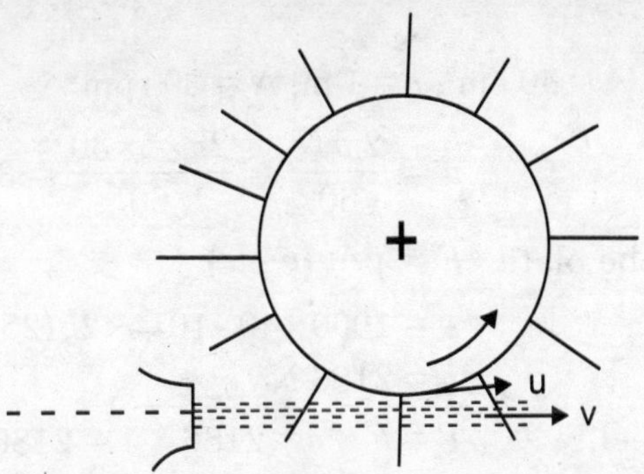


**Fig. 1.10 Fluid jet impinging on a series of moving plates.**

Mass of water striking the plate, $m = \rho A v$

Force exerted by the jet, $F = m(v - u)$

$$= \rho A v(v - u)$$

Work done on the plates, $W = F \times u = \rho A v(v - u)u$

$$= \rho Q(v - u)u$$

Kinetic energy of jet, $KE = \frac{1}{2} m v^2 = \frac{1}{2}\rho A v^3$

Efficiency of system, $\eta = \frac{W}{KE} = \frac{\rho A v(v - u)u^2}{\frac{1}{2}\rho A v^3} = \frac{2(v - u)u}{v^2}$

For efficiency to be maximum, $\frac{d\eta}{du} = 0$

$$\therefore \quad \frac{d}{du}\left(vu - u^2\right) = 0$$

$$v - 2u = 0$$

Or $$u = \frac{v}{2}$$

And $$\eta_{max} = \frac{2\left(v - \frac{v}{2}\right)\frac{v}{2}}{v^2} = \frac{1}{2} \text{ or } 50\ \%$$

**Example 1.7** A water jet of velocity 25 m/s having a cross – sectional area of 40 cm$^2$ strikes a series of flat plates mounted on the periphery of a wheel such that each plate appears successively before the jet. Determine:

(c) The total force exerted by the jet on the plate,

(d) The torque on the wheel,

(e) The work done per second on the plate, and

(f) The efficiency of the jet.

The jet strikes the plates at a radial distance of 1m from the wheel axis and the wheel rotates at a speed of 30 rpm.

**Solution.**

Given: $v = 25$ m/s, $A = 40$ cm$^2$, $r = 1$ m, $N = 30$ rpm

$$u = \frac{2\pi rN}{60} = \frac{2\pi \times 1 \times 30}{60} = 3.14 \text{ m/s}$$

Force exerted on the plate, $F = \rho Av(v-u)$

$$= 1000 \times 40 \times 10^{-4} \times 25(25-36.14)$$

$$= 2186 \text{ N}$$

Torque on the wheel, $T = F \times r = 2186 \times 1 = 2186$ Nm

Work done, $W = F \times u = 2186 \times 3.14 = 6867$ Nm/s

Efficiency of jet, $$\eta = \frac{2(v-u)u}{v^2}$$

$$= \frac{2(25-3.14)\times 3.14}{25^2} = 0.2196 \text{ or } 21.96\ \%$$

**Example 1.8** A nozzle of 50 mm diameter delivers a stream of water of 25 m/s perpendicular to a plate that moves away from the jet at 5 m/s. Calculate: (a) the force on the plate, (b) the work done, and (c) the efficiency of the jet.

**Solution.**

Given: $v = 25$ m/s, $u = 5$ m/s, $d = 50$ mm

Force on the plate, $F = \rho A(v-u)^2$

$$= 1000 \times \frac{\pi}{4}\left(50 \times 10^{-3}\right)^2 (25-5)^2$$

$$= 785.4 \text{ N}$$

Work done, $W = F \times u = 785.4 \times 5 = 39.27$ N. m/s

Kinetic energy of water jet, $KE = \frac{1}{2}mv^2 = \frac{1}{2}(\rho Av)\times v^2 = \frac{1}{2}\rho Av^3$

$$= \frac{1}{2} \times 1000 \times \frac{\pi}{4}\left(50 \times 10^{-3}\right)^2 \times (25)^2$$

$$= 15339.8 \text{ N. m/s}$$

Efficiency of jet, $$\eta_{jet} = \frac{W}{KE} = \frac{3927}{15339.8} \times 100$$

$$= 25.6\ \%$$

**Example 1.9** A 60 mm diameter jet of water having a velocity of 25 m/s strikes a flat plate, the normal of which is inclined at 30° to the axis of the jet. Calculate the normal force on the plate, when (a) the plate is stationary, and (b) the plate is moving with a velocity of 10 m/s in the direction of the jet, away from the jet. Also determine the power and efficiency of the jet when the plate is moving.

**Solution.**

Given: $d = 60$ mm, $v = 25$ m/s, $\theta = 90° - 30° = 60°$, $u = 10$ m/s

(a)
$$F_n = \rho A v^2 \sin\theta$$
$$= 1000 \times \frac{\pi}{4}\left(60 \times 10^{-3}\right)^2 \times (25)^2 \times \sin 60°$$
$$= 1530.4 \text{ N}$$

(b)
$$F_n = \rho A (v-u)^2 \sin\theta$$
$$= 1000 \times \frac{\pi}{4}\left(60 \times 10^{-3}\right)^2 \times (25-10)^2 \times \sin\theta$$
$$= 550.94 \text{ N}$$

Work done by the jet when plate is moving, $W = F_n \sin\theta \times u$
$$= 550.94 \sin 60° = 4771.3 \text{ Nm/s}$$

Kinetic energy of jet, $KE = \frac{1}{2}\rho A v \times v^2$
$$= \frac{1}{2}\rho A v^3 = \frac{1}{2} \times 1000 \times \frac{\pi}{4}\left(60 \times 10^{-3}\right)^2 \times (25)^3$$
$$= 22089.3 \text{ Nm/s}$$

Efficiency of jet, $\eta_{jet} = \frac{W}{KE} = \frac{4771.3}{22089.3} \times 100 = 21.6\ \%$

### 1.6.4 Jet Impinging on a Moving Curved Plate

***(a) Curved plate normal to jet***

Consider a fluid jet striking normal to a moving curved plate, as shown in Fig. 1.11.

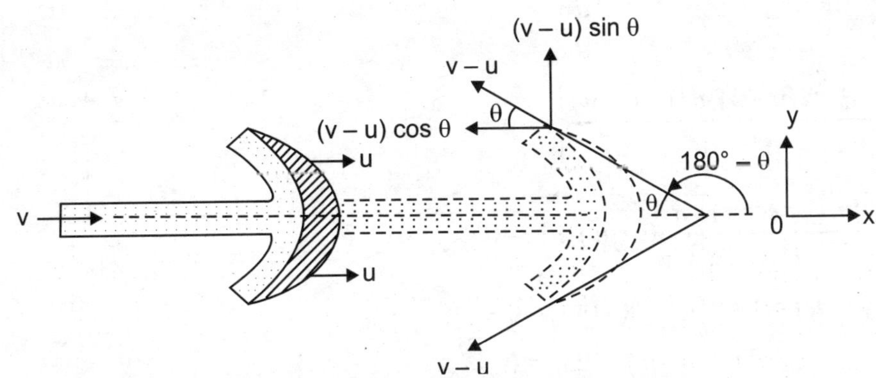


**Fig. 1.11 Fluid jet impinging normal to moving curved plate.**

Let $v$ = absolute velocity of jet
$u$ = velocity of plate

$A$ = area of cross – section of jet

Relative velocity of jet w.r.t. plate = $v - u$

Its components are:

$$v_x = -(v-u)\cos\theta$$
$$v_y = (v-u)\sin\theta$$

Mass of fluid striking the plate per second $\dot{m} = \rho A(v-u)$

Force exerted by jet on the plate,

$$F_x = \dot{m} \times \text{change in velocity}$$
$$= \dot{m}\left[(v-u)-\{-(v-u)\}\cos\theta\right]$$
$$= \rho A(v-u)^2(1+\cos\theta)$$

Work done on the curved plate per second,

$$W = F_x \times u$$
$$= \rho A(v-u)^2(1+\cos\theta)u$$

Kinetic energy of jet, $KE = \frac{1}{2}mv^2$

$$= \frac{1}{2}\rho Av \times u^2 = \frac{1}{2}\rho Av^3$$

Efficiency, $\eta = \frac{W}{KE} = \frac{\rho A(v-u)(1+\cos\theta)u}{\frac{1}{2}\rho Av^3}$

$$= \frac{2(v-u)^2 u(1+\cos\theta)}{v^3}$$

For a given jet velocity, efficiency will be maximum, when,

$$\frac{d\eta}{du} = 0$$

$$\therefore \quad \frac{d}{du}\left[\frac{2(v-u)^2 u(1+\cos\theta)}{v^3}\right] = 0$$

$$\left[\frac{2u(1+\cos\theta)}{v^3}\right]\frac{d}{du}(v-u)^2 u] = 0$$

Or $(v-u)^2 \times 1 + 2(v-u)(-1)u = 0$

Or $(v-u)^2 - 2u = 0$

$$u = \frac{v}{3}$$

Or $(v-u) = 0$

$\therefore$ $v = u$

When $v = u$, the work done becomes zero.

Thus, for maximum efficiency, $u = \frac{v}{3}$, i.e., vane velocity is one – third of the jet velocity.

$$\eta_{max} = \frac{2\left(v - \frac{v}{3}\right)^2 \times (1+\cos\theta)}{v^3}$$

$$= 2 \times \frac{4}{9} v^2 \times \frac{v}{3} \times (1+\cos\theta) \times \frac{1}{v^3}$$

$$= \frac{8}{27}(1+\cos\theta)$$

***(b) Jet impinging series of curved vanes***

In this case, there is always one vane or the other facing the jet and the entire fluid is utilized.

The mass of fluid striking the vanes per second,

$$\dot{m} = \rho A v$$

Thrust exerted by the jet on the vanes,

$$F_x = \rho A v\left[(v-u) - \{-(v-u)\}\cos\theta\right]$$

$$= \rho A v (v-u)(1+\cos\theta)$$

Kinetic energy supplied by the jet,

$$KE = \frac{1}{2}\rho A v \times v^2 = \frac{1}{2}\rho A v^3$$

Efficiency,
$$\eta = \frac{W}{KE} = \frac{\rho A v (v-u)(1+\cos\theta) u}{\frac{1}{2}\rho A v^3}$$

$$= \frac{(v-u) u (1+\cos\theta)}{v^2}$$

For efficiency to be maximum for a given value of $v$,

$$\frac{d\eta}{du} = 0$$

$$\left[\frac{2(1+\cos\theta)}{v^2}\right]\frac{d}{du}\left[(v-u)u\right] = 0$$

$$(v-u)\times 1 + u(-1) = 0$$

$$v - 2u = 0$$

$$u = \frac{v}{2}$$

Hence, the efficiency is maximum when the peripheral speed of wheel is half of the jet velocity

$$\eta_{max} = \frac{2}{v^2}\left[\left(v - \frac{v}{2}\right)\frac{2}{v}(1+\cos\theta)\right]$$

$$= \frac{1}{2}(1+\cos\theta)$$

For $\theta = 0°$, i.e., semi – circular vanes, $\eta_{max} = 1$ or 100 %.

**Example 1.10** A jet of water 60 mm diameter strikes a curved cane at its centre with a velocity of 20 m/s. The curved vane is moving with a velocity of 5 m/s in the direction of the jet. The jet is deflected through an angle of 160°. Assuming the plate to be smooth, find

(a) Thrust on the vane in the direction of jet.

(b) Power of the jet, and

(c) Efficiency of the jet,

**Solution.**

Given: $d = 60$ mm, $v = 20$ m/s, $u = 5$ m/s, $\theta = 180° - 160° = 20°$

*(a) Thrust on the vane,*

$$F_x = \rho A(v-u)^2(1+\cos\theta)$$

$$= 1000 \times \frac{\pi}{4}(0.060)^2(20-5)^2(1+\cos 20°)$$

$$= 1233.98 \text{ N}$$

*(b)* Power of jet $= F_x \times u$

$$= \frac{1233.98 \times 5}{10^3} = 6.17 \text{ kW}$$

*(c)* Efficiency of jet, $\eta = \dfrac{\text{Work done}}{K.E. = \frac{1}{2}\rho A v^3}$

$$= \frac{6169.9}{\frac{1}{2} \times 1000 \times \frac{\pi}{4}(0.60)^2 \times 20^3}$$

$$= 0.5455 \text{ or } 54.55\ \%$$

### 1.6.5 Jet Striking a Moving Curved Vane Tangentially at One Tip and Leaving at the Other Tip

The jet striking and leaving a curved moving vane tangentially is shown in Fig. 1.12 The velocity triangles at inlet and outlet are also shown in the Fig. 1.12

Let $v_1$ = absolute velocity of jet at inlet

$v_2$ = absolute velocity of jet at outlet

$u_1$ = peripheral velocity of vane at inlet

$u_2$ = peripheral velocity of vane at outlet
$v_{r1}$ = relative velocity of jet at inlet
$v_{r2}$ = relative velocity of jet at outlet
$v_{w1}$ = velocity of whirl at inlet
$v_{w2}$ = velocity of whirl at outlet
$v_{f1}$ = velocity of flow at inlet
$v_{f1}$ = velocity of flow at outlet
$\alpha_1, \alpha_2$ = angles which the absolute velocities make with the direction of motion of vane at inlet and outlet respectively.
$\beta_1, \beta_2$ = tip angles at inlet and outlet respectively
= angles which the relative velocity make with the direction of motion of vane at inlet and outlet respectively.

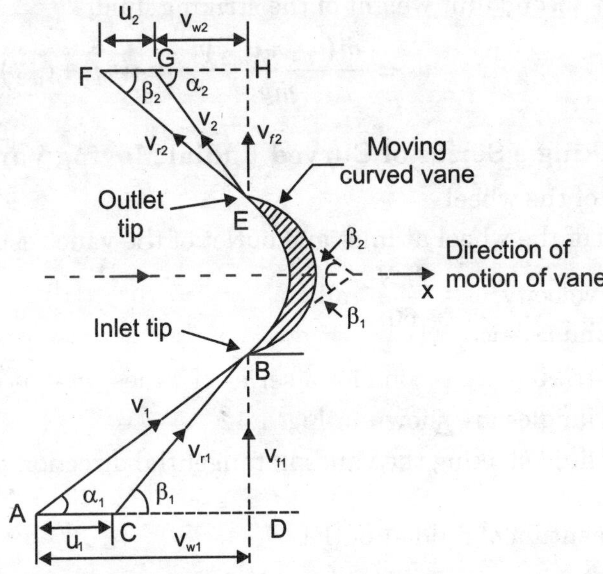


**Fig. 1.12 Velocity triangles for jet striking and leaving a moving curved vane tangentially.**

The velocity if whirl is the component of absolute velocity in the direction of motion of vane.

$$\therefore \quad v_{w1} = v_1 \cos \alpha_1 \text{ and } v_{w2} = v_2 \cos \alpha_2$$

The velocity of flow is the component of absolute velocity perpendicular to the direction of motion of vane.

$$\therefore \quad v_{f1} = v_1 \sin \alpha_1 \text{ and } v_{f2} = v_2 \sin \alpha_2$$

The triangles ABC and EFG are called the velocity triangles at inlet and outlet respectively.

For a smooth vane, $u_1 = u_2 = u$

Mass of fluid striking the vane per second, $\dot{m} = \rho A v_{r1}$

Where $A$ = area of fluid jet

Force exerted by fluid jet in the direction of motion,

$$F_x = \dot{m}\left[v_{r1}\cos\beta_1 - (-v_{r2}\cos\beta_2)\right]$$

$$= \dot{m}\left[v_{r1}\cos\beta_1 + v_{r2}\cos\beta_2\right]$$

$$v_{r1}\cos\beta_1 = v_{w1} \text{ and } v_{r2}\cos\beta_2 = v_{w2} + u_2$$

$$\therefore \quad F_x = \dot{m}(v_{w1} - u_2 + v_{w2} + u_2)$$

$$= \dot{m}(v_{w1} + v_{w2}) \text{ for } u_1 = u_2$$

$$= \rho A v_{r1}(v_{w1} + v_{w2})$$

Work done per second by the jet on the vane,

$$W = F_x \times u = \rho A v_{r1}(v_{w1} + v_{w2})u$$

Work done per second unit weight of the striking fluid,

$$\dot{w} = \frac{\dot{m}(v_{w1} + v_{w2})u}{\dot{m}g} = (v_{w1} + v_{w2})\frac{u}{g}$$

### 1.6.6 Jet Striking a Series of Curved Radial Moving Vanes

Let $N$ = rpm of the wheel

$r_1, r_2$ = radii of the wheel at inlet and outlet of the vane respectively.

Then angular velocity, $\omega = \dfrac{2\pi N}{60}$ rad/s

$u_1 = \omega r_1$ and $u_2 = \omega r_2$

Mass of fluid striking per second for a series of vanes, $\dot{m} = \rho A v_1$

The velocity triangles are shown in Fig. 1.12.

Momentum of fluid striking the vanes in tangential direction per second at inlet = $\dot{m} \times v_{w1}$

Similarly momentum of fluid at outlet

$$= \dot{m}(-v_{w2}) = -\dot{m}\, v_{w2}$$

–ve sign is taken as $v_{w2}$ is in opposite direction to motion of vanes.

Angular momentum rate at inlet = $\dot{m}\, v_{w1} r_1$

Angular momentum rate at outlet = $\dot{m}\, v_{w2} r_2$

Torque exerted by fluid on the wheel,

$T$ = rate of charge of angular momentum

$$= \dot{m}\left[v_{w1}r_1 - (-v_{w2}r_2)\right]$$

$$= \dot{m}\left[v_{w1}r_1 + v_{w2}r_2\right]$$

Work done per second on the wheel, $W = T \times \omega$

$$= \dot{m}(v_{w1}r_1 + v_{w2}r_2)\omega$$

$$= \rho A v_1(v_{w1}u_1 + v_{w2}u_2)$$

Efficiency of radial curved vanes,

$$\eta_{\text{vane}} = \frac{\text{work done per second}}{\text{kinetic energy per second}}$$

$$= \frac{\dot{m}(v_{w1}u_1 + v_{w2}u_2)}{\frac{1}{2}\dot{m}v_1^2} = \frac{2(v_{w1}u_1 + v_{w2}u_2)}{v_1^2}$$

**Example 1.11** A jet of water having a velocity of 50 m/s impinges without shock on a series of vanes moving at 15 m/s, the direction of motion of the vanes is inclined at 20° to that of jet. The relative velocity at outlet is 0.9 of that at inlet. The absolute velocity of water at exit is to be normal to the motion of vanes. Determine:

(a) Vane angles at inlet and outlet.

(b) Work done on vanes per $N$ (Newton) of water supplied by the jet, and

(c) Hydraulic efficiency.

**Solution.**

The velocity triangles are shown in Fig.1.13.

Given: $u$ = 15m/s, $v_1$ = 50 m/s, $\alpha_1$ = 20°, $v_{r2}$ = 0.9 $v_{r1}$, $\alpha_2$ = 90°

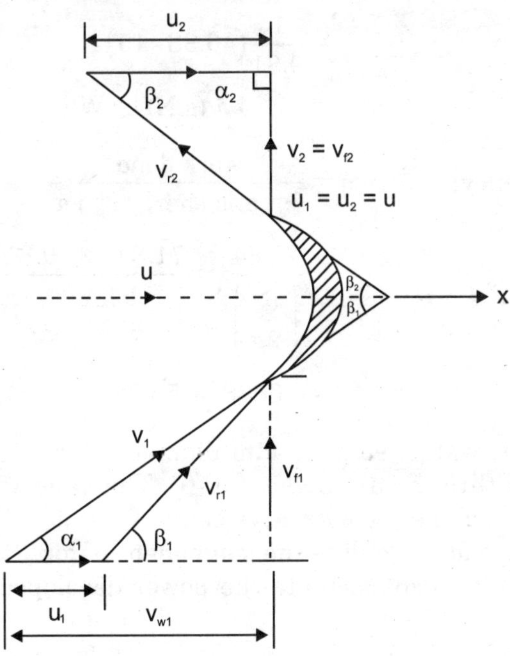


**Fig. 1.13 Velocity triangles.**

***(a) From inlet velocity triangle, we have***

$$\tan\beta_1 = \frac{v_{r1}}{v_{w1} - u_1} = \frac{v_1 \sin\alpha_1}{v_1\cos\alpha_1 - u_1}$$

$$= \frac{50\sin 20°}{50\cos 20° - 15} = 0.53466$$

$$\beta_1 = 28.13°$$

$$v_{w1} = v_1 \cos \alpha_1 = 50 \sin 20° = 46.98 \text{ m/s}$$

$$v_{f1} = v_1 \sin \alpha_1 = 50 \sin 20° = 17.1 \text{ m/s}$$

$$\sin \beta_1 = \frac{v_{f1}}{v_{r1}}, v_{r1} = \frac{17.1}{\sin 28°.13} = 36.27 \text{ m/s}$$

$$v_{r2} = 0.9\, v_{r1} = 0.9 \times 36.27 = 32.64 \text{ m/s}$$

From outlet velocity triangle,

$$\cos \beta_2 = \frac{u_2}{v_{r2}} = \frac{15}{32.64} = 0.4595$$

$$\beta_2 = 62.64°$$

Work done per second per $N$ of water striking

$$= \frac{1}{g}(v_{w1}u_1 + v_{w2}u_2)$$

$$= \frac{1}{g}(v_{w1}u_1) \text{ as } v_{w2} = 0$$

$$= \frac{1}{9.81}(46.98 \times 15)$$

$$= 71.84 \text{ J/(s.N) or W/N}$$

Hydraulic efficiency,

$$\eta = \frac{\text{work done}}{\text{kinetic energy of jet}}$$

$$= \frac{71.84}{\left(\frac{v_1^2}{2g}\right)} = \frac{71.84 \times 2 \times 9.81}{50^2}$$

$$= 0.5638 \text{ or } 56.38\%$$

**Example 1.12** A water jet of 30 mm diameter strikes on a curved vane and is deflected through an angle of 170°, The vane moves in the same direction as that of the jet with a velocity of 30 m/s. The rate of flow is 120 litres per second. Calculate the component of force on the vane in the direction of motion. Also calculate the power developed by the vane and water efficiency.

**Solution.**

Given: $d$ = 30 mm, $\beta_2$ = 180° – 170° = 10°, $u$ = 30 m/s, $Q$ = 120 lps = 0.12 m³/s, $\alpha_1 = 0$, $\beta_1 = 0$.

The velocity diagrams are shown in Fig. 1.14

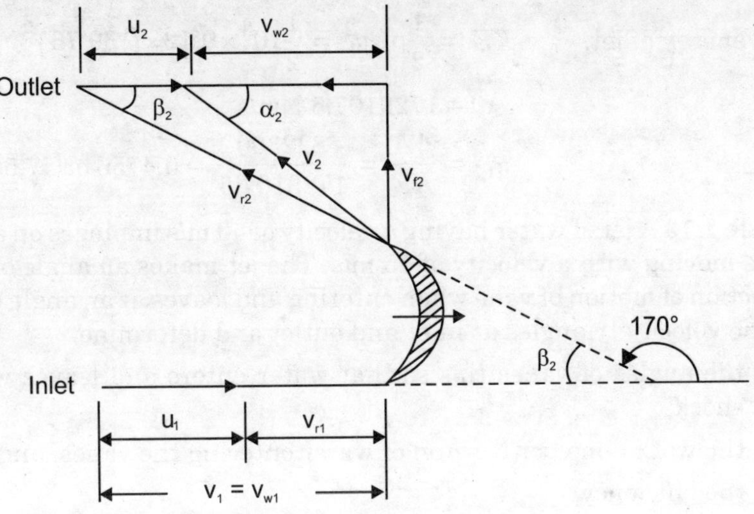


**Fig. 1.14 Velocity triangle**

Area of jet, $$A = \frac{\pi}{4}d^2 = \frac{\pi}{4}\times(0.03)^2 = 7.07\times10^{-4}\,m^2$$

$$u_1 = u_2 = u = 30 \text{ m/s}$$

$$v_1 = \frac{Q}{A} = \frac{0.12}{7.07\times10^{-4}} = 169.76 \text{ m/s}$$

From inlet velocity triangles, we have

$$v_{r1} = v_1 - u_1 = 169.76 - 30 = 139.76 \text{ m/s}$$

$$v_{w1} = v_1 = 169.76 \text{ m/s}$$

From outlet velocity triangle, we have

For no friction loss over the vane, $v_{r2} = v_{r1}$ = 139.76 m/s

$$v_{w2} = v_{r2}\cos\beta_1 - u_2$$

$$= 139.76 \cos 10^\circ - 30 = 107.64 \text{ m/s}$$

Component of force on the vane in the direction of motion,

$$F = \rho A v_{r1}\left(v_{w1} + v_{w2}\right)$$

$$= 10^3 \times 7.07\times10^{-4}\times139.76\left(169.76+107.64\right)$$

$$= 27410 \text{ N}$$

Work done, $$W = F\times u = 27410 \times 30$$

$$= 822300 \text{ N. m/s or } 822.3 \text{ kW}$$

Hence, power developed by the vane, $P$ = 822.3 kW

Efficiency of vane or water efficiency,

$$\eta_w = \frac{\text{work done on the vane}}{\text{kinetic energy of jet}}$$

Kinetic energy of jet, $KE = \frac{1}{2}\rho Q v_1^2 = \frac{1}{2}10^3 \times 0.12 \times (169.76)^2$

$$= 1729107.5 \text{ Nm/s}$$

$$\eta_w = \frac{W}{KE} = \frac{822300}{1729107.5} = 0.4756 \text{ or } 47.56\ \%$$

**Example 1.13** A jet of water having a velocity of 40 m/s impinges on a series of vanes moving with a velocity of 15 m/s. The jet makes an angle of 30° to the direction of motion of vane when entering and leaves at an angle of 120°. Draw the velocity triangles at inlet and outlet and determine.

(a) the angles of vanes tips so that water enters and leaves without shock,

(b) the work done per Newton of water entering the vanes, and

(c) the efficiency.

**Solution.**

Given: $v_1 = 40$ m/s, $u = 15$ m/s, $\alpha_1 = 30°$, $\alpha_2 = 180° - 120° = 60°$

***(a) The velocity triangles are shown in Fig. 1.15***

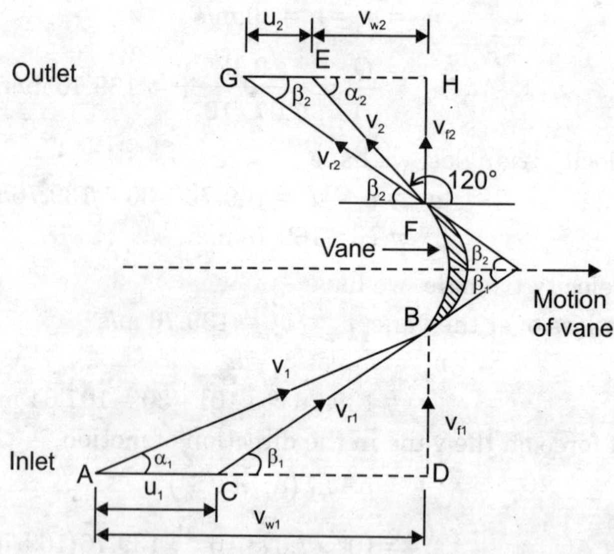


**Fig. 1.15 Velocity triangles.**

From inlet velocity triangles, we have

$$v_{w1} = v_1 \cos \alpha_1 = 40 \cos 30° = 34.64 \text{ m/s}$$

$$v_{f1} = v_1 \sin \alpha_1 = 40 \cos 30° = 20 \text{ m/s}$$

$$\tan \beta_1 = \frac{v_{f1}}{v_{w1} - u_1} = \frac{20}{34.64 - 15} = 1.018$$

$$\beta_1 = 45.52°$$

$$\sin \beta_1 = \frac{v_{f1}}{v_{r1}}$$

$$v_{r1} = \frac{20}{\sin 45.52°} = 28.03 \text{ m/s}$$

Without any friction loss on the vane, $v_{r2} = v_{r1} = 28.03$ m/s

From outlet velocity triangle, we have

$$\angle GEF = 180° - 60° = 120°$$

Applying sine formula, we have

$$\frac{v_{r2}}{\sin \angle GEF} = \frac{u_2}{\sin \angle EFG}$$

$$\frac{28.03}{0.866} = \frac{15}{\sin(60° - \beta_2)}$$

$$\sin(60 - \beta_2) = 0.4634 = \sin 27.6°$$

$$\therefore \quad 60° - \beta_2 = 27.6°$$

$$\beta_2 = 32.39°$$

Work done per Newton of water entering the vanes,

$$W = (v_{w1} + v_{w2})u$$

$$v_{w2} = v_2 \cos \alpha_2 = v_{r2} \cos \beta_2$$

$$= 28.03 \cos 32.39° - 15 = 8.64 \text{ m/s}$$

$$W = \frac{1}{9.81}(34.64 + 8.64) \times 15 = 66.81 \text{ N.m}$$

$$\text{Efficiency} = \frac{\text{work done}}{\text{energy supplied by jet} = \dfrac{v_1^2}{2g}}$$

$$= \frac{66.18 \times 2 \times 9.81}{40^2}$$

$$= 0.8115 \text{ or } 81.15\ \%$$

**Example 1.14** A wheel consists of radial blades with inner and outer radii of 380 mm and 760 mm respectively. Water enters the blades at the outer periphery with a velocity of 50 m/s and supply jet makes an angle of 30° with the tangent to wheel at inlet tip. Water leaving the blade has a flow velocity of 10 m/s. The blade angle at entrance and exit are 40° and 30° respectively. Determine.

(a) Work done per Newton of water,

(b) Speed of wheel, and

(c) Efficiency of blading

**Solution.**

Given: $R_1 = 760$ mm, $R_2 = 380$ mm, $v_1 = 50$ m/s, $\alpha_1 = 30°$, $v_{f2} = 10$ m/s, $\beta_1 = 40°$, $\beta_2 = 30°$

The velocity triangles are shown in Fig. 1.16.

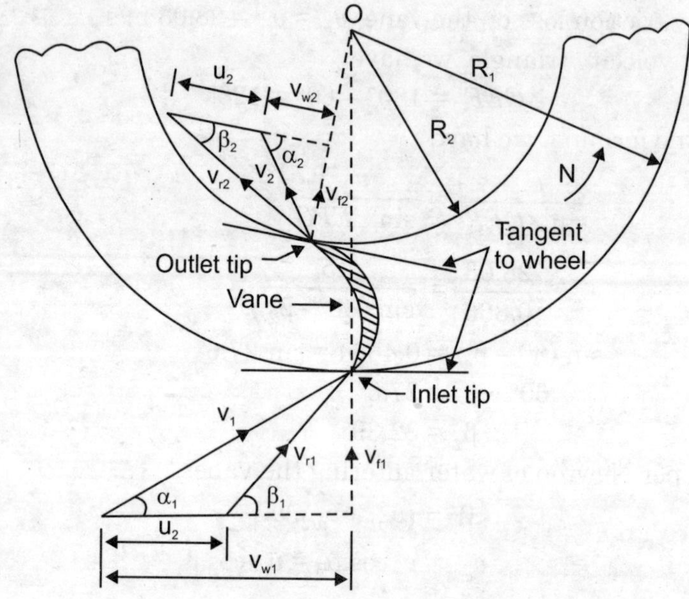


**Fig. 1.16 Velocity triangles.**

From inlet velocity triangle, we have

$$v_{f1} = v_1 \sin \alpha_1 = 50 \sin 30° = 25 \text{ m/s}$$

$$v_{w1} = v_1 \cos \alpha_1 = 50 \sin 30° = 43.3 \text{ m/s}$$

$$u_1 = v_{w1} - v_{f1} \cot \beta_1$$

$$= 43.3 - 25 \cot 400 = 13.5 \text{ m/s}$$

From inlet velocity triangle, we have

$$\frac{u_2}{u_1} = \frac{R_2}{R_1}$$

$$u_2 = \frac{380}{760} \times 13.5 = 6.75 \text{ m/s}$$

$$v_{w2} = \frac{v_{f2}}{\tan \beta_2} - u_2 = \frac{10}{\tan 30°} - 6.75 = 10.57 \text{ m/s}$$

(a) Work done per Newton of water,

$$W = \frac{1}{g}(u_1 v_{w1} + u_2 v_{w2})$$

$$= \frac{1}{9.81}[13.5 \times 43.3 + 6.75 \times 10.57]$$

$$= 66.86 \text{ N. m}$$

Angular velocity, $\omega = \frac{u_1}{R_1} = \frac{13.5}{0.760} = 17.76 \text{ rad/s}$

Now $\omega = \frac{2\pi N}{60}$

$$N = \frac{60 \times 17.76}{2\pi} = 169.6 \text{ rpm}$$

Efficiency of blading,

$$\eta_b = \frac{\text{work done}}{\text{energy supplied by jet}}$$

$$= \frac{66.86 \times 2 \times 9.81}{(50)^2} = 0.5247 \text{ or } 52.47\ \%$$

**Example 1.15** A jet of water moving at 10 m/s impinges on a concave shape vane to deflect the jet through 120° when stationary. The vane is moving at 4 m/s. Calculate:

(a) The angle of jet so that there is no shock at inlet,

(b) The absolute velocity of the jet at exit both in magnitude and direction, and

(c) The work done per second per Newton of water. Assume that the blade is smooth.

**Solution.**

Given: $v_1 = 10$ m/s, $u = 4$ m/s

Assuming the blade to be symmetrical,

$$\beta_1 = \beta_2$$

Now $\beta_1 + \beta_2 = 180° - 120° = 60°$

$\therefore$ $\beta_1 = \beta_2 = 30°$

The velocity triangles are shown in Fig. 1.17

$$u_1 = u_2 = u = 5 \text{ m/s}$$

Applying sine formula to inlet velocity triangle ABC, we have

$$\frac{v_1}{\sin(180° - \beta_1)} = \frac{u_1}{\sin(\beta_1 - \alpha_1)}$$

$$\frac{10}{\sin(180° - 30°)} = \frac{4}{\sin(30° - \alpha_1)}$$

$$\sin(30° - \alpha_1) = \frac{4}{10}\sin 150° = 0.20 = \sin 11.54°$$

$\therefore$ $30° - \alpha_1 = 11.54°$

$$\alpha_1 = 18.46°$$

Also
$$\frac{v_{r1}}{\sin\alpha_1} = \frac{u_1}{\sin(\beta_1 - \alpha_1)}$$

$$v_{r1} = \sin 18.46° \times \left(\frac{4}{\sin(30° - 18.46°)}\right)$$

$$= 6.33 \text{ m/s}$$

$$v_{w1} = v_1 \cos\alpha_1 = 10 \cos\alpha_1\ 18.46° = 9.48 \text{ m/s}$$

Since the vane is smooth, $v_{r2} = v_{r1} = 6.33$ m/s

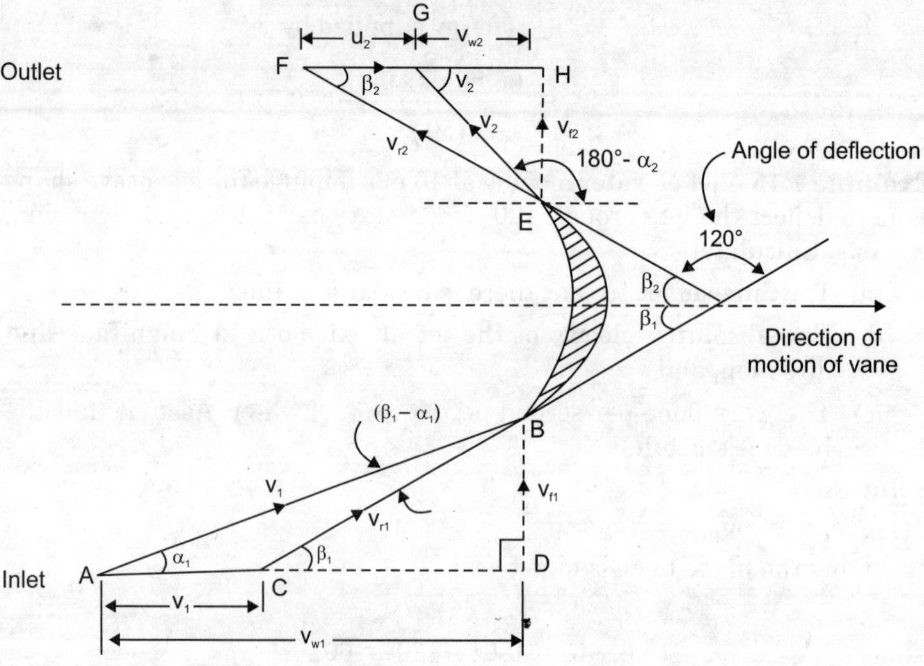


**Fig. 1.17 Velocity triangles.**

From outer velocity triangles, we have

$$v_{r2} \cos\beta_2 = u_2 + v_{w2}$$

$$6.33 \cos 30° = 4 + v_{w2}$$

$$v_{w2} = 1.48 \text{ m/s}$$

$$v_{f2} = v_{r2} \sin\beta_2$$

$$= 6.33 \sin 30° = 3.16 \text{ m/s}$$

$$\tan\alpha_2 = \frac{v_{f2}}{v_{w2}} = \frac{3.16}{1.48} = 2.4324$$

$$\alpha_2 = 67.65°$$

Hence, angle made by $v_2$ at outlet with direction of motion of vane = 180° – 67.65° = 112.35°.

Now $$v_2 = \sqrt{v_{w2}^2 + v_{f2}^2}$$

$$= \sqrt{(1.48)^2 + (3.16)^2} = 3.49 \text{ m/s}$$

***(c) Work done per second or Newton of water***

$$= \frac{1}{g}(v_{w1} + v_{w2})u$$

$$= \frac{1}{9.81}[9.48 + 1.48] \times 4 = 4.469 \text{ Nm/s}$$

## 1.7 Jet Propulsion

### Working Principle

Consider an orifice of area of cross – section A fixed to a tank full of water shown in Fig. 1.18. The jet issues out from the orifice with a velocity $v$. Let $Q$ be the discharge from the orifice per second. Applying liner – momentum equation, the force exerted upon the fluid to change its velocity from zero to $v$ is:

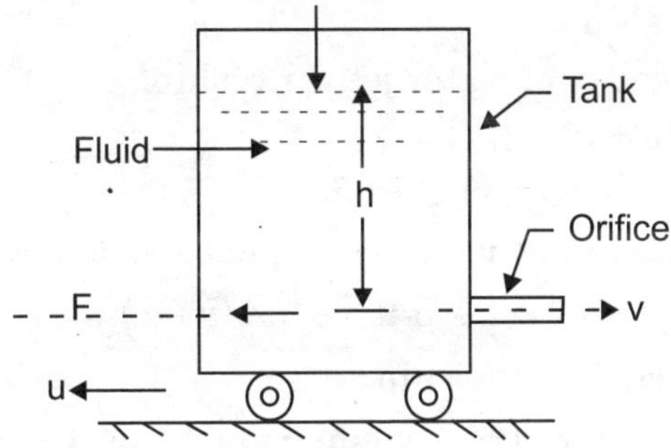


**Fig. 1.18 Principle of jet propulsion.**

$$F = \rho Q(v - 0) = \rho Q v$$

Where

$Q = Av$

$v = K_v\sqrt{2gh}$

$K_v$ = coefficient of velocity

$h$ = head of fluid over the centre line of orifice.

According to Newton's third law of motion, action and reaction are equal and opposite. Therefore, an equal and opposite force – $F$ is exerted on the tank. If the tank is provided with wheels, it will start moving towards the left with a velocity $v$ (say). Thus $F$ is the propulsive force which may be used to propel the crafts such as air craft, ships or submarines to which the orifice or nozzle is fitted.

### 1.7.1 Jet Propulsion of Ships

The principle explained above may be applied to propel the ship through water. The ship carries pumps which takes water from the surroundings of the ship and discharges by forcing through the orifice at the back of the ship.

**Case I:** Inlet orifices are at night angles to the direction of motion of the ship.

Consider the ship shown in Fig. 1.19 having inlet orifices at right angles to the direction of motion of the ship.

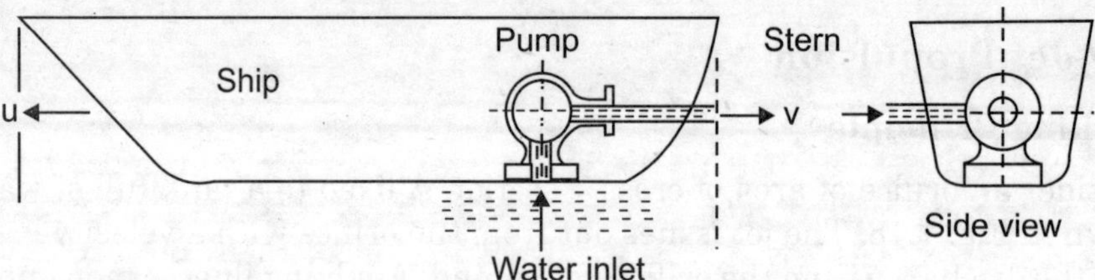


**Fig. 1.19 Jet propulsion.**

Let $v$ = velocity of issuing water jet

$u$ = velocity of ship

$v_r$ = relative velocity of water jet w.r.t. ship

$= v - (-u) = v + u$

$A$ = area of cross – section of orifices

Mass of water issuing from the orifice at the back of ship,

$$\dot{m} = \rho A v_r = \rho A(v+u)$$

Propulsive force exerted on the ship,

$$F = \text{mass flow rate of water} \times \text{change in jet velocity}$$

$$= \dot{m}(v_r + u)$$

$$= \rho A\,(v+u)\,[(v+u) - u]$$

$$= \rho A\,(v+u)v$$

Thrust or propulsive work done per second = propulsive force × velocity of ship

$$= F \times u = \rho A\,(v+u)vu$$

For right angle inlet of water to the direction of motion of the ship, inlet velocity of water is zero. Hence the outlet velocity equals $v_r$.

Kinetic energy supplied by the jet per second,

$$KE = \frac{1}{2}\dot{m}v_r^2 = \frac{1}{2}(\rho A v_r)v_r^2$$

$$= \frac{1}{2}\rho A v_r^3 = \frac{1}{2}\rho A(v+u)^3$$

Propulsion efficiency, $\eta_p = \dfrac{\text{Propulsive work}}{\text{KE supplied by the jet}}$

$$= \frac{\rho A(v+u)vu}{\frac{1}{2}\rho A(v+u)^3}$$

$$= \frac{2uv}{(v+u)^2}$$

For a given jet velocity $v$, the propulsion efficiency shall be maximum when $\dfrac{d\eta_p}{du} = 0$.

$$\frac{d}{du}\left[\frac{2uv}{(v+u)^2}\right] = 0$$

$$\frac{(v+u)^2 \times 2v - 2uv \times 2(v+u)}{(v+u)^4} = 0$$

Or $2v(v+u) - 4uv = 0$ $(v = -u)$

$$2v^2 + 2uv - 4uv = 0$$

$$v^2 - uv = 0$$

$$v - u = 0$$

∴ $u = v$ $(v \neq 0)$

$$(\eta_p)_{max} = \frac{2v^2}{4v^2} = \frac{1}{2} \text{ or } 50\ \%$$

**Case II:** Inlet orifices face the direction of motion of the ship Fig. 1.20 shows a ship in which water enters through a pipe having an open end at the front. As the water is at rest and the ship moves with velocity $u$, the water will enter the pipe with velocity $u$ relative to the motion of the ship.

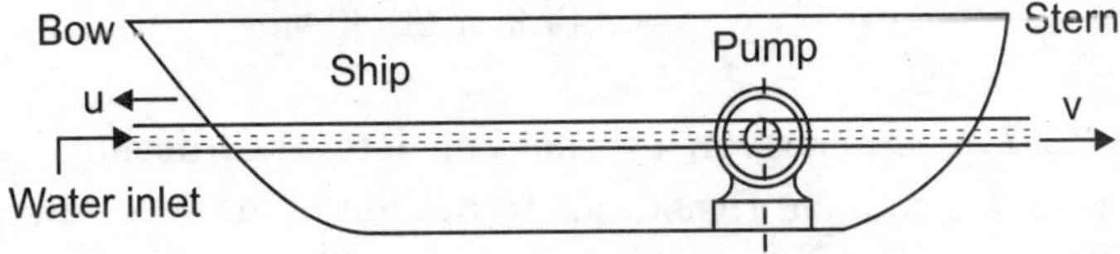


**Fig. 1.20 Jet propulsion.**

Propulsive work done per second = $\rho A(v+u)uv$

$$KE \text{ supplied by the jet} = \frac{1}{2}\dot{m}\left(v_r^2 - u^2\right)$$

$$= \frac{1}{2}\rho A v_r\left(v_r^2 - u^2\right)$$

$$= \frac{1}{2}\rho A(v+u)\left[(v+u)^2 - u^2\right]$$

$$= \frac{1}{2}\rho A(v+u)(v+2u)v$$

Propulsion efficiency, $$\eta_p = \frac{\rho A(v+u)uv}{\frac{1}{2}\rho A(v+u)(v+2u)}$$

$$= \frac{2u}{v+2u}$$

For $v = u$, $$\eta_p = \frac{2u}{3u} = \frac{2}{3} \text{ or } 66.67\ \%$$

**Example 1.16** A jet propelled boat moves at 30 km/h. The cross – sectional are of jet is 220 sq. cm. The relative velocity of boat is 10 m/s. Find the propulsive power required.

Assume that the orifices face the direction of motion of boat. Also find the propulsive efficiency of jet.

**Solution.**

Given: $u = \frac{30 \times 10^3}{3600} = 8.33$ m/s, $A = 220$ cm² or $220 \times 10^{-4}$ m², $v_r = 10$ m/s,

$v_r = v + u$, $v = 10 - 8.33 = 1.67$ m/s

$$\text{Propulsive power} = \rho A(v+u)vu/10^3$$

$$= 10^3 \times 220 \times 10^{-4} \times 10 \times 1.67 \times 8.33/10^3$$

$$= 3.06 \text{ kW}$$

$$\text{Propulsive efficiency} = \frac{2u}{v+2u}$$

$$= \frac{2 \times 8.33}{1.67 + 2 \times 8.33}$$

$$= 0.9089 \text{ or } 90.89\ \%$$

**Example 1.17** A ship driven by reaction jets and discharging water at the stern is found to have resistance to motion of 3000 N when moving at 30 km/h. The velocity of jet relative to ship is 20 m/s. Determine:

(a) Number of jets if each jet has an area of cross – section of 80 cm².

(b) Power required to drive the pump.

(c) Propulsive efficiency when (i) inlet orifices are at right angles to ship motion, and (ii) inlet orifices face the direction of ship motion.

**Solution.**

Given: $R = 3000\ N$, $u = \dfrac{30\times10^3}{3600} = 8.33$ m/s, $v_r = 20$ m/s, $A = 80$ cm²

Absolute velocity of jet, $v = v_r - u$

$= 20 - 8.33 = 11.67$ m/s

Thrust force exerted on the ship,

$$F = pA(v+u)\times v\times n$$

$$= 1000\times80\times10^{-4}\times20\times11.67\times n$$

$$= 1867.2 \text{ nN}$$

For equilibrium of ship, $F = R$

∴ Number of jets required, $n = \dfrac{3000}{1867.2} = 1.6 \approx 2$

Propulsive work $= Fu$

$= 3000 \times 8.33 = 25{,}000$ N. m/s

(i) *K.E.* supplied by jets $= \dfrac{1}{2}\rho An(v+u)^3$

$$= \frac{1}{2}\times1000\times80\times10^{-4}\times2\times20^3$$

$$= 64{,}000 \text{ N. m/s}$$

Propulsive efficiency, $\eta_p = \dfrac{\text{Propulsive work}}{K.E. \text{ of jets}}$

$$= \frac{25{,}000}{64{,}000}\times10 = 39.06\ \%$$

(ii) *K.E.* supplied by jets $= \dfrac{1}{2}\times\rho An(v+u)\left[(v+u)^2 - u^2\right]$

$$= \frac{1}{2}\times1000\times80\times10^{-4}\times2\times20\left[20^2-(8.33)^2\right]$$

$$= 52{,}897.8 \text{ N. m/s}$$

$$\eta_p = \frac{25{,}000}{52{,}897.8}\times100 = 47.26\ \%$$

**Example 1.18** A small ship is fitted with jets of total area 0.65 m². The relative velocity of water through the jet is 9 m/s and speed of the ship is 18 km/h in sea water. The efficiencies of engine and pump are 85 % and 65 % respectively. If water is taken at amid ship, determine:

(a) Propelling force, and
(b) Overall efficiency.

Assume pipe losses to be 10 % of the kinetic energy of jets.

**Solution.**

Given: $A = 0.65\ m^2$, $v_r = 9$ m/s, $u = \dfrac{18\times10^3}{3600} = 5$ m/s, $\eta_{engine} = 0.85$, $\eta_{pump} = 0.65$

Pipe losses, $h_f = 10\ \%$ of *K.E.* of jets

$$= 0.1\times\frac{1}{2}v_r^2 = \frac{v_r^2}{20} \text{ per kg of water}$$

Absolute velocity of jet, $v = v_r - u = 9 - 58 = 4$ m/s

Propelling force, $F = \rho A(v+u)v$

$$= 1000\times0.65(4+5)\times4 = 23{,}400 \text{ N}$$

Work done by jets per second, $W = F \times u = 23{,}400 \times 5 = 117{,}000$ N. m/s

Output of pump per kg of water = *K.E.* of jet + pipe losses

$$= \frac{v_r^2}{2}+\frac{v_r^2}{20} = 0.55\ v_r^2$$

$$\text{Input to pump per kg of water} = \frac{0.55v_r^2}{\eta_{pump}} = \frac{0.55v_r^2}{0.65} = 0.846\ v_r^2$$

$$\text{Output of engine} = \frac{\text{input to pump}}{\eta_{engine}}$$

$$= \frac{0.846v_r^2}{0.85} = 0.9955\ v_r^2$$

$$= 0.9955\times9^2 = 80.63 \text{ N.m/kg of water}$$

$$\text{Total input to engine} = (\rho A v_r)\times80.63$$

$$= (1000\times0.65\times9)\times80.63 = 471706 \text{ N. m/s}$$

Overall efficiency, $$\eta_{overall} = \frac{\text{work done by jets}}{\text{total input to engine}}$$

$$= \frac{117{,}000}{471{,}706}\times100 = 24.58\ \%$$

**Example 1.19** (a) A jet propelled ship discharging astern is having a relative velocity of 10 mps when moving at 30 km/h speed. The cross – sectional area of the jets at discharge is 230 cm$^2$.

Calculate resistance to the ship and propulsive work.

(b) If water enters through orifices facing the direction of motion of ship and pump is 80 percent efficient and frictional losses in the pipe are equivalent to 3.5 m of water head,

Calculate (i) power requires to drive the pump, and (ii) overall efficiency of propulsion.

**Solution.**

Given: $v_r = 10$ m/s, $u = \dfrac{30\times10^3}{3600} = 8.33$ m/s, $A = 230$ cm², $\eta_{pump} = 0.80$,

$h_f = 3.5$ m of water.

Propulsive force,
$$F = \rho A\,(v + u)v$$
$$v_r = v + u$$
$$v = 10 - 8.33 = 1.67 \text{ m/s}$$
$$F = 1000\times230\times10^{-4}\,(1.67+8.33)\times1.67$$
$$= 384.1 \text{ N}$$

$\therefore$ Resistance to motion $= 384.1$ N

Propulsive work $= F\times u = 384.18.33$

$= 3200$ N. m/s or 3.2 kW

$$\text{Energy supplied by jet} = \frac{1}{2}(\rho A v_r)\left(v_r^2 - u^2\right) + (\rho A v_r)gh_f$$

$$= \frac{1}{2}\times1000\times230\times10^{-4}\times10\left[10^2 - (8.33)^2\right] + 1000\times230\times10^{-4}\times10\times9.81\times3.5$$

$= 11417.3$ N. m/s or 11.42 kW

$=$ Pump output

$$\text{Power required to drive the pump} = \frac{\text{pump output}}{\eta_{pump}} = \frac{11.42}{0.8} = 14.28 \text{ kW}$$

Overall efficiency of propulsion,

$$\eta_{overall} = \frac{\text{propulsive power}}{\text{power supplied to pump}}$$

$$= \frac{3.2\times100}{14.28} = 22.4\ \%$$

## 1.8 Forces Due to Deviated Flow

***Flow round a pipe – bend and reducer***

Consider pipe in which a fluid is flowing with velocity $v_1$ and the pipe is inclined at an angle $\theta_1$ with x – axis. Let the pipe bend and now the velocity of flow is $v_2$ inclined at angle $\theta_2$ (Fig.1.21).

Dynamic force component in positive x–direction exerted on flowing fluid by the bend,

$$F_x = \rho Q\left(v_2\cos\theta_2 - v_1\cos\theta_1\right)$$

Where $Q$ = rate of flow in the pipe

$\rho$ = density of fluid

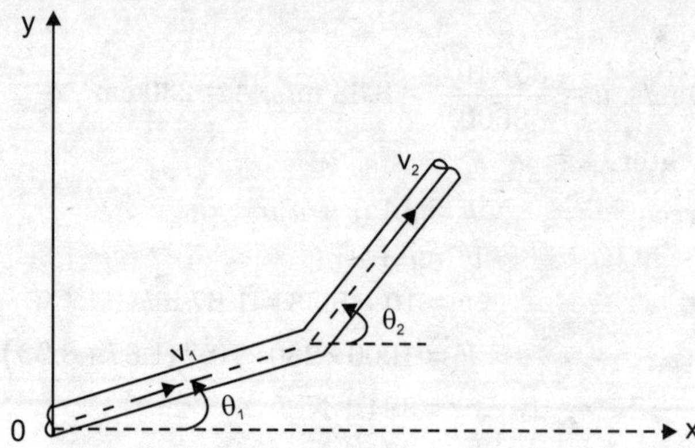


**Fig. 1.21 Flow in a pipe bend and reducer.**

Dynamic force in positive y-direction exerted on flowing fluid by The bend,

$$F_y = \rho Q\left(v_2 \sin\theta_2 - v_1 \sin\theta_1\right)$$

Resultant force acting on the flowing fluid,

$$F = \sqrt{F_x^2 + F_y^2}$$

Direction of resultant force with the horizontal,

$$\tan\phi = \frac{F_y}{F_x}$$

According to Newton's third law of motion, the flowing fluid will also exert a force equal in magnitude but opposite in direction on the bend.

## Summary

1. A fluid jet is a stream of fluid issuing from a nozzle with a high velocity thereby possessing high kinetic energy.
2. The force exerted by a jet of water on a stationary plate:

$F_x = \rho A v^2$ for a vertical plate

$= \rho A v^2 \sin^2\theta$ for an inclined plate

$= \rho A v^2 (1 + \cos^2\theta)$ for a curved plate when jet strikes at the centre

$= \rho A v^2 \cos\theta$ for a curved plate when jet strikes at the tip of vane

3. The force exerted by a jet of water on a moving plate:

$F_x = \rho A (v - u)^2$ for a moving vertical plate

$= \rho A (v - u)^2 \sin^2\theta$ for an inclined moving plate

$= \rho A (v - u)^2 (1 + \cos\theta)$ when jet strikes the vane at the centre

$= \rho A v(v - u)$ for series of plates mounted on a rotating wheel.

4. When a jet of water strikes a curved vane at one tip and leaves at the other tip:

Force exerted, $F_x = \rho A v_{r1} (v_{w1} \pm v_{w2})$

Work done per second $= \rho A v_{r1} (v_{w1} \pm v_{w2})u$

Take +ve sign for $\alpha_2 < 90°$ and –ve sign for $\alpha_2 > 90°$

Work done per second per newton of fluid

$$= \frac{1}{g}(v_{w1} + v_{w2})$$

For series of vanes: $F_x = \rho A v_1 (v_{w1} \pm v_{w2})$

Work done $= F_x \times u$

Work done per second per Newton of fluid $= \frac{1}{2}(v_{w1} + v_{w2})u$

5. For series of radial curved vanes:

Work done per second on the wheel

$$= \rho A v_1 (v_{w1} u_1 + v_{w2} u_2)$$

Efficiency of radial curved vanes $= \dfrac{2(v_{w1} u_1 + v_{w2} u_2)}{v_1^2}$

6. Jet propulsion of ships:

(a) Orifices at right angles to the direction of motion of the ship.

Propulsive force on the ship, $F = \rho A(v + u)v$

Thrust or propulsive work done per second, $W_p = \rho A(v + u)v \times u$

Propulsion efficiency, $\eta_p = \dfrac{2uv}{(v + u)^2}$

For propulsion efficiency to be maximum, $v = u$

$$(\eta_p)_{max} = \frac{1}{2} \text{ or } 50\ \%$$

(b) Orifices face the direction of motion of the ship.

Propulsion work, $W_\rho = \rho A(v + u)uv$

Propulsion efficiency, $\eta_p = \dfrac{2u}{v + 2u}$

For $u = \upsilon$, $\eta_p = \dfrac{2}{3}$ or 66.67 %

## Fill in the Blanks

1. Work done by a fluid jet impinging on a stationary plate is .................... .
2. When a fluid jet strikes a stationary inclined plate, the ratio of discharge is .................... .
3. The efficiency of a fluid jet impinging on a moving curved plate will be maximum when the jet velocity is .................. the vane velocity.
4. The maximum efficiency of a fluid jet impinging on a series of curved vanes occure when the peripheral speed of the wheel is .................. of the jet velocity.
5. Efficiency of a fluid jet striking moving curved vane is given by ................. divided by .................... .
6. Maximum efficiemc6y of jet propulsion of ships with inlet orifices at right angles to direction of motion of ship is ................ .
7. The condition for maximum propulsion efficiency of jet propulsion with inlet orifices at right angles to direction of motion of ship is .................. .
8. The efficiency of jet propulsion with inlet orifices at right angles to direction of motion of ship is .................... .
9. The efficiency of jet propulsion with inlet orifices facing the direction of motion of ship is .................... .
10. The maximum possible efficiency of jet propulsion with inlet orifices facing the direction of motion of ship is .................... .

## Answers

| | | | |
|---|---|---|---|
| 1. | Zero | 2. | $\dfrac{1+\cos\theta}{1-\cos\theta}$ |
| 3. | Three times | 4. | One – half |
| 5. | Work done, kinetic energy of jet | 6. | 50 % |
| 7. | $u = v$ | 8. | $\dfrac{2uv}{(v+u)^2}$ |
| 9. | $\dfrac{2u}{v+2u}$ | 10. | 66.67 % |

## Multi-Choice Questions

1. The force exerted by a jet of water striking a stationary vertical plate in the direction of jet is given by

   (a) $\rho Av$ (b) $\rho Av^2$

   (c) $\rho A^2v$ (d) $\rho Av^3$

2. The force exerted by a jet of water striking a moving vertical plate in the direction of motion of the plate is given by

   (a) $\rho Av^2$ (b) $\rho Av^3$

   (c) $\rho A(v-u)^2$ (d) $\rho A(v-u)^3$

3. When a steady jet impinges on a fixed inclined surface

   (a) The flow is divided into two parts which are proportional to the angle of inclination of the surface

   (b) No force is exerted on the inclined surface

   (c) The momentum component is unchanged parallel to the surface.

   (d) The flow is divided into two equal ports.

4. For maximum efficiency of a series of curved moving vanes, the speed is

   (a) Equal to the jet speed (b) $\frac{3}{4}$th of the jet speed

   (c) Half of the jet speed (d) $\frac{1}{3}$rd of the jet speed

5. The normal force exerted by a jet on a fixed inclined plate is

   (a) $\rho Av \sin\theta$ (b) $\rho Av^2 \sin\theta$

   (c) $\rho Av \cos\theta$ (d) $\rho Av^2 \cos\theta$

6. The force exerted by a jet of water of area $A$, velocity $v$ impinging on a series of flat plates moving with speed $u$ is

   (a) $\rho Av(v-u)$ (b) $\rho Av(v-u)^2$

   (c) $\rho Avu$ (d) $\rho A(v-u)^2$

7. The tangential force due to a jet of water of area A, velocity striking a fixed curved vane of inlet angle $\alpha_1$ and outer angle $\alpha_2$ is

   (a) $\rho Av^2(\cos\alpha_1 + \cos\alpha_2)$

   (b) $\rho Av^2(\cos\alpha_1 - \cos\alpha_2)$

   (c) $\rho Av^2(\sin\alpha_1 + \sin\alpha_2)$

   (d) $\rho Av^2(\sin\alpha_1 - \sin\alpha_2)$

8. The tangential force due to a jet of water of area $A$, velocity $\upsilon$, passing over a moving curved vane is

(a) $\rho Av(v_{w1}+v_{w2})$ (b) $\rho Av(v_{w1}-v_{w2})$

(c) $\dfrac{\rho Av^2(v_{w1}+v_{w2})}{v_1^2}$ (d) $\rho Av^2(v_{w1}-v_{w2})$

9. When a fluid jet impinges an inclined stationary plate, the jet divides into two parts whose ratio of discharge is

(a) $\dfrac{\cos\theta}{\sin\theta}$ (b) $\dfrac{1+\cos\theta}{1-\cos\theta}$

(c) $\dfrac{1+\cos\theta}{1+\cos\theta}$ (d) $\dfrac{1-\cos\theta}{1-\cos\theta}$

10. The efficiency of jet propulsion with orifices at right angles to the direction of motion of ship is

(a) $\dfrac{2u}{v+u}$ (b) $\dfrac{2u}{(v+u)^2}$

(c) $\dfrac{2uv}{(v+u)^2}$ (d) $\dfrac{uv}{(v+u)^2}$

11. The maximum efficiency of jet propulsion with inlet orifices at right angles to direction of motion of ship is

(a) 50 % (b) 60 %

(c) 66.67 % (d) 70 %

12. The efficiency of jet propulsion with inlet orifices facing the direction of motion of the ship is

(a) $\dfrac{2v}{v+u}$ (b) $\dfrac{2v}{v+2u}$

(c) $\dfrac{2uv}{v+u}$ (d) $\dfrac{uv}{v+u}$

---

## Answers

1. (b) 2. (c) 3. (a) 4. (c) 5. (b) 6. (a) 7. (a) 8. (a) 9. (b) 10. (c) 11. (a) 12. (b)

---

## Review Questions

1. Define fluid machinery. What are the types of fluid machinery?
2. What is a turbomachine? What are its types?
3. Explain the impulse-momentum principle. What are its applications?
4. Derive an expression for the force exerted by a jet of water on a fixed vertical plate in the direction of the jet.
5. Show that when a jet of water impinges on a series of curved vanes, maximum efficiency is obtained when the vane is semi-circular in section and the velocity of the vane is half that of the jet.
6. Derive an expression for the force exerted by a jet of water on moving inclined plate in the direction of jet.
7. Derive an expression for the efficiency of curved radial vanes.
8. What is jet propulsion? How do the ships move with this principle?
9. Derive an expression for the efficiency of propulsion when the inlet orifices face the direction of motion of the ship.
10. How is the force on the bend of a pipe-line running full of water determined?

## Numerical Problems

1. A jet of water 50 mm in diameter having a velocity of 10 m/s impinges on a stationary vertical flat plate. Find the force exerted by the jet on the plate and the work done.

   [**Ans.** 196.35 N, Zero]

2. A jet of water of diameter 75 mm having a velocity of 20 m/s strikes a fixed plate such that the angle between the jet and the plate is 60°. Calculate the force exerted by the jet on the plate in the direction (a) normal to the plate, and (b) of jet.

   [**Ans.** 1530 N, 1325N]

3. A jet of water 100 mm diameter and having a velocity of 30 m/s strikes tangentially on a wheel which deflects the jet through an angle of 120°. Calculate the hydrodynamic thrust on the vane when

   (a) The axis of symmetry of the vane is horizontal, and

   (b) The tangent at inlet tip is horizontal.

   [**Ans.** 1958 N, 30°; 1958 N, 30°]

4. A nozzle of 50 mm diameter delivers a jet of water at 20 m/s perpendicular to a plate that moves away from the jet at 5 m/s. Calculate

(a) The force on the plate

(b) The work done, and

(c) The efficiency of jet.

[**Ans.** 441.45 N, 2207 N. m/s, 28.1%]

5. A wheel consists of radial blades with inner and outer radii of 0.3 m and 0.6 m respectively. Water enters the blades at the outer periphery with a velocity of 50 m/s and the supply jet makes an angle of 25° with tangent to wheel at inlet tip. Water leaving the blades has a flow velocity of 10 m/s. If the blade angles at entrance and exit are 40° and 30° respectively, calculate: (a) work done per Newton of water, (b) speed of the wheel, and (c) efficiency of blading.

[**Ans.** 100.43 N.m, 320.4 rpm, 78.8%]

6. A jet of water 80 mm in diameter moving with a velocity of 12 m/s strikes a flat vane which is normal to the axis of the stream. Calculate:

   (a) The force exerted by the jet if the vane moves with a velocity of 5 m/s.

   (b) The force exerted by the jet if there is a series of vanes before the jet moving with velocity of 5 m/s.

[**Ans.** 246.23 N, 421.83 N]

7. A circular water jet having a cross-sectional area of 25 $cm^2$ moves with a velocity of 35 m/s and strikes tangentially a curved plate. The angle of curvature of plate at outlet is 120° with the x-axis. Assuming the plate to be frictionless, calculate the force exerted by the jet on the plate in x-direction, when:

   (a) the plate is stationary, and

   (b) the plate is moving in the direction of jet with velocity of 15 m/s.

[**Ans.** 4601 N, 1501 N]

8. A jet of water 50 mm in diameter impinges on a curved vane and is deflected through an angle of 175°. The vane moves in the same direction as that of the jet with a velocity of 35 m/s. The rate of flow is 170 litres/s. Determine the component of force on the vane in the direction of motion. Neglecting friction, calculate the power developed by the vane and the water efficiency.

[**Ans.** 10545.8 N,369.1 Kw, 57.6%]

9. A 75 mm diameter water jet having a velocity of 15 m/s impinges on the bucket of a wheel. The axis of the jet coincides with the axis of the bucket. The bucket is a part of a sphere of radius 200 mm, the depth being 100 mm. Determine the forces exerted by the jet on bucket, power developed and maximum efficiency, when:

(a) the bucket is fixed,

(b) the bucket is moving in the same direction as the jet with a velocity such that the work done per second by the jet on bucket would be maximum,

(c) there is a series of buckets in place of only one, and moving with a velocity such that the efficiency is maximum.

[**Ans.** 1857.1 N, Zero, Zero; 819.1 N, 4.09 kW, 55 %; 927 N, 6.95 kW, 93.5 %]

10. A water jet issues from a nozzle 50 mm diameter, strikes a plate with a velocity of 30 m/s. the plate makes an angle of 60° to the horizontal. If frictional loss is such as to reduce the velocity of stream leaving the plate to 25 m/s, calculate:

    (a) The component of force in the direction of jet,

    (b) The component of force normal to jet, and

    (c) The magnitude and direction of resultant force exerted by water jet.

    [**Ans.** 1030 N; 1275.3 N; 1638.3 N at 51°-12′ with direction of jet]

11. A jet of water 50 mm diameter having a velocity of 20 m/s, glides without shock into a series of smooth curved vanes moving in a direction parallel to the jet and away from it with a velocity of 8 m/s. The water is turned round through an angle of 15° as it passes over the vane. Calculate the pressure exerted on the vanes in the direction of rotation, and power developed.

    If for any reason the blade is held stationary, what will be the pressure on it?

    [**Ans.** 873.1 N, 6.985 kW; 1461.7 N]

12. A series of curved vanes (entrance angle 30° and exit angle 5°) deflect a jet of water 10 sq. cm in area of cross-section, moving at 50 m/s and inclined at 15° to the line of motion of the vane. Determine:

    (a) The velocity of vane to avoid shock at entry,

    (b) The magnitude and direction of the resultant force on the vane and force in the direction of motion, and

    (c) The magnitude and direction of velocity at exit.

    [Ans. 25.8 m/s; 2383.8 N at 7.5° and 2364.2 Nin the direction of motion; 6.7 m/s at 82.5°]

13. A jet of water having a velocity of 35 m/s impinges on a series of vanes moving with a velocity of 20 m/s. The jet makes an angle of 30° to the direction of motion of vanes when entering and leaves at an angle of 120°. Find:

    (a) The angles of vane tips so that water enters and leaves without shock,

(b) The work done per kg of water entering the vanes, and

(c) The efficiency.

[**Ans.** 120.1°, 1.6°; 600 N.m/s; 98.6 %]

14. A motor-boat with jet propulsion draws 300 lit/s through orifices amid ships and discharges it astern through orifices having an effective area of cross-section of 0.05 $m^2$. If the boat travels at 4.5 m/s, calculate the propelling force.

[**Ans.** 449.3 N]

15. In a jet propelled boat, the resistance to the motion is 3924 N. The boat is provided with two jets each having an area of 0.02 $m^2$. The water is drawn through orifices amid ship. If the boat travels at a speed of 6 m/s, determine the quantity of water to be pumped per second and the efficiency of jet propulsion.

[**Ans.** 0.534 $m^3$/s, 49.5 %]

CHAPTER

# 2 Whirling of Fluids

## 2.1 Types of Fluid Motion

A fluid may possess at any instant one or more types of motion as given below in Fig. 2.1 .

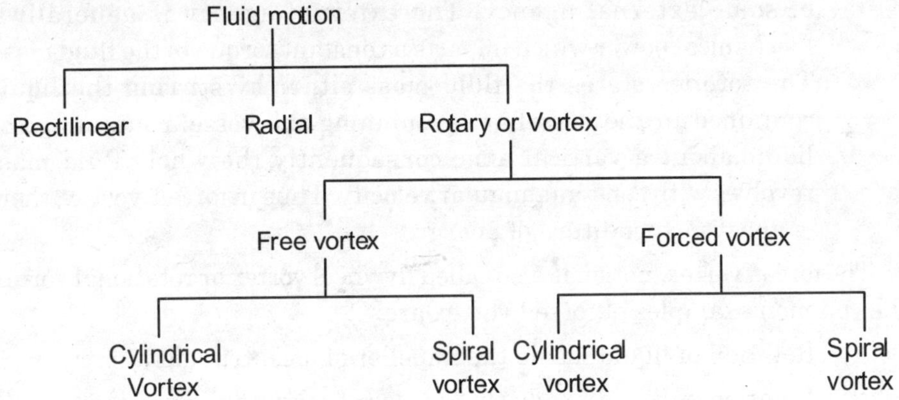


**Fig. 2.1 Types of fluid motion.**

1. **Rectilinear motion**: This type of motion of the fluid is one – dimensional in which stream lines are parallel.
2. **Radial motion**: It is defined as one in which the fluid flows in radial direction in such a way that pressure and velocity at any point change with respect to distance of that point from the central axis only. In this case, the velocity of flow is in radial direction. The flow may take place radially inward to or outward from the centre. The flow is two – dimensional and streamlines are not parallel.
3. **Rotary or vortex motion**: The vortex motion is defined as a motion in which the whole fluid mass rotates about an axis. A mass of fluid in rotation about a fixed axis is called vortex. Rotary motion of fluid is also called vortex motion. In this case, the rotating fluid particles have velocity in tangential direction. The flow lines are curved stream lines. The vortex motion is of two types: free vortex and forced vortex.

(i) *Free vortex* – is that in which the fluid mass rotates without any external impressed contact force. The whole fluid mass rotates either due to fluid pressure itself or the gravity or due to rotation previously imparted. Energy is not expended by any outside source.

The free vortex motion is also called potential vortex or irrotational vortex. The common examples of free vortex are:

1. Flow of liquid in a centrifugal pump casing after it has left the impeller.
2. Flow of water in a turbine casing before it enters the guide vanes.
3. Flow around a circular bend.
4. Whirlpool in a river.
5. Rotary flow observed in a shallow vessel, such as wash basin or bath tub while draining liquid through the outlet at the bottom.

(ii) *Forced vortex* – is one in which fluid mass is made to rotate by means of some external agency. The external agency is generally a mechanical power which imparts a constant torque to the fluid mass. The torque rotates the fluid mass either by stirring the liquid contained in the vessel or by spinning the vessel containing the liquid,about a vertical axis; consequently the whole fluid mass revolves with constant angular velocity. Thus in forced vortex, there is always expenditure of energy.

The forced vortex motion is also called flywheel vortex or rotational vortex. The common examples of forced vortex are:

1. Rotation of liquid inside the impeller of a centrifugal pump.
2. Rotation of liquid inside the runner of a hydraulic turbine.
3. Rotation of liquid in a vertical cylindrical vessel.

The free or forced vortex is further classified as cylindrical or spiral vortex.

(a) *Cylindrical Vortex* – is one in which the fluid mass rotates in concentric circles. Hence, it is also known as circulatory flow.

(b) *Spiral Vortex* – When cylindrical vortex is superimposed on the radial flow, the resulting vortex is known as spiral vortex. Its common example is motion of liquid in spiral or volute casing of a centrifugal pump.

## 2.2 Radial Flow

Radial flow can be produced by constraining a stream of fluid to flow between two parallel circular plates, the opening being at the centre and the periphery. Fig. 2.2 shows two parallel circular plates in which the lower plate has an opening at the centre of radius $r_1$. A pipe of same radius $r_1$ is connected to the opening. If liquid is inserted in the pipe then it will flow radially outward between the two parallel plates. If the liquid is pushed in at the periphery between the two plates, it will flow radially inward to the centre of the pipe.

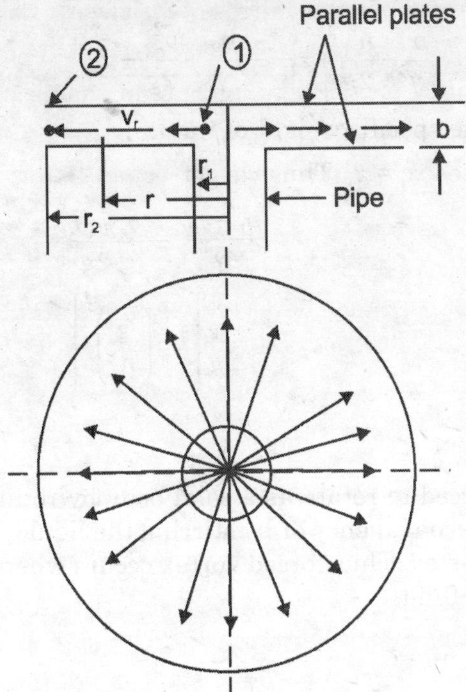


**Fig. 2.2 Radial flow.**

Let $b$ = distance between the two parallel circular plates.

$r_1$ = radius of pipe

$r_2$ = outer radius of plates

$v_r$ = radial velocity at a distance r from the centre

$Q$ = quantity of fluid flowing per second.

Then area across flow, $A_f = 2\pi rb$

Applying continuity equation, we have

$$Q = A_f v_r = 2\pi rb v_r$$

For constant value of $Q$,

$$rbv_r = \text{constant}$$

or

$$v_r \propto \frac{1}{r}$$

i.e. radial velocity of flow is inversely proportional to the radius of circular parallel plates through which the radial flow takes places.

Consider two points 1 and 2 at the entrance and exit from the plates. Let $p_1$ and $p_2$ be the pressure and $v_{r1}$ and $v_{r2}$ be the radial velocity at points 1 and 2 respectively.

Applying Bernoulli's theorem between the two points, we have

$$\frac{p_1}{\gamma}+\frac{v_{r1}^2}{2g}+z_1 = \frac{p_2}{\gamma}+\frac{v_{r2}^2}{2g}+z_2$$

Where $\gamma = \rho g$, is the specific weight of fluid.

For horizontal plates, $z_1 = z_2$. Thus we get

$$\frac{p_1 - p_2}{\gamma} = \frac{v_{r2}^2 - v_{r1}^2}{2g}$$

$$= \frac{v_{r2}^2}{2g}\left[1-\left(\frac{r_2}{r_1}\right)^2\right] \qquad \left[\because v_r \propto \frac{1}{r}\right]$$

## 2.3 Forced Vortex

When a fluid is forced to rotate as a solid body by rotating the containing vessel by some external agency or by stirring the liquid, it will get a motion known as forced vortex. Thus forced vortex occurs when constant torque is maintained on the fluid.

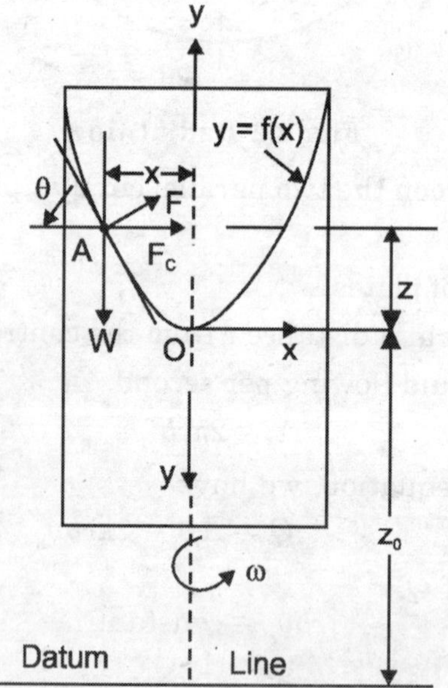


Fig. 2.3 Forced cylindrical vortex.

### 2.3.1 Cylindrical forced vortex

In a cylindrical forced vortex, the liquid is rotated in a closed cylindrical container by supplying energy from an external source. Let the section of liquid surface be any curve $y = f(x)$ and let $A$ be any point on it given by $(x, y)$ measured from the origin at the lowest point on the curve.

Consider the equilibuium of a small mass $m$ of the liquid at a point $A$ on free surface. The forces acting on this elementary mass are:

1. Weight of liquid mass acting vertically downward,

$$W = mg$$

2. Centripetal force acting horizontally,

$$F_c = m\omega^2 x$$

Where $\omega = \dfrac{2\pi N}{60}$ is the angular speed of cylindrical vessel

and $N = rpm$

3. Fluid pressure force acting normal to the surface or tangent at point $A$.

Let $\theta$ = angle made by the tangent at $A$ with $x$-axis.

Resolving forces along the tangent, we have

$$m\omega^2 x \cos\theta = mg \sin\theta$$

Or

$$\tan\theta = \frac{\omega^2 x}{g}$$

Now for the assumed curve $y = f(x)$, $\tan\theta = \dfrac{dy}{dx}$

$$\because \quad \frac{dy}{dx} = \frac{\omega^2 x}{g}$$

Integrating, we get

$$y = \frac{\omega^2 x^2}{2g} + C$$

where C = constant of integration

At $x = 0$, $y = 0$, therefore, $C = 0$

Or

$$y = \frac{\omega^2 x^2}{2g}$$

Or

$$x^2 = \frac{2g}{\omega^2} \cdot y$$

This is the equation of a parabola touching the origin and symmetrical about $y$-axis. The surface of liquid is generated by the revolution of parabola about the $y$-axis. This surface is called the paraboloid of revolution.

If $x$ and $y$ are replaced by $r$ and $z$ respectively, then

$$r^2 = \frac{2g}{\omega^2} \cdot z$$

Or

$$z = \frac{\omega^2 r^2}{2g} = \frac{v^2}{2g}$$

Where $v = \omega r$ is the tangential velocity of particle.

Let $z_o$ = height of origin O from some datum line

Then $$z = \frac{\omega^2 r^2}{2g} + z_o = \frac{v^2}{2g} + z_o$$

Or $$z_2 - z_1 = \frac{v_2^2 - v_1^2}{2g}$$

### 2.3.2 Forced vortex in open vessel

In this case, liquid is free to form a paraboloid of revolution as shown in Fig. 2.4.

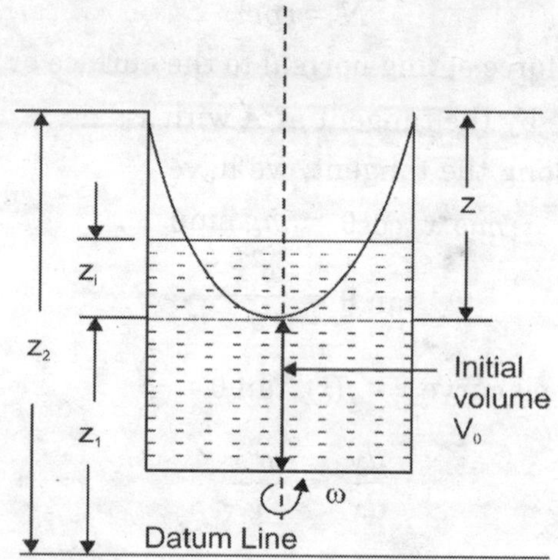


**Fig. 2.4 Forced vortex in open vessel.**

Let $z_i$ = depression of level at the centre of vessel due to formation of forced vertex.

$z$ = height of the paraboloid.

$V_o$ = initial volume of liquid in the vessel upto centre of depression

$r$ = radius of the vessel

For volume constancy, we have

Volume of liquid in vessel before rotation = Volume of liquid in vessel after rotation.

$$V_o + \pi r^2 z_i = V_o + \frac{1}{2}\pi r^2 z$$

$\because$ $$z_i = \frac{z}{2} = \frac{1}{2}(z_2 - z_1)$$

Now $$r^2 = \frac{2g}{\omega^2} z = \frac{2g}{\omega^2} \times 2z_i = \frac{4g}{\omega^2} z_i$$

$\therefore$ $$z_i = \frac{\omega^2}{4g} \cdot r^2 = k\omega^2$$

Where $k = \frac{r^2}{4g}$ is a constant for a vessel.

Or $$z_i \alpha \omega^2$$

Thus the depression of vortex below the original level of still liquid is proportional to square of angular speed of rotation.

### 2.3.3 Forced vortex in closed vessel

The forced vortex in a closed vessel is shown in Fig. 2.5. Let the liquid touch the lid when the angular speed is *w*.

Then $$r^2 = \frac{2g}{\omega^2} \cdot z$$

If the speed is increase further to $\omega^1$, then

$$(r')^2 = \frac{2g}{(\omega')^2} \cdot z'$$

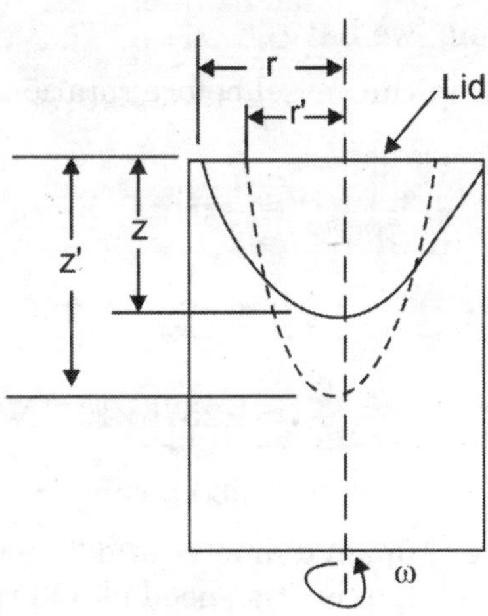


**Fig. 2.5 Forced vortex in closed vessel.**

Since volume of liquid remains the same, the volume of air enclosed also remains the same.

$\therefore$ $$\frac{1}{2}\pi r^2 z = \frac{1}{2}\pi (r')^2 (z')$$

Or $$r^2 z = (r')^2 \, . \, z' = \frac{2g}{(\omega')^2} \cdot (z')^2$$

Or $$(z')^2 = \frac{r^2 z}{2g}\left(\omega^1\right)^2 = k^2 \, . \, (\omega')^2$$

Or $$z' = k.\omega'$$

Where $k = \sqrt{\frac{r^2 z}{2g}}$, is a constant

**Example 2.1** An open cylindrical vessel 0.3m internal diameter and 1m long is filled with water upto a height of $\frac{2}{3}m$. Calculate the speed of rotation about its vertical axis when the water is just going to spill out of vessel.

**Solution.**

Given: $r = 0.15\text{m}$, $z_{empty} = 1-\frac{2}{3} = \frac{1}{3}m$

$$\text{Angular speed of rotation, } \omega = \frac{2\pi N}{60} \text{ rad/s}$$

$$\text{Centrifugal head, } z = \frac{v^2}{2g} = \frac{\omega^2 r^2}{2g}$$

$$= \frac{\left(\frac{2\pi N}{60}\right)^2 \times (0.15)^2}{2\times 9.81} = 1.2576 \times 10^{-5} N^2 \text{ m}$$

At the instant of spilling, we have

Volume of empty space in the vessel before rotation = volume of paraboloid of revolution

$$\pi r^2 z_{empty} = \frac{1}{2}\pi r^2 z$$

$$z = 2\, z_{empty} = 2\times\frac{1}{3} = \frac{2}{3}m$$

$$\therefore \quad \frac{2}{3} = 1.2576\times 10^{-5} N^2$$

$$N = 230.2 \text{ rpm}$$

**Example 2.2** A cylinder 15m in diameter and 37.5m high containing water is rotated about its vertical axis at a speed of 320 rpm, so that a portion of water spills out.

(a) If the cylinder is now brought to rest, what would be the depth of water in it?

(b) If now the speed is increased to 600 rpm, how much water will be left in the vessel?

**Solution.**

Given: $r = 7.5$ cm, $h = 37.5$ cm, $N = 320$ rpm

(a) Angular velocity of vessel, $\omega = \frac{2\pi N}{60} = \frac{2\pi\times 320}{60} = 33.51 \text{ rad/s}$

Peripheral velocity of vessel, $v = \omega r = 33.51\times 7.5\times 10^{-2} = 2.513 \text{ m/s}$

Centrifugal head generated, $z = \frac{v^2}{2g} = \frac{(2.513)^2}{2\times 9.81} = 0.322 \text{ m}$

Volume of water spilled out, $V$ = volume of paraboloid of revolution

$$= \frac{1}{2}\pi r^2 z = \frac{1}{2}\times\pi\times\left(7.5\times10^{-3}\right)\times0.322$$

$$= 2.845 \times 10^{-3} \text{ m}^3$$

Initial volume of water in cylinder when full,

$$V_o = \pi r^2 h = \pi\left(7.5\times10^{-2}\right)^2 \times37.5\times10^{-2}$$

Remaining water, $V_f = V_o - V = (6.626-2.845)\times10^{-3}$

$$= 6.626 \times 10^{-3} \text{ m}^3$$

$$= 3.781 \times 10^{-3} \text{ m}^3$$

Depth of water after the vessel comes to rest $= \dfrac{V_f}{\pi r^2}$

$$= \frac{3.781\times10^{-3}}{\pi\left(7.5\times10^{-2}\right)^2} = 0.214 \text{ m or } 21.4 \text{ cm}$$

(b) Centrifugal head generated at 600 rpm,

$$z = \frac{\left(\dfrac{2\pi rN}{60}\right)^2}{2g} = \frac{\left(\dfrac{2\pi\times7.5\times10^{-2}\times600}{60}\right)^2}{2\times9.81}$$

$$= 1.132 \text{ m}$$

The shape of free surface will be as shown in Fig. 2.6.

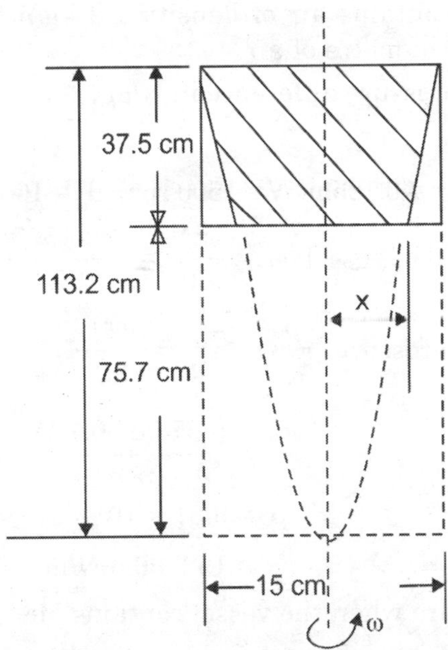


**Fig. 2.6 Forced vortex in open vessel.**

Volume of water spilled out = Hatched area in Fig. 2.6

Let $x$ = radius of parabola at the bottom of vessel

$$h = 113.2 - 37.5 = 75.7 \text{ cm}$$

$$x = \frac{\sqrt{2gh}}{\omega} = \frac{\sqrt{2\times 9.81\times 75.7\times 10^{-2}}}{\left(\frac{2\pi\times 600}{60}\right)}$$

$$= 0.0613 \text{ m}$$

$$\text{Volume of water spilled out} = (\text{total volume of paraboloid}) - (\text{volume of paraboloid below the bottom of vessel})$$

$$= \left[\frac{1}{2}\times\pi\times\left(7.5\times 10^{-2}\right)^2\times 1.132\right] - \left[\frac{1}{2}\times\pi\times(0.0613)^2\times 0.757\right]$$

$$= 0.010002 - 0.004468$$

$$= 5.534 \times 10^{-3} \text{ m}^3$$

$$\text{Remaining water} = \pi(7.5\times 10^{-2})^2 \times 37.5\times 10^{-2} - 5.534\times 10^{-3}$$

$$= 1.0928 \times 10^{-3} \text{ m}^3$$

**Example 2.3** A cylindrical vessel 0.25m internal diameter is completely filled with water and is rotated about its vertical axis at 1500 rpm. Calculate the difference of pressure between its circumference and the centre in metre of water of density 1000 kg/m³.

If the vessel now contains air of density 1.2 kg/m³, what would be the pressure difference in metre of air?

Also calculate the pressure difference in N/m².

**Solution.**

Given: $d = 0.25$m or $r = 0.125$m, $N = 1500$ rpm, $\rho_\omega = 1000$ kg/m³, $\rho_a = 1.2$ kg/m³

$$\text{Angular speed of vessel, } \omega = \frac{2\pi N}{60} = \frac{2\pi\times 1500}{60} = 157.08 \text{ rad/s}$$

$$\text{Difference of pressure, } \frac{p}{\gamma_w} = \frac{v^2}{2g} = \frac{(\omega r)^2}{2g}$$

$$= \frac{(157.08\times 0.125)^2}{2\times 9.81} = 19.65\text{m of } H_2O$$

Or

$$p = 9.81\times 10^3\times 19.65$$

$$= 192766.5 \text{ N/m}^2$$

Difference of pressure when the vessel contains air,

$$\frac{p}{\gamma_a} = \frac{v^2}{2g} = 19.65 \text{ m of air}$$

Or $$P = 1.2 \times 10^3 \times 19.65$$

$$= 23580 \text{ N/m}^2$$

**Example 2.4** A closed cylindrical vessel 15 cm internal diameter and 1m long, stands vertically. It is filled with water upto a height of $\frac{2}{3}$ m and is rotated about its vertical axis. Calculate the speed of rotation of the vessel, when:

(a) The top of the paraboloid formed by water just touches the top lid of the vessel, and

(b) The bottom of the paraboloid formed by water just touches the bottom of the vessel.

**Solution.**

Given: $r = 0.075\text{m}$, $l = 1\text{m}$, $z_i = 1 - \frac{2}{3} = \frac{1}{3} m$

(a) When top of paraboloid just touches the lid (Fig. 2.7)

Volume of hollow paraboloid = volume of empty space in the vessel before rotation

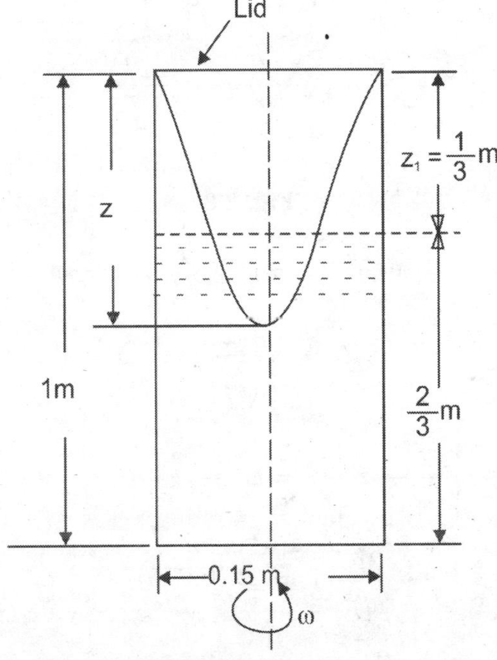


**Fig. 2.7 Forced vortex in closed vessel.**

$$\frac{1}{2}\pi r^2 z = \pi r^2 z_i$$

$$z = 2z_i = 2\times\frac{1}{3} = \frac{2}{3}m$$

Centrifugal head developed by rotation,

$$z = \frac{\omega^2 r^2}{2g} = \frac{\left[\frac{2\pi N}{60}\times 0.075\right]^2}{2\times 9.81}$$

$$= 3.144 N^2 \times 10^{-5} = \frac{2}{3}$$

$$N = 460.5 \text{ rpm}$$

(b) When bottom of paraboloid touches the bottom of vessel (Fig. 2.8)

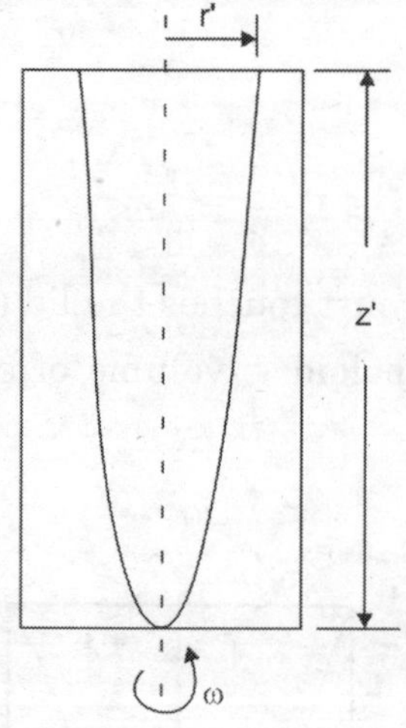


**Fig. 2.8**

$$\frac{1}{2}\pi(r')^2 . z' = \pi r^2 . z_i , \quad z_i = \frac{1}{3}m$$

$$(r')^2 = \frac{2g}{(\omega')^2}\cdot z' , \quad z' = 1m$$

$$\therefore \quad \frac{1}{2}\times\pi\left[\frac{2g}{(\omega')^2}\cdot z'\right]\cdot z = \pi r^2 z_i$$

$$\frac{2g}{(\omega')^2}\cdot 1 = 2\times(0.075)^2\times\frac{1}{3}$$

$$(\omega')^2 = \frac{2\times 9.81\times 3}{2\times(0.075)^2} = 5232$$

$$\omega' = 72.33 \text{ rad/s}$$

$$N = \frac{60 \times 72.83}{2\pi} = 690.7 \text{ rpm}$$

**Example 2.5** Find the maximum speed of an open circular cylinder having 200 mm diameter, 1.25m long and containing water upto a height of 1m, at which it should be rotated about its vertical axis so that no water spills out. Also find the speed at which axial depth becomes zero.

**Solution.**

Given: $r$ = 100mm or 0.1m, $l$ = 1.25m, $h$ = 1m, $z_i$ = 1.25 – 1.0 = 0.25m

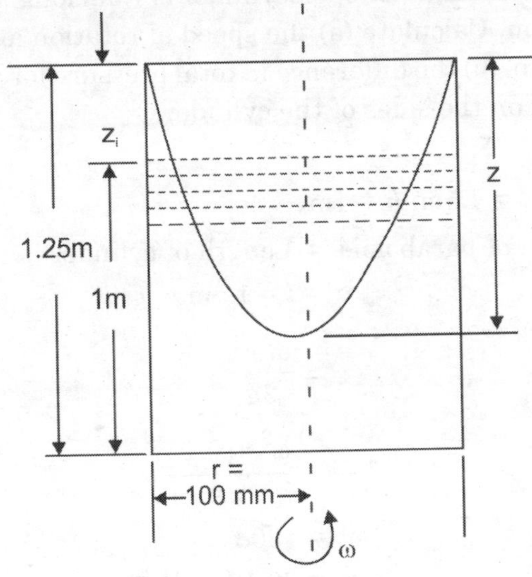


**Fig. 2.9**

From Fig. 2.9, we have

Hollow volume of paraboloid = Empty volume in cylinder before rotation

$$\frac{1}{2}\pi r^2 z = \pi r^2 z_i$$

$$z = 2z_i = 2 \times 0.25 = 0.50\text{m}$$

Now $$z = \frac{\omega^2 r^2}{2g} = \frac{\left(\frac{2\pi N r}{60}\right)^2}{2g}$$

$$= \frac{\left(\frac{2\pi N \times 0.1}{60}\right)^2}{2 \times 9.81} = 0.50$$

$$N = 299.1 \text{ rpm}$$

When axial depth becomes zero then depth of paraboloid = 1.25m

$$z = \frac{\omega^2 r^2}{2g} = \frac{\left(\frac{2\pi N r}{60}\right)^2}{2g}$$

$$1.25 = \left(\frac{2\pi r}{60}\right)^2 \times \frac{N^2}{2g} = \left(\frac{2\pi \times 0.1}{60}\right)^2 \times \frac{N^2}{2 \times 9.81}$$

$$N^2 = 223641.2$$

$$N = 472.9 \text{ rpm}$$

**Example 2.6** A 0.3m diameter open cylinder is 1.5m long. It contains water upto a height of 1m. Calculate (a) the speed of rotation at which the axial depth becomes zero, (b) the difference in total pressure force due to rotation at the bottom and on the sides of the cylinder.

**Solution.**

Given: $r = 0.15$m, $l = 1.5$m, $h = 1$m

(a) Depth of paraboloid = Length of cylinder

$$z = l = 1.5\text{m}$$

$$z = \frac{\omega^2 r^2}{2g}$$

$$1.5 = \frac{\omega^2 \times (0.15)^2}{2 \times 9.81}$$

$$\omega^2 = 1308$$

$$\omega = 36.17 \text{ rad/s}$$

$$N = 345.4 \text{ rpm}$$

(b) Total pressure force at the bottom of cylinder before rotation,

$$F_i = \gamma A \bar{h}$$

Where $\gamma = \rho g = 10^3 \times 9.81 = 9810$ N/m$^3$

$$A = \text{area of bottom} = \pi r^2$$

$$= \pi (0.15)^2 = 0.07068 \text{ m}^2$$

$$\bar{h} = 1\text{m}$$

$$F_i = 9810 \times 0.07068 \times 1 = 693.43 \text{ N}$$

After rotation, the depth of water at the bottom is not constant. Therefore, the pressure force is not constant at the bottom. To calculate the pressure force, consider an elementary ring of water of radius $r$ and width $dr$ at a height $z$ from the bottom (Fig. 2.10).

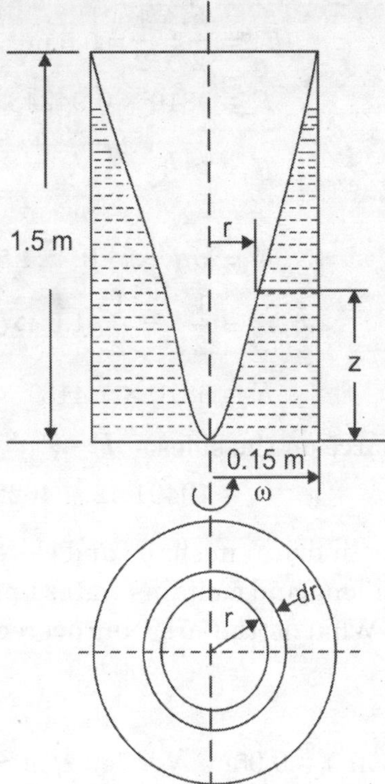


Fig. 2.10

Hydrostatic force of the ring on the bottom,

$$dF = \gamma \times 2\pi r.\, dr \times z$$

$$= 9810 \times 2\pi r.dr \times \frac{\omega^2 r^2}{2g}$$

$$= \left[\frac{9810 \times 2\pi \times 1308}{2 \times 9.81}\right] \times r^3\, dr$$

$$= 4109203.2 \times r^3 .\, dr$$

$$\text{Total pressure force, } F = \int_0^{0.15} 4109203.2 \times r^3\, dr$$

$$= \frac{4109203.2}{4} \times \left| r^4 \right|_0^{0.15} = 520.07 \text{ N}$$

Difference in pressure force at the bottom = $F_i - F$

$$= 693.43 - 520.07 = 173.36 \text{ N}$$

Total pressure force on the sides before rotation,

$$F_i = \gamma A \bar{h}$$

Where $\gamma = 9810$ N/m³

$$A = \pi dh = \times 0.3 \times 1.0 = 0.94248 \text{ m}^2$$

$$\bar{h} = \frac{h}{2} = \frac{1}{2} = 0.5 \text{ m}$$

$$F = 9810 \times 0.94248 \times 0.5 = 4622.85 \text{ N}$$

After rotation, $F = \gamma A \bar{h}$

Where $\gamma = 9810 \text{ N/m}^3$

$$A = \pi dl = \pi \times 0.3 \times 1.5 = 1.4137 \text{ m}^2$$

$$\bar{h} = \frac{1}{2} l = \frac{1}{2} \times 1.5 = 0.75 \text{ m}$$

$$F = 9810 \times 1.4137 \times 0.75 = 10401.42 \text{ N}$$

Difference in pressure force on the sides = $F - F_i$

$$= 10401.42 - 4622.85 = 5778.57 \text{ N}$$

**Example 2.7** A vessel cylindrical in shape and closed by a lid at the top is 20 cm in diameter, 1.4m long and contains water upto a height of 0.95m. It is rotated at 740 rpm. What is the area uncovered at the bottom of the vessel.

**Solution.**

Given: $r = 0.1\text{m}$, $l = 1.4\text{m}$, $h = 0.95\text{m}$, $N = 740$ rpm

Angular speed of vessel, $\omega = \frac{2\pi N}{60}$

$$= \frac{2\pi \times 740}{60} = 77.5 \text{ rad/s}$$

Height of paraboloid if the vessel is open at the top,

$$z = \frac{\omega^2 r^2}{2g} = \frac{(77.5)^2 \times (0.1)^2}{2 \times 9.81} = 3.06 \text{ m}$$

In Fig. 2.11, we have $z_1 + 1.4 + z_2 = 3.06$

Or $z_1 + z_2 = 1.66$ m ...(1)

For the paraboloid GOH, we have

$$1.4 + z_1 = \frac{\omega^2 r_1^2}{2g}$$

$$= \frac{(77.5)^2 \times r_1^2}{2 \times 9.81} = 306.13 \times r_1^2 \quad \text{...(2)}$$

For the paraboloid IOJ, we have

$$z_1 = \frac{\omega^2 r_2^2}{2g}$$

$$= \frac{(77.5)^2 \times r_2^2}{2 \times 9.81} = 306.13 \times r_2^2 \quad \text{...(3)}$$

Also volume of air before rotation = volume of air after rotation

$$\pi r^2 (l-h) = \text{volume of paraboloid GOH} - \text{volume of paraboloid IOJ}$$

$$\pi\times(0.1)^2\times(1.4-0.95) = \frac{1}{2}\pi r_1^2(1.4+z_1)-\frac{1}{2}\pi r_2^2 z_1$$

$$0.0045\times 2 = r_1^2(1.4+z_1)-r_2^2 z_1$$

Or

$$0.009 = 1.4r_1^2+\left(r_1^2-r_2^2\right)z_1 \quad \text{...(4)}$$

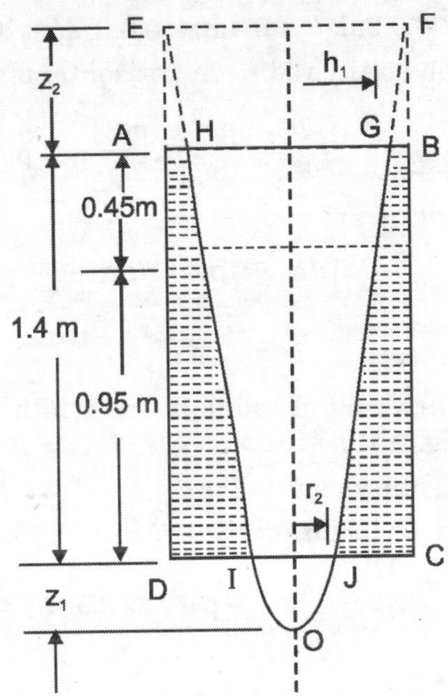


**Fig. 2.11**

From Eq. (2), we have
$$r_1^2 = \frac{1.4+z_1}{306.13}$$

From Eq. (3), we have
$$r_2^2 = \frac{z_1}{306.13}$$

Substituting in Eq. (4), we have

$$0.009 = \frac{1.4(1.4+z_1)}{306.13}+\left[\frac{(1.4+z_1)-z_1}{306.13}\right]z_1$$

$$0.009\times 306.13 = 1.96 + 1.4\,z_1 + 1.4\,z_1$$

$$2.8\,z_1 = 0.7952$$

$$z_1 = 0.284 \text{ m}$$

$$r_2^2 = \frac{0.284}{306.13} = 9.276\times10^{-4}$$

$$r_2 = 0.03045 \text{ m}$$

Area uncovered at the base $= \pi r_2^2 = \pi \times 9.276 \times 10^{-4}$

$$= 2.914 \times 10^{-3} \text{ m}^2 \text{ or } 0.0029 \text{ m}^2$$

**Example 2.8** A closed vertical cylinder 0.4m in diameter and 0.4m in height is completely filled with an oil of specific gravity 0.80. If the cylinder is rotated about its vertical axis at 200 rpm, calculate the thrust of oil on the top and bottom covers of the cylinder.

**Solution.**

Given: $D = 0.4$m or $R = 0.2$m, $h = 0.4$m, $s = 0.8$, $N = 200$ rpm

The pressure variation at any radius in horizontal plane is given as:

$$\frac{\partial p}{\partial r} = \frac{\rho v^2}{r} = \frac{\rho \omega^2 r^2}{r} = \rho \omega^2 r$$

Integrating both sides, we get

$$\int dp = \int \rho \omega^2 r \, dr$$

Or

$$p = \frac{1}{2} \rho \omega^2 r^2$$

Consider an elementary ring of radius $r$ and width $dr$ on the top of the vessel as shown in Fig. 2.12.

Area of elementary ring $= 2\pi r \, dr$

Force on the elementary ring, $dF = p \times 2\pi r \, dr$

$$= \frac{1}{2} \rho \omega^2 r^2 \times 2\pi r \, dr = \pi \rho \omega^2 r^3 \, dr$$

Total force on top of the vessel, $F_{top} = \pi \rho \omega^2 \int_0^R r^3 dr$

$$= \frac{\pi}{4} \rho \omega^2 \Big|_2^0 r^4 \Big|_0^R = \frac{\pi}{4} \rho \omega^2 R^4$$

Total pressure force on the bottom of vessel,

$$F_{bottom} = F_{top} + \text{weight of water in cylinder}$$

$$= \frac{\pi}{4} \rho \omega^2 R^4 + \gamma \times \pi R^2 h$$

Now

$$\omega = \frac{2\pi N}{60} = \frac{2\pi \times 200}{60} = 20.944 \text{ rad/s}$$

$$F_{top} = \frac{\pi}{4} \rho \omega^2 R^4$$

$$= \frac{\pi}{4} \times 1000 \times 0.8 \times (20.944)^2 \times (0.2)^4$$

$$= 440.98 \text{ N}$$

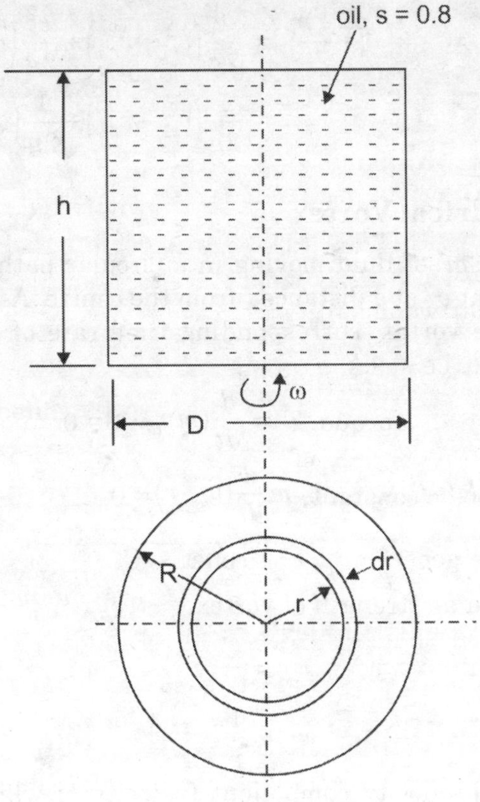


**Fig. 2.12**

$$F_{bottom} = F_{top} + \gamma \times \frac{\pi D^2}{4} \times h$$

$$= 440.98 + 9.81 \times 1000 \times 0.8 \times \frac{\pi}{4} (0.4)^2 \times 0.4$$

$$= 440.98 + 394.48$$

$$= 835.46 \text{ N}$$

## 2.4 Spiral Vortex

The flow of fluid in runners of turbines and pumps is usually a combination of (i) Circulatory or cylindrical forced flow, i.e. flow in concentric circles, and (ii) Radial flow i.e. flow involving a change of distance from the axis of rotation. The path resulting from the superimposition of these two flows is of the form of spiral.

The pressure difference between any two points in spiral forced vortex taken on the same horizontal plane, is found by adding the pressure difference due to radial flow and that due to cylindrical forced vortex, both considered separately, i.e.

$$\frac{p_2 - p_1}{\gamma} = \frac{1}{2g}\left(v_{r1}^2 - v_{r2}^2\right) + \frac{w^2}{2g}\left(r_2^2 - r_1^2\right)$$

$$= \frac{1}{2g}\left(v_{r1}^2 - v_{r2}^2\right) + \frac{1}{2g}\left(v_2^2 - v_1^2\right)$$

## 2.5 Free Cylindrical Vortex

Consider a mass '*m*' of fluid moving in a circular path with a tangential velocity component $v_w$ at a distance *r* from the centre. As no external torque is applied in free vortex, corresponding time rate of change of angular momentum is zero, i.e.,

$$\text{Torque, } T = \frac{d}{dt}(mv_w \cdot r) = 0$$

For mass 'm' to be constant, $m\frac{d}{dt}(v_w \cdot r) = 0$

Integrated, we get $v_w \cdot r = \text{const.} = \text{C}$

Where *C* is known as strength of vortex.

$$\therefore \quad v_w = \frac{C}{r}$$

$$\text{Or} \quad v_w \propto \frac{1}{r}$$

i.e. tangential velocity component (velocity of whirl) $v_w$ is inversely proportional to radius *r*.

In case of forced vortex, tangential velocity,

$$v = \omega r = \text{const.} \times r = C_1 \times r$$

And in radial flow, $v_w \cdot r = \text{const.} = C_2$

$$\therefore \quad v_w = \frac{C_2}{r}$$

Therefore, in case of free vortex, the velocity is in tangential direction, i.e. velocity of whirl, while in radial motion the velocity is in radial direction, i.e. velocity of flow.

In free vortex no energy is imparted to or extracted from the fluid, the total head at any point is constant throughout. Therefore, Bernoulli's theorem holds good, if there are no losses. Hence,

$$\frac{p}{\gamma} + \frac{v_w^2}{2g} + z = \text{const.}$$

$$\text{Or} \quad \frac{p_1}{\gamma} + \frac{v_{w1}^2}{2g} + z_1 = \frac{p_2}{\gamma} + \frac{v_{w2}^2}{2g} + z_2$$

Discharge, $Q = \int v_w \cdot dA = \int \frac{c}{r} dA$

Difference of pressure, $\frac{p_2 - p_1}{\gamma} = \frac{v_{w1}^2 - v_{w2}^2}{2g} + (z_1 - z_2)$

$$= \frac{c^2}{2g}\left(\frac{1}{r_1^2} - \frac{1}{r_2^2}\right) + (z_1 - z_2)$$

$$= \frac{v_{w1}^2}{2g}\left[1 - \left(\frac{r_1}{r_2}\right)^2\right] \quad \text{for } z_1 = z_2$$

## 2.6 Free Spiral Flow

The spiral flow in the case of free vortex can be obtained by the super – imposition of radial flow over circulatory flow. For such a flow the distance '$b$' between the two plates will vary. Thus

$$R.b.v = \text{const.} = C$$

Or

$$v = \frac{C}{R.b}$$

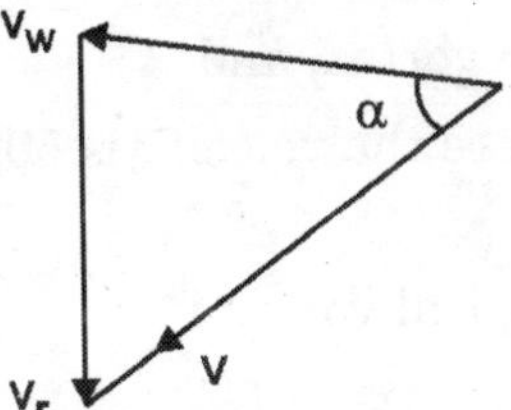


**Fig. 2.13 Spiral flow.**

Since both $v_w$ and $v_r$ vary inversely as $R$, angle α shown in Fig. 2.13 must remain constant. Then the resultant path is an equiangular or logarithmic spiral.

$$\tan \alpha = \frac{v_r}{v_w}$$

**Example 2.9** In a free cylinderical vortex flow, at a point in the fluid at a radius of 250 mm and a height of 150 mm, the velocity and pressure are 12.5 m/s and 0.15 MPa respectively. If the fluid is air having specific weight 0.012 kN/m³, calculate the pressure at a radius of 500 mm and a height if 300 mm.

**Solution.**

Given: $r_1 = 250$ mm, $z_1 = 150$ mm, $v_1 = 12.5$ m/s, $p_1 = 0.15$ MPa, $\gamma_a = 0.012$ kN/m³, $r_2 = 500$ mm, $z_2 = 300$ mm

For a free vortex flow, $vr = \text{const}$

or $$v_1 r_1 = v_2 r_2$$

$$v_2 = \frac{12.5 \times 250}{500} = 6.25 \text{ m/s}$$

**Bernoullis's equation:**

$$\frac{p_1}{\gamma_a} + \frac{v_1^2}{2g} + z_1 = \frac{p_2}{\gamma_a} + \frac{v_2^2}{2g} + z_2$$

$$\frac{0.15 \times 10^6}{0.012 \times 10^3} + \frac{(12.5)^2}{2 \times 9.81} + 0.15 = \frac{p_2}{0.012 \times 10^3} + \frac{(6.25)^2}{2 \times 9.81} + 0.3$$

$$p_2 = (5.97286 + 12500 - 0.15) \times 0.012 \times 10^3$$

$$= 0.15007 \text{ MPa}$$

**Example 2.10** A cylindrical drum has an opening in the centre of its bottom. It is initially filled with water with the opening plugged. Then the opening is unplugged and water is allowed to flow out in the form of a free spiral vortex. If the tangential velocity at a radial distance of 35 cm is found to be 25 cm/s, determine:

(a) the velocity at a radial distance of 15 cm,

(b) the velocity law for the vortex, and

(c) the radial distance where the velocity is 50 cm/s.

**Solution.**

The velocity-law for free spiral vortex is,

$$vr = C$$

Where $v$ = tangential velocity at a radial distance $r$.

(a) $$C = 25 \times 35 = 875$$

At $$r = 15\text{cm}$$

$$v = \frac{C}{r} = \frac{875}{15} = 58.33 \text{ cm/s}$$

(b) Velocity law is, $$vr = 875$$

(c) $$r = \frac{C}{v} = \frac{875}{50} = 17.5 \text{ cm}$$

## Summary

1. A fluid may have rectilinear, radial and rotary or vortex motion.
2. In a free vortex, the fluid mass rotates without any external impressed contact force.
3. In a forced vortex, the fluid mass is made to rotate by some external agency.
4. A cylindrical vortex is one in which the fluid mass rotates in concentric circles.
5. A cylindrical vortex superimposed on radial flow gives rise to spiral vortex.
6. In a radial flow, radial velocity, $v_r \propto \frac{1}{r}$
7. In a cylindrical forced vortex, the centrifugal head developed is, $z = \frac{\omega^2 r^2}{2g}$.
8. The shape of the surface generated in cylindrical forced vortex is a paraboloid of revolution.
9. The depression of vortex below the original level of still liquid due to forced vortex in an open vessel is proportional to the square of angular speed of rotation.
10. The depression of vortex in a closed vessel due to forced vortex is proportional to the angular speed of rotation.
11. In a free cylindrical vortex, velocity of whirl, $v_w \propto \frac{1}{r}$
12. In a forced cylindrical vortex, tangential velocity, $v \propto r$.
13. In radial flow, radial velocity of flow, $v_r \propto \frac{1}{r}$.
14. In a free spiral flow, $v \propto \frac{1}{R}$.
15. If a closed cylindrical vessel completely filled with water is rotated about its vertical axis, the total pressure force acting on the top and bottom are:

$$F_{top} = \frac{l}{4}\omega^2 \pi R^4$$

$$F_{bottom} = F_{top} + \gamma \times \pi R^2 h$$

## Multi-Choice Questions

1. Which of the following is an example of free vortex flow?
   (a) Whirlpool in a river
   (b) Flow of liquid inside the impeller of a centrifugal pump
   (c) Flow of liquid inside the runner of a Francis turbine
   (d) All of the above.
2. In the case of a forced vortex, the rise of liquid level at the sides of the vessel as compared to fall of liquid level at the axis of rotation is
   (a) Less than
   (b) more than
   (c) equal to
   (d) depends on speed of rotation
3. For a free vortex flow the governing equation is:
   (a) $\frac{p}{\gamma}+\frac{v^2}{2g}+z=const.$ (b) $\frac{p}{\gamma}+\frac{v}{2g}+z=const.$
   (c) $\frac{p}{\rho}+\frac{v^2}{g}+z=const.$ (d) $\frac{p}{\rho}+\frac{v^2}{2g}+z=const.$
4. If a closed cylindrical vessel completely filled with water is rotated about its vertical axis, the total pressure force acting on the top surface is equal to
   (a) $\frac{\pi}{4}\omega^2 R\rho$ (b) $\frac{\pi}{4}\omega R^2\rho$
   (c) $\frac{\pi}{4}\omega^2 R^3\rho$ (d) $\frac{\pi}{4}\rho\omega^2 R^4$
5. For a cylindrical vortex the centrifugal head generated is equal to
   (a) $z=\frac{\omega^2 r^2}{g}$ (b) $\frac{\omega^2 r^2}{2g}$
   (c) $\frac{\omega^2 r}{2g}$ (d) $\frac{\omega r^2}{2g}$
6. For a free cylindrical vortex, the whirl velocity $v_\omega$ and radial distance $r$ are related as:
   (a) $v_w \times r = const$ (b) $v_w \times \frac{1}{r} = const$
   (c) $v_w \times r^2 = const$ (d) $v_w \times r^3 = const$

7. For a forced cylindrical vortex, the tangential velocity $v$and radius $r$ are related as:

(a) $v \propto r$ (b) $v \propto \frac{1}{r}$

(c) $v \propto \sqrt{r}$ (d) $v \propto r^2$

## Answers

1. (a) 2. (c) 3. (a) 4. (d) 5. (b) 6. (a) 7. (a)

## Review Questions

1. Distinguish between radial, vortex and spiral motions.
2. Differentiate between free vortex and forced vortex.
3. Prove that if a liquid in a deep vessel is stirred it forms two types of vortices, the forced vortex in the centre and free vortex on the outside.
4. Prove that in an open cylindrical vessel the depression of vortex below the original level of still water is proportional to the square of angular speed of rotation.
5. What is a spiral vortex? Give examples.
6. Prove that in a free vortex, the tangential velocity as well as the radial velocity is inversely proportional to the distance from the centre.
7. Derive an expression for the difference of pressure between two points in a free vortex flow.

## Numerical Problems

1. A glass tube 5 am in diameter, open at the top, containing a liquid, rotates about its axis which is vertical, at 700 rpm. What is the depression of the lowest point of the surface below the surface of liquid when at rest?

[**Ans.** 8.55cm]

2. The closed cylindrical container shown in Fig. 2.14 is filled with oil having a specific gravity of 0.85. By means of paddles, the oil is rotated so that each particle has an angular velocity of 100 rpm about $O$. To what height above $O$ will the oil rise in the open tube?

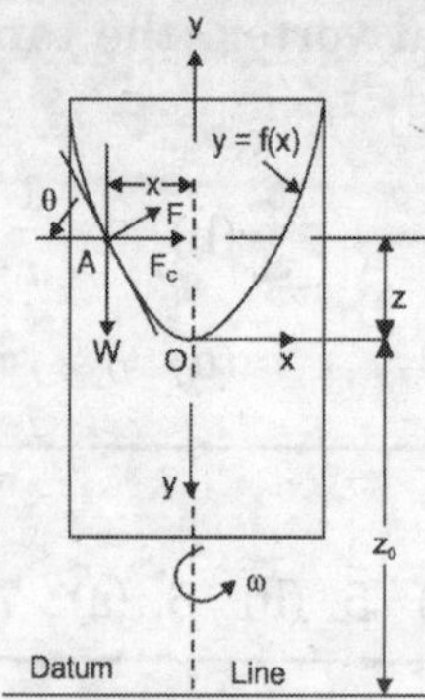


**Fig. 2.14**

3. An open cylindrical tank 1 m in diameter and 1.5 m deep is filled with water and rotated about its axis at 60 rpm. How much liquid is spilled and how deep is the water at the axis?

   At what speed should the tank be rotated in order that the centre of the bottom of the tank has zero depth of water?

   [**Ans.** 0.2 $m^3$; 1.997 m; 103.5 rpm]

4. A closed cylinder, 1 metre in diameter and 2 m high is mounted on a vertical rotating shaft. The cylinder is filled with water upto three – fourth of its height. Show that the ratio of speed to *y*, where *y* is the distance from the lowest point of the paraboloid to the top of the cylinder, is constant between certain limits of speed. Find out these limiting speeds.

   [**Ans.** 84.5rpm; 169 rpm]

5. A closed cylindrical vessel with 0.4 metre diameter, is rotated about its axis which is vertical. The vessel is 2 m long and is filled, with liquid upto a height of 1.5 m. Find the maximum speed in rpm at which the vessel can be rotated so that the bottom of the vessel is wholly covered with thin layer of the liquid.

   [**Ans.** 422.8 rpm]

6. A closed cylinder 30 cm diameter and 0.25 cm deep is completely filled with water. It is rotated about its axis which is vertical, at 240 revolutions per minute. Calculate the total pressure force of water on each end.

   [**Ans.** 251.1N, 252.9N]

7. In a free cylindrical cortex of water, the tangential velocity at a radius of 10 cm from the axis of rotation is found to be 7 m/s and the intensity of pressure is 29.9 N/$cm^2$. Find the intensity of pressure at a radius of 20 cm from the axis.

   [**Ans.** 31.26 N/$cm^2$]

8. A pump impeller is 37.5 cm diameter and it discharges water with component velocities of 2 and 12 m/s in the radial and tangential directions respectively. The impeller is surrounded by a concentric cylindrical chamber with parallel sides, the outer diameter being 45 cm. If the flow in the chamber is a free spiral vortex, find the component velocities of the water on leaving and increase in pressure if there is no loss.

[**Ans.** 1.67m/s; 10m/s; 2.29m of water]

9. Find the maximum speed of an open circular cylinder having 150 mm diameter, 1 m length and containing water up to a height of 800 mm at which it should be rotated about its vertical axis so that no water spills.

[**Ans.** 356.67 r.p.m.]

10. A 150 mm diameter open circular cylinder is 1 m long and contains water up to a height of 0.7 m. Estimate the speed at which the cylinder may be rotated about its vertical axis so that the axial depth becomes zero.

[**Ans.** 563.88 r.p.m.]

11. An open cylindrical vessel 120 mm in diameter and 300 mm deep is filled with water up to the top. Estimate the volume of water left in the vessel when it is rotated about its vertical axis:

(i) With a speed of 300 r.p.m and

(ii) With a speed of 600 r.p.m.

[**Ans.** (i) 2369.37 $cm^3$ (ii) 7.2.67 $cm^3$]

12. A cylindrical vessel closed at the top is 0.2 m in diameter, 1.2 m long and contains water up to a height of 0.8 m.

(i) Find the height of the paraboloid formed, if it is rotated at 400 r.p.m. about its vertical axis.

(ii) Find the speed of rotation of the vessel when axial depth is zero.

[**Ans.** (i) 0.845 m (ii) 567.2 r.p.m.]

13. A vessel cylindrical in shape and closed at the top and bottom is 300 mm in diameter, 1 m long and contains water up to a depth of 0.8 m. The air above the water surface is at a pressure of 60 $kN/m^2$. If the vessel is rotated at a speed of 250 r.p.m. about its vertical axis find the pressure head at the bottom of the vessel

(i) At the centre, and

(ii) At the edge.

[**Ans.** (i) 6.439 m of water, (ii) 7.225 m of water.]

14. A vessel cylindrical in shape and closed at the top and bottom is 200 mm in diameter and 0.15 m in height. The vessel is completely filled with the water. If it is rotated about its vertical axis with a speed of 200 r.p.m., what is the total pressure force exerted by water on the top and bottom of the vessel?

**[Ans.** 0.0344 kN, 0.0806 kN]

CHAPTER 3

# Hydro Power Generation Fundamentals

## 3.1 Introduction

Energy is the primary input to every activity of development. Electrical energy is the most convenient form. The per capita electrical energy consumed is considered to be an index of development and progress of a nation. The purpose of hydropower generation is to harness power from water flowing under pressure. Hydro-power is generated in a hydro-electric plant which incorporates a number of water driven prime-movers known as water turbine or hydraulic turbines.

Water flowing under pressure has two forms of energy: potential and kinetic. The hydraulic turbine converts the energy of flowing water into mechanical energy. The mechanical energy is used in running an electric generator which is directly coupled to the shaft of the hydraulic turbine. The electric generator generates electric power which is transmitted over long distances by means of transmission lines.

## 3.2 Essential components of Hydro-Electric Power Plant

The essential components of hydro-electric power plant are:

1. Storage reservoir
2. Dam with control works
3. Waterways with control works
4. Power house with turbine and other machinery.
5. Tail race
6. Generation and transmission of electric power.

**1. Storage reservoir:** The water available from the catchment area is stored in a large lake called reservoir. The stored water is utilised to run the turbine for generating electric power as per the requirement throughout the year. The lake may be natural or artificial. Natural lake is found in high and huge mountains. Artificial lake is made by constructing dam on one or two openings of a valley in order to store water.

**2. Dam with control works:** Dam is a structure erected on suitable site to provide for the storage of water and to create head. Dam may be built to make an artificial reservoir from a valley or it may be erected in a river to control flowing water. Different types of dams are: earth dam, masonry dam, or steel dam. Masonry dams are very common.

Structures and appliances to control the supply of water from the storage reservoir through the dam are known as control or Head works. The principal parts of control works are:

(i) Gates and valves and their operating mechanism.

(ii) Structures necessary for their operation.

(iii) Devices for the protection of gates and hydraulic machines.

**Trashracks:** are made up of a row of rectangular cross-sectional structural steel bars placed across the entire intake opening in an inclined position. They are used to obstruct debris from going into the intake.

3. **Waterways with control works:** Waterways is a passage through which the water is carried from the storage reservoir to the power house. It may consist of tunnels, canals, flumes, fore bays, penstocks. Tunnel is a water passage made by cutting the mountain to save distance. Forebay is an enlarged section of a canal spread out to accommodate the required width of intake. The function of fore bay is to store temporarily the rejected water by the plant when its load is reduce and to meet the instantaneous increased demand of water due to sudden increase of load.

Penstocks are pipes of large diameter carrying water under pressure from the storage reservoir to the turbine. A surge tank is a safety device for the penstock. A surge tank is a reservoir fitted near the valves to receive the rejected water when the pipe line is suddenly closed. The surge tank, therefore, controls the pressure variations due to sudden closing of valves thus eliminating water hammer effects.

4. **Power house:** It is a building to house the turbines, generators and other accessories for operating the machines.

5. **Tail race:** It is a waterway to carry the water discharged from the turbine to a suitable point where it can be safely disposed off or stored to be pumped back into the original reservoir.

6. **Generation and transmission of electric power:** It consists of electrical generating machines, transformers, switching equipment and transmission lines.

### 3.3 Classification of Hydro-Electric plants

The hydro-electric plants may be classified as follows:

1. **Operating head:** High head, medium head, and low head plants.
2. **Functions:** Run – off river plants, storage plants, and pumped storage plants.
3. **Load capacity:** Base load plants, and peak load plants.
4. **Source of energy:** River (or reservoir) power plants and tidal power plants.

### 3.3.1 High Head Water Power Plants

Such plants work under heads from 100 m and above. Water is usually stored up in lakes on high mountains during the rainy season or when the snow melts. From one end of the lake, tunnels are constructed which lead the water into smaller reservoir known as fore bays. The fore bays distribute the water to penstocks through which it is lead to the turbines. Fig. 3.1 shows the layout of high head power plant. The water is discharged after passing through turbines into the tail-race for further use for irrigation purposes.

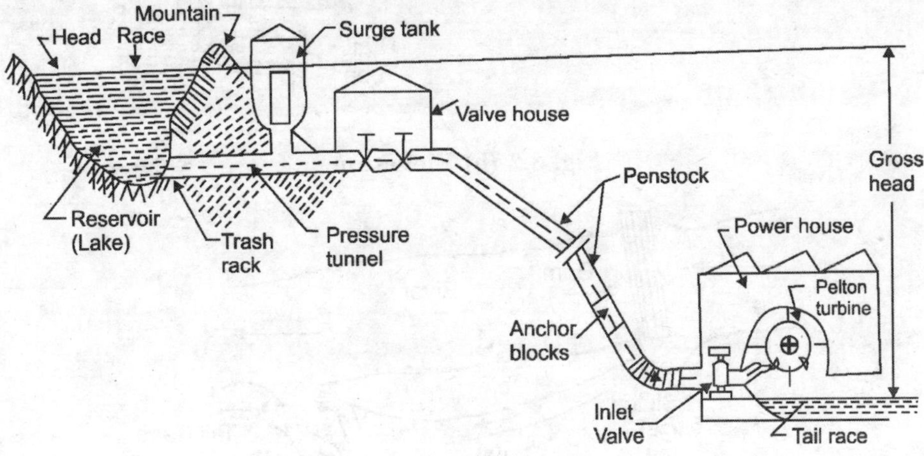


**Fig. 3.1 High head water power plant layout.**

### 3.3.2 Medium Head Water Power Plants

A medium head power plant works between 30-500 m. The plant layout of medium head plant is shown in Fig. 3.2 (a) and (b).

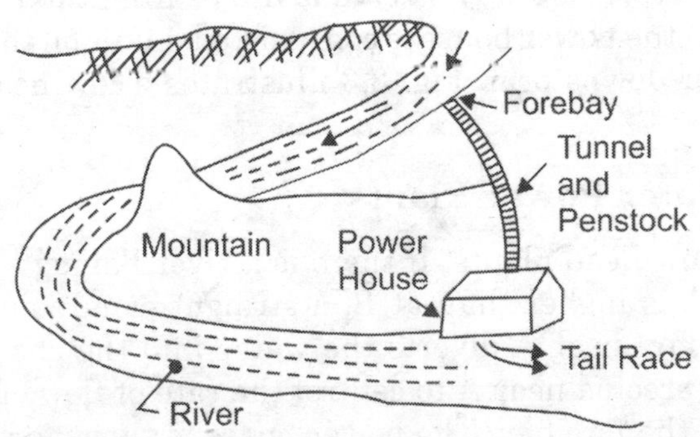


**Fig. 3.2 (a) Plan**

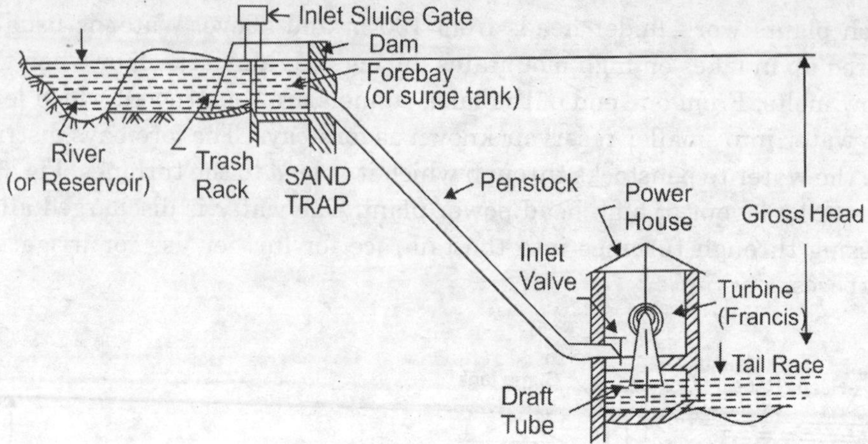


**Fig. 3.2 (b) Elevation**

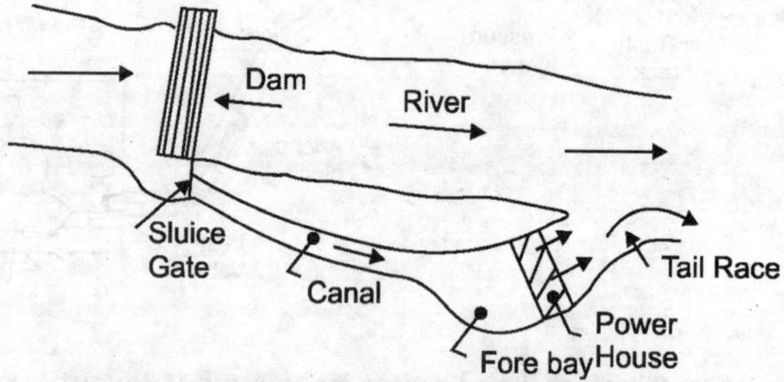


**Fig. 3.3 Low Head Water Power Plant Layout.**

### 3.3.3 Low Head Water Power Plants

These plants work from 25-80 m head. These plants unusually consisit of a dam across a river. A side way stream diverges from the river of the dam. Over this stream the power house is constructed. Later on this channel joins the river further downstorm. Fig. 3.3 illustrates a low head water power plant.

### 3.3.4 Canal Water Power Plants

These are also low head plants. If there is a river having a bend (Fig. 3.4) whose level at A is higher than at B, a straight canal joining A and B is constructed. A dam near A diverts the water into the channel. The main sluice gates are erected near A to control the rate of flow of water into the canal as well as the fore bay. The power house is shown on the right hand side of the fore bay. On leaving the turbine the water is lead back into the river by the canal at B.

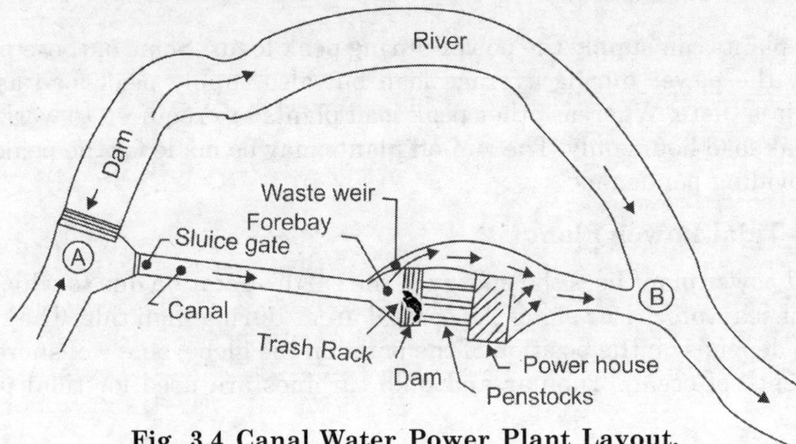


**Fig. 3.4 Canal Water Power Plant Layout.**

### 3.3.5 Run – Off River Water Plants

These plants do not store any water and uses the water as it flows. The utility of such a plant is limited. Run – off river plants may be without water pondage or with water pondage. The plants with pondage are provided with a weir or barrage to accommodate sufficient storage to take care of load fluctuations.

### 3.3.6 Storage Plants

The storage plants have reservoir provided by constructing a dam across the river if the plant is low head or by closing the opening of a valley by a dam in case of a high head plant. The storage is collection of water which takes care of fluctuations of water supply during flood and draught periods, as well as load fluctuations, thus increasing the utility of plant. Such a plant is also useful as peak load plant.

### 3.3.7 Pumped Storage Plant

This plant caters for the peak load of the power station. It stores power by pumping back a portion of water from the tail race back to the reservoir during off – peak hours, when excess power can be available for this purpose. The pumping is done by the same turbine-generator (producing power) which now acts as a pump-motor set. In certain cases where it is not possible to use turbine as jump, a separate pump is coupled to the same shaft. The stored water is utilised during peak load period on demand. Such plants are built where there is a continuous shortage of water.

### 3.3.8 Base-Load Plants

These plants are those which cater for the base load of the system. Such plants are required to supply a constant power when connected to the grid. Thus they run without stop and are often remote controlled.

### 3.3.9 Peak-load Plants

These plants can supply the power during peak loads. Some of these plants supply the power during average load but also supply peak load as and when it is there. Whereas other peak load plants are required to work during peak load hours only. The run-off plants may be made for the peak load by providing pondage.

### 3.3.10 Tidal Power Plant

A tidal power plant uses the energy of the earth's rotation due to which the level of sea water rises along the coastal areas during high tide. The range of tide depends on the position of the point on the globe, shape of shore line and depth of ocean. Tubular and bulb turbines are used for tidal power plant.

## 3.4 Classification of Water Turbines

The water turbines may be classified as follows:

1. Based upon the head and quantity of water available.
   (a) Impulse turbine – requires high head and small quantity of flow of water.
   (b) Reaction turbine – requires low head and high rate of flow of water.
2. Based after the name of originator.
   (a) Pelton turbine – named in honour of Lester Allen Pelton, is an impulse turbine used for high head and low discharge.
   (b) Francis turbine – named after James Bichens Francis, is reaction type of turbine used for medium head and medium discharge.
   (c) Kalpan turbine – named in honour of Dr. Victor Kaplan, is a reaction type of turbine for low head and large quantity of flow.
3. According to action of water on moving blades.

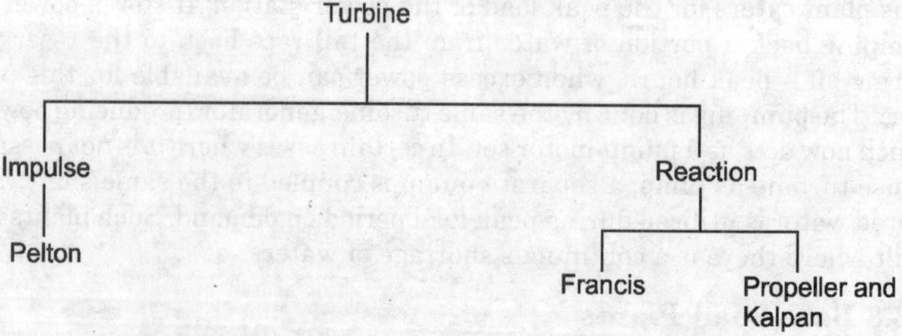


4. According to direction of flow of water in the runner.

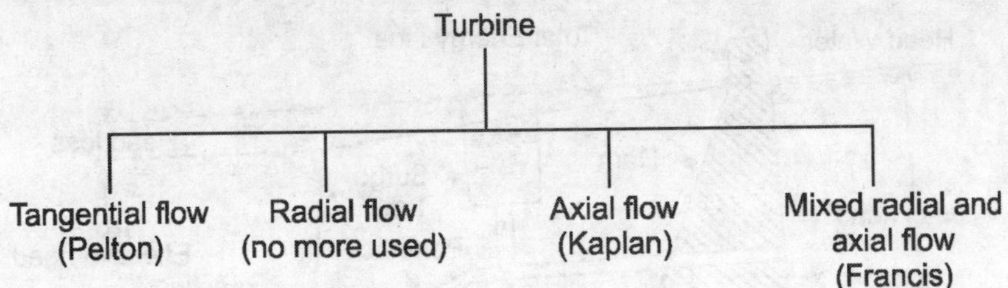


5. According to direction of flow of shaft.
   Horizontal shaft – Pelton turbine
   Vertical shaft – Francis and Kaplan turbines.
6. According to specific speed.

The specific speed of a turbine is the speed of a geometrically similar turbine that would develop one brake power under a head of one metre.

$$\text{Specific speed, } N_s = \frac{N\sqrt{P_t}}{H^{5/4}}$$

Where $N$ = speed of turbine, rpm.
$P_t$ = turbine output in brake power, kW
$H$ = effective head in metres.

**Table 3.1 Specific speed of turbines.**

| Type of turbine | Specific speed in S.I.Units |
|---|---|
| 1. Pelton: | |
| (a) Single jet | 8.5 – 30 |
| (b) Multiple jets | 30 – 50 |
| 2. Francis | 50 – 340 |
| 3. Kaplan | 250 – 1025 |

## 3.5 Heads and Discharge

Head is the difference in elevation between two levels of water. Head can be characterised by:

1. Gross head, and
2. Net or effective head.

A typical installation of a reaction turbine is shown in Fig. 3.5.

**1. Gross Head:** It is defined as the difference in elevation between the head race level at the intake and the tail race level at the discharge side. Both the elevations have to be measured simultaneously. The normal head would be that for which the plant works most of the time. In rainy season, the head race and tail race level increase. Whereas at the time of draught both these levels decrease.

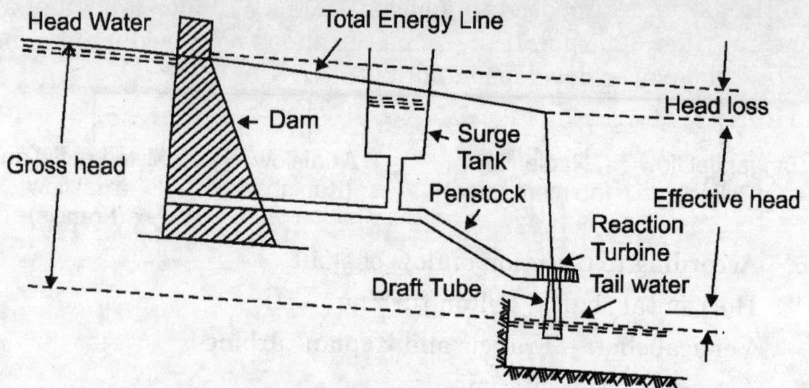


**Fig. 3.5 Typical installation of a reaction turbine.**

**2. Net or Effective Head:** It is the head obtained by subtracting from gross head all the losses in carrying water from the head race to the entrance of turbine. The losses are due to friction occuring in tunnels, canals, and penstocks which lead the water in to the turbine. Net or effective head is, therefore, the true pressure difference between the entrance to the turbine casing and the tail race water elevation. The measurement of the pressure head at the entrance to the turbine casing must be corrected for the velocity head at that point and the tail – water elevation corrected for residual velocity head.

**Discharge:** It is the quantity of water used by the water turbine in a unit time. It is measured in $m^3/s$ or litres/s.

**Example 3.1** A turbine develops 14,000 kW under a head of 500 m, when running at 300 rpm. Calculate the specific speed. What type of turbine is this?

**Solution.**

Given: $P_t = 14{,}000$ kW, $H = 500$ m, $N =$ rpm

$$\text{Specific speed, } N_s = \frac{N\sqrt{P_t}}{H^{5/4}} = \frac{300\sqrt{14{,}000}}{500^{5/4}} = 15$$

According to Table 3.1, it is a Pelton turbine.

**Example 3.2** A turbine develops 25,000 kW under a head of 30 m when running at 166.7 rpm.

Calculate its specific speed and type of turbine.

**Solution.**

Given: $P_t = 25{,}000$ kW, H = 30 m, N = 166.7 rpm.

$$\text{Specific speed, } N_s = \frac{N\sqrt{P_t}}{H^{5/4}} = \frac{166.7\sqrt{25{,}000}}{30^{5/4}} = 375.4$$

According to Table 3.1, it is a Kaplan turbine.

**Example 3.3** It is proposed to develop 1500 kW under a head of 150 m, while running at 300 rpm. If the same turbine is now used under a head of 30 m, find the power developed and its rpm.

**Solution:**

Given: $P_{t1} = 1500$ kW, $H_1 = 150$ m, $N_1 = 300$ rpm, $H_2 = 30$ m

$$P_{t2} = P_{t1}\left(\frac{H_2}{H_1}\right)^{3/2} = 1500\left(\frac{30}{150}\right)^{3/2} = 134.164 \text{ kW}$$

$$N_2 = N_1\left(\frac{H_2}{H_1}\right)^{1/2} = 300\left(\frac{30}{150}\right)^{1/2} = 134.164 \text{ rpm}$$

## Review Questions

1. List the essential components of hydro – electric power plants.
2. Define head works. What is their purpose?
3. What are trash racks? State their uses.
4. What are waterways? List their control works.
5. What is fore bay?
6. What is a surge tank?
7. Describe the basis under which the water power plants can be classified.
8. What are the ranges of water head for high, medium, and low head power plants?
9. Why it is difficult to classify water power plants according to heads under which they work?
10. Draw the line diagram of a medium head water power plant.
11. Different between storage and pumped storage plants.
12. What is the difference between bass load and peak load plants?
13. What is water hummer in a penstock? How this can be decreased?
14. How water turbines can be classified?
15. Differentiate between impulse and reaction turbine/
16. Define specific speed. Give its ranges for impulse and reaction turbines.
17. Define head of water on a turbine.
18. What are the various types of heads in a water turbine?

## Numerical Problems

1. It is desired to build a turbine to operate at 100 rpm and develop 10,000 kW under a head of 100 m. What type of turbine would you suggest?

   [**Ans.** $N_s = 31.6$, multi – jet impulse turbine]

2. A discharge of 300 lit /s of water is to be used under a head of 75 m to operate a turbine at a speed of 600 rpm. What is the solution?

   [**Ans.** $N_s = 40.0$, Pelton turbine either two nozzles with one wheel or two units with a single nozzle]

# 4 Unit and Specific Quantities

## 4.1 Introduction

Two similar hydraulic machines designed for different specifications are required to be compared. Concept of 'Unit and Specific Quantities' is a pre-requisite for the comparison of hydraulic machines.

## 4.2 Unit and Specific Quantities

The rate of flow, speed, power, etc. of hydraulic machines are all functions of the working head. To facilitate correlation, comparison and use of experimental data, these quantities are usually reduced to unit head. These reduced quantities are known as unit quantities.

A specific quantity is obtained by reducing any quantity to a value corresponding to unit head and some unit size. The unit size may be the inlet diameter of runner in case of reaction turbines and least jet diameter in Pelton turbine.

**Specific speed of a turbine**:– It is defined as the speed of a geometrically similar turbine working under a unit head and developing unit power.

**Specific speed of a pump**: It is defined as the speed of a geometrically similar pump working against a unit head and raising unit quantity of water.

## 4.3 Expressions for Unit Quantities

### 1. Unit rate of flow

Rate of flow = cross – sectional area × velocity of flow

$$Q \propto v_f$$

Where $\quad v_f = K_f\sqrt{2gH}$

$K_f$ = flow velocity coefficient

$H$ = head

$$\therefore \quad Q \propto \sqrt{H}$$

$$\text{Or} \quad Q = k_1\sqrt{H}$$

By definition, unit rate of flow $Q_1$ is the value of $Q$ when $H = 1$.

$$Q_1 = k_1\sqrt{1} = k_1 = \frac{Q}{\sqrt{H}}$$

Unit rate of flow $Q_1 = \dfrac{Q}{\sqrt{H}}$

**2. Unit speed**

Peripheral speed of runner,

$$u = \frac{\pi DN}{60}$$

Where $N = rpm$

Also $u = K_u\sqrt{2gh}$

$\therefore$ $N \propto u \propto \sqrt{H}$

Or $N = k_2\sqrt{H}$

Where $k_2 =$ a constant

By definition, unit speed, $N_1 = k_2\sqrt{1} = k_2 = \dfrac{N}{\sqrt{H}}$

**3. Unit power**

Available power of a turbine,

$$P_a = \frac{\gamma QH}{10^3}\, kW$$

Brake power, $P_t = \eta_t p_a = \dfrac{\eta_t \gamma QH}{10^3}$

Where $\eta_t$ = turbine overall efficiency

In general, $P \propto QH$

But $Q \propto \sqrt{H}$

$\therefore$ $P \propto H^{\frac{3}{2}} = k_3 H^{\frac{3}{2}}$

Where $k_3$ = a constant

By definition, unit power, $P_1 = k_3\,(1)^{\frac{3}{2}} = k_3$

$$= \frac{P}{H^{\frac{3}{2}}}$$

**4. Unit force**

The force exerted by jet on the Pelton turbine runner at its periphery is,

$$F = \frac{\gamma Q}{g}(v_{w1} - v_{w2})$$

Or $F \propto Q.v_w$

But $Q \propto \sqrt{H}$

and $v_w \propto \sqrt{H}$

$\therefore$ $F \propto H = k_4 H$

By definition, unit Force, $F_1 = k_4.1 = k_4$

Or $F_1 = \frac{F}{H}$

**5. Unit torque**

Torque on runner $=$ Force at periphery $\times$ Radius

$$T = F \times R$$

Or $T \propto F \propto H$

Or $T = k_5 H$

By definition, unit torque, $T_1 = K_5.1 = K_5$

$$= \frac{T}{H}$$

## 4.4 Expressions for Specific Quantities

**1. Specific flow in a reaction turbine**

Rate of flow, $Q = \pi DB.v_f$

Now $v_f \propto \sqrt{H}$

$$Q \propto D^2\sqrt{H}$$

$$= k_6 D^2 \sqrt{H}$$

By definition, specific rate of flow,

$$Q_{11} = k_6.1^2.\sqrt{1} = k_6$$

$$= \frac{Q}{D^2\sqrt{H}}$$

For a Pelton turbine,

$$Q = \frac{\pi}{4} d^2 v \propto d^2 \sqrt{H}$$

Where $d =$ least diameter of water jet

$$\therefore \quad Q_{11} = \frac{Q}{d^2 \sqrt{H}}$$

**2. Specific power**

Power, $\quad P \propto Q.H$

Now $\quad Q \propto D^2 \sqrt{H}$

$$\therefore \quad P \propto D^2 H^{\frac{3}{2}} = k_7 D^2 H^{\frac{3}{2}}$$

By definition, specific power, $P_{11} = k_7.1^2.1 = k_7$

$$= \frac{P}{D^2 H^{\frac{3}{2}}}$$

Similarly for a Pelton turbine,

$$P_{11} = \frac{P}{d^2 H^{\frac{3}{2}}}$$

**3. Specific force of jet or periphery of runner**

$$\boldsymbol{F} = \frac{\gamma Q}{g}(v_{w1} - v_{w2})$$

Or $\quad F \propto Q.v_w$

But $\quad Q \propto d^2 \sqrt{H}$

and $\therefore \quad F \propto d^2 H$

$$= k_8 d^2 H$$

By definition, specific force,

$$F_{11} = k_8.1^2.1 = k_8$$

$$= \frac{F}{d^2 H}$$

**4. Specific torque**

$$T = F.R$$

Or $\quad T \propto F \propto d^2 H = k_9 d^2 H$

By definition, specific torque, $T_{11} = k_9.1^2.1 = k_9$

$$= \frac{T}{d^2 H}$$

## 4.5 Specific Speed of a Turbine

Specific speed is the speed of a geometrically similar turbine working under unit head and delivering unit brake power.

Now $\quad u = \pi DN$

and $\quad u_1 \propto \sqrt{H}$

$\therefore \quad D \propto \frac{\sqrt{H}}{N}$

Also brake power, $P_t \propto Q.H$

Where $\quad Q \propto D^2.\sqrt{H}$

$\therefore \quad P_t \propto D^2 H^{\frac{3}{2}} \times \frac{H}{N^2} \cdot H^{\frac{3}{2}} \times \frac{H^{\frac{5}{2}}}{N^2}$

Or $\quad N \propto \left[\frac{H^{\frac{5}{2}}}{P_t}\right]^{\frac{1}{2}} \propto \frac{H^{\frac{5}{4}}}{\sqrt{P_t}}$

Or $\quad N = N_S \frac{H^{\frac{5}{4}}}{\sqrt{P_t}}$

Where $N_S = \frac{N\sqrt{P_t}}{H^{\frac{5}{4}}}$ is the specific speed of turbine.

## 4.6 Specific Speed of a Pump

Specific speed of a pump is the speed of a geometrically similar pump delivering unit quantity of water against a unit head.

Now $\quad u = \pi DN$

and $\quad u \propto \sqrt{H}$

$\therefore \quad D \propto \frac{\sqrt{H}}{N}$

Also $Q \propto D^2 \sqrt{H} \propto \frac{H}{N^2} \cdot \sqrt{H} \propto \frac{H^{\frac{3}{2}}}{N^2}$

Or $$N \propto \left[\frac{H^{\frac{3}{2}}}{Q}\right]^{\frac{1}{2}} \propto \frac{H^{\frac{3}{4}}}{\sqrt{Q}} = N_S \frac{H^{\frac{3}{4}}}{\sqrt{Q}}$$

$\therefore \quad N_S = \frac{N\sqrt{Q}}{H^{\frac{3}{4}}}$ is the specific speed of pump.

**Example 4.1** A Francis turbine works under a maximum head of 24 m generating 2100 kW at 250 rpm. Calculate:

(a) The unit power and rpm of this turbine.

(b) The unit speed and producing one kW, and

(c) The power of the turbine working under 1 m head and running at 1 rpm.

**Solution.**

Given: $H = 24m$, $P_t = 2100kW$, $N = 250rpm$

(a) $$P_1 = P_t\left(\frac{1}{H}\right)^{\frac{3}{2}} = 2100\left(\frac{1}{24}\right)^{\frac{3}{2}} = 17 \cdot 86kW$$

$$N_1 = \frac{N}{\sqrt{H}} = \frac{250}{\sqrt{24}} = 51rpm$$

(b) $$N_S = \frac{N\sqrt{P_t}}{H^{\frac{5}{4}}} = \frac{250\sqrt{2100}}{24^{\frac{5}{4}}} = 215 \cdot 7rpm$$

(c) $$P_t = \left[\frac{N_s \cdot H^{\frac{5}{4}}}{N}\right]^2 = \left[\frac{215_7 \times (1)^{\frac{5}{4}}}{1}\right]^2 = 46 \cdot 512kW$$

**Example 4.2** A turbine develops 3750 kW under a head of 100 m at 110 rpm. Calculate the specific speed. What would be its normal speed and output under a head of 80 m?

**Solution.**

**Given:** $P_t = 3750kW$, $H = 100m$, $N = 110rpm$, $H_1 = 80m$

Specific speed, $$N_S = \frac{N\sqrt{P_t}}{H^{\frac{5}{4}}} = \frac{110\sqrt{3750}}{100^{\frac{5}{4}}} = 21 \cdot 3$$

$$N \propto H^{\frac{1}{2}}$$

$$N_1 = N\left(\frac{H_1}{H}\right)^{\frac{1}{2}} = 110\left(\frac{80}{100}\right)^{\frac{1}{2}} = 98 \cdot 4rpm$$

$$P_t \propto H^{\frac{3}{2}}$$

$$P_{t1} = 3750\left(\frac{80}{100}\right)^{\frac{3}{2}} = 2683 \cdot 3kW$$

**Example 4.3** Calculate the specific speed of a centrifugal pump delivering $1.5m^3/s$ against a net head of 15 m when running at 740 rpm.

**Solution.**

Given: $N = 740rpm, Q = 1.5m^3/s, H = 15m$

Specific speed of a pump, $N_s = \frac{N\sqrt{Q}}{H^{\frac{3}{4}}} = \frac{740\sqrt{1.5}}{15^{\frac{3}{4}}} = 118 \cdot 9$

## Review Questions

1. Define a unit quantity and a specific quantity.
2. What is the necessity of defining unit and specific quantities?
3. Derive expressions for specific flow, specific power, and specific torque of a hydraulic turbine.
4. Define specific speed of a hydraulic turbine and derive an expression for it.
5. Define specific speed of a pump and derive an expression for it.
6. Derive expressions for unit flow, unit speed, and unit power of a hydraulic turbine.

## Numerical Problems

1. A Pelton turbine develops 60 MW under a head of 720 m. Determine its specific speed, if it runs at a synchronous speed of 50 Hz.

   [**Ans.** 197]

2. A turbine develops 1800 kW under a head of 160 m when running at 370 rpm.

   Calculate: (a) The specific speed of turbine, and (b) The power developed and its rpm under a head of 30 m.

   [**Ans.** 27.586, 146.142 kW, 160.2 rpm]

# Pelton Turbine

## 5.1 Introduction

The Pelton turbine is an impulse turbine which operates under a high head of running water and requires comparatively less quantity of water. The water is conveyed from a reservoir in the mountains to the turbine in the power house through a long pipe called penstock. A nozzle is fitted at the end of the penstock through which water comes out in the form of a free and compact jet. The water jet having velocity is made to impinge on the buckets fixed round the circumference of a wheel mounted on a shaft. The impact of water on the surface of the bucket produces a force which causes the wheel to rotate, thus, supplying a torque or mechanical power on the shaft.

## 5.2 Main Components of Pelton Turbine

A Pelton turbine consists of the following main components:

**1. Guide mechanism:** This mechanism controls the quantity of water passing through the nozzle and striking the buckets. It maintains the speed of the turbine constant even when the head varies. It mainly consists of a defelctor and a spear operated by a governor.

**2. Buckets and Runner:** Each bucket is divided vertically into two parts by a splitter giving the shape of a double hemispherical cup. The buckets are bolted to the circumference of a round disc forming the runner of the Pelton turbine.

**3. Casing:** It is necessary to prevent splashing of water and to lead the water to the tail race and also as a safeguard against accidents. It has no hydraulic function to perform.

**4. Hydraulic brake:** It consists of water of a small nozzle fitted in such a way that on being opened, it directs a jet of water on the back of the buckets to bring the revolving runner quickly to rest. The single jet and double jet Pelton turbines are shown in Fig.5.1.

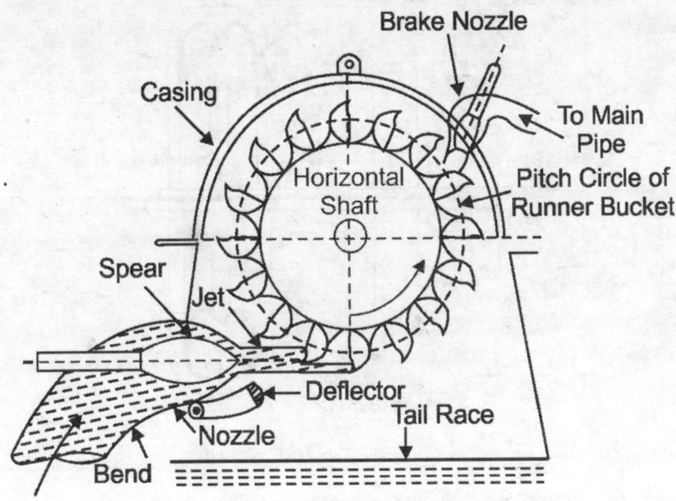


**Fig, 5.1 (a) Single jet, Horizontal Shaft Pelton Turbine.**

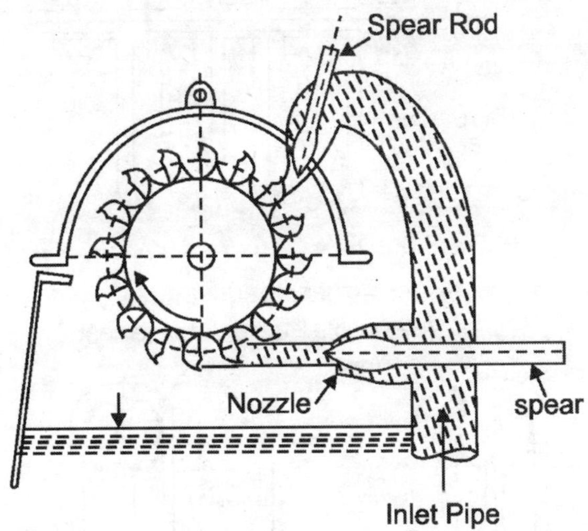


**Fig. 5.1 (b) Double jet, Horizontal Shaft Pelton Turbine.**

The arrangements of runners for Pelton Turbines are shown in Fig.5.2.

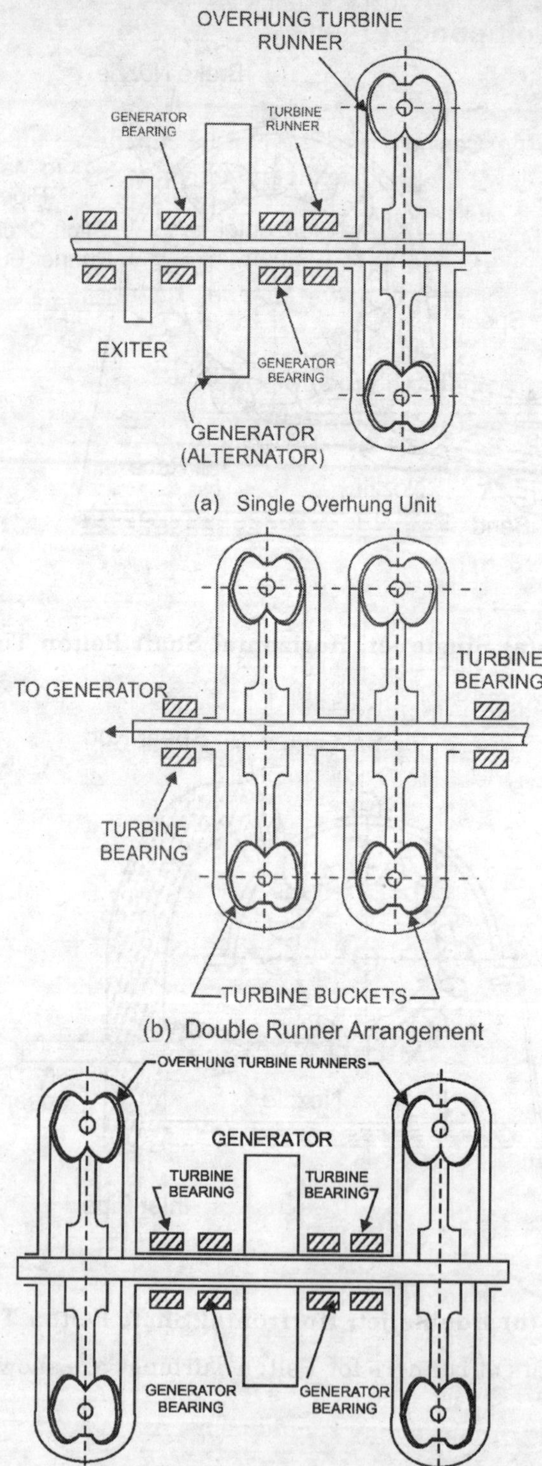


(a) Single Overhung Unit

(b) Double Runner Arrangement

(c) Double Overhung Runner Arrangement

**Fig. 5.2 Arrangements of runners for Pelton turbines.**

## 5.3 Design of Component Parts

### 1. Turbine Power

The power available from the turbine can be calculated for a specified net head and rate of flow of water, as given below:

Power available, $P_a = \dfrac{\gamma QH}{10^3}$ kW

wherer $\gamma$ = specific weight of water = 9810 N/m$^3$

$Q$ = rate of flow of water, m$^3$/s

$H$ = net head acting at the turbine inlet, m

Power supplied on the turbine shaft $P_t = P_a \times \eta_t$

Where $\eta_t$ = overall efficiency of turbine.

= 0.85 to 0.90

### 2. Nozzle and jet diameters

Discharge through jet, $Q = \dfrac{\pi}{4} d_1^2 v$ m$^3$/s

Where $d_1$ = least diameter of jet (at vena contracts), m

$v_1$ = velocity of jet, m/s

$= K_{v1}\sqrt{2gH}$

$k_{v1}$ = coefficient of velocity

= 0.98-0.99

$H$ = net head, m

$$\therefore \quad d_1 = \sqrt{\frac{4Q}{\pi K_{v1}\sqrt{2gH}}}$$

Nozzle diameter, $d = 1.25d_1$

### 3. Multiple Jets

Specific speed of Pelton turbine,

$$N_s = \frac{N\sqrt{P_t}}{H^{5/4}}$$

For a Pelton turbine having n number of jets,

$$N_s = \frac{N\sqrt{nP_t}}{H^{\frac{5}{4}}}$$

*When* $N_s > 30$, multiple jets should be used.

$n_{max}$ = 2 for a horizontal turbine

= 6 for a vertical turbine.

### 4. Mean diameter of Pelton runner

Circumferential velocity of bucket,

$$u_1 = K_{u1}\sqrt{2gH}$$

Where $K_{u1}$ = velocity coefficient or speed ratio

$$= \frac{\text{peripheral velocity of buckets at their mean diameter}}{\text{theoretical spouting velocity of water under the head acting on turbine}}$$

$$\approx 0.45$$

Now $$u_1 = \frac{\pi D_1 N}{60} \text{ m/s}$$

Where $D_1$ = mean or pitch circle core pitch circle diameter of Pelton runner, m

$N$ = rpm

$$\therefore \quad \frac{\pi D_1 N}{60} = K_{u1}\sqrt{2gH}$$

$$D_1 = \frac{60}{\pi}.\frac{K_{ui}}{N}.\sqrt{2gH}$$

$$= \frac{84.6 K_{u1}\sqrt{H}}{N}$$

If an electrical generator is directly coupled to the turbine, then

Frequency of alternator, $$f = \frac{pN}{60} \text{ Hz}$$

Where $p$ = number of pairs of poles.

For $f = 50Hz$, $$N = \frac{3000}{p} \text{ rpm}$$

### 5. Selection of speed

The selection of speed of a Pelton turbine is very important because maximum efficiency is obtained at a definite speed. If the speed of the turbine is made higher, then

(a) The specific speed will increase.

Its advantages are:

(i) The size of the turbine will become smaller and hence it will be less costly.

(ii) The jet diameter will decrease. It will raise the jet ratio and increase the turbine efficiency.

The disadvantage of having higher specific speed is that it requires multi-jets with which the governing becomes complicated and more expensive.

(b) The speed of directly coupled generator will increase which will require smaller number of pair of poles. Hence the generator will be less costly.

(c) The material required for high speed turbine and generator will be costly.

**6. Jet Ratio**

Jet ratio,
$$m = \frac{\text{Mean diameter of runner}, D_1}{\text{Least diameter of jet}, d_1}$$
$$= 10 \text{ to } 30$$
$$\simeq 11\text{-}14$$

**7. Minimum numbers of buckets**

Number of buckets, $z = 0.5m + 15$ for $m = 6$ to $35$

**Example 5.1** A Pelton turbine develops 15MW when working under a head of 850 m and running at 600 rpm. Calculate:

(a) the least diameter of runner
(b) the diameter of nozzle orifice tip
(c) the mean diameter of runner
(d) jet ratio
(e) number of buckets

Take overall efficiency of turbine = 89.2%, $K_{v1} = 0.985$, $K_{u1} = 0.45$.

**Solution.**

Given:

$P_t = 15$ MW, $H = 850$m, $N = 600$rpm, $\eta_t = 0.892$
$K_{u1} = 0.45$, $K_{v1} = 0.985$

(a)
$$P_t = \frac{\gamma Q H}{10^3} \times \eta_t$$
$$15 \times 10^6 = 9.81 \times 10^3 \times Q \times 850 \times 0.892$$
$$Q = 2.017 \text{m}^3/\text{s}$$

Now
$$Q = \frac{\pi}{4} d_1^{\,2}\, v_1 = \frac{\pi}{4} d_1^{\,2} \times K_{v1} \sqrt{2gh}$$
$$2.017 = \frac{\pi}{4} d_1^{\,2} \times 0.985 \times \sqrt{2 \times 9.81 \times 850}$$
$$d_1 = 0.142 \text{m or } 142\text{mm}$$

(b) Diameter of nozzle orifice, $d = 1.25\ d_1 = 1.25 \times 142 = 177.5 mm$

(c)
$$u_1 = \frac{\pi D_1 N}{60} = K_{u1} \sqrt{2gH}$$

$$D_1 = \frac{60}{\pi} \times \frac{0.45}{600} \times \sqrt{2 \times 9.81 \times 850}$$

$$= 1.85\text{m or } 1850\text{mm}$$

(d) Jet ratio, $m = \frac{D_1}{d_1} = \frac{1850}{142} = 13$

(e) Number of buckets, $z = 0.5m + 15 = 0.5 \times 13 + 15 = 21.5 = 22$

**Example 5.2** A Pelton turbine is to operate a 25MW generator under an effective head of 450m.

*Determine:* (a) size of jet, (b) mean diameter of runner, (c) synchronous speed, and (d) specific speed of turbine.

Assume $\eta_{gen} = 94\%$, $\eta_t = 85\%$, $K_{u1} = 0.46$, and $m = 12$

**Solution:**

Given: $P_g = 25MW$, $H = 450m$

(a)Turbine brake power, $P_t = \frac{P_t}{\eta_{gen}} = \frac{25}{0.94} = 26.596MW$

Available turbine power, $P_\alpha = \frac{P_t}{\eta_t} = \frac{26.596}{0.85} = 31.289MW$

Now, $P_\alpha = \frac{\gamma QH}{10^3} kW$

$$31.289 \times 10^6 = 9.81 \times 10^3 \times Q \times 450$$

$$Q = 7.088 \text{ m}^3/\text{s}$$

Velocity of jet, $\upsilon_1 = K_{\upsilon 1}\sqrt{2gH}$

$$= 0.97\sqrt{2 \times 9.81 \times 450} = 91.14 \; m/s$$

$$\frac{\pi}{4} d_1^{\,2} \times \upsilon_1 = Q$$

$$d_1 = \sqrt{\frac{4 \times 7.088}{\pi \times 91.14}} = 0.3147 \text{ m}$$

or 314.7 mm = $315mm$

(b) Mean diameter of runner, $D_1 = m \times d_1$

$$= 12 \times 315 = 3780mm$$

(c) Peripheral velocity of runner, $u_1 = K_{u1}\sqrt{2gH}$

$$= 0.46\sqrt{2 \times 9.81 \times 450}$$

$$= 43.2 \; m/s$$

Also $$u_1 = \frac{\pi D_1 N}{60}$$

$$N = \frac{43.2\times 60}{\pi\times 3.780} = 218.27\,rpm$$

Now $$f = \frac{pN}{60}$$

Let $$p = 14$$

$$50 \times 60 = 14 \times N_{syn}$$

$$N_{syn} = \frac{3000}{14} = 214.3\,rpm$$

Revised diameter of runner, $$D_1 = \frac{218.27}{214.3}\times 3780 = 3850\,mm$$

(d) Specific speed, $$N_s = \frac{N\sqrt{P_t}}{H^{5/4}} = \frac{214.3\sqrt{26596}}{450^{5/4}} = 16.86$$

## 5.4 Velocity Triangles

In a Pelton turbine, the water jet issuing from the nozzle strikes a number of buckets mounted on the wheel simultaneously. In drawing the velocity triangles, the following points should be kept in mind:

(i) $u_1 = u_2$, since $r_1 = r_2$

(ii) $v_{r1} = v_{r2}$, assuming there is no friction at the blades. In actual practice, $v_{r2} < v_{r2}$.

(iii) $v_2 < v_1$

**1. Ideal velocity triangles:**

The actual and assumed velocity triangle are shown in Fig.5.3

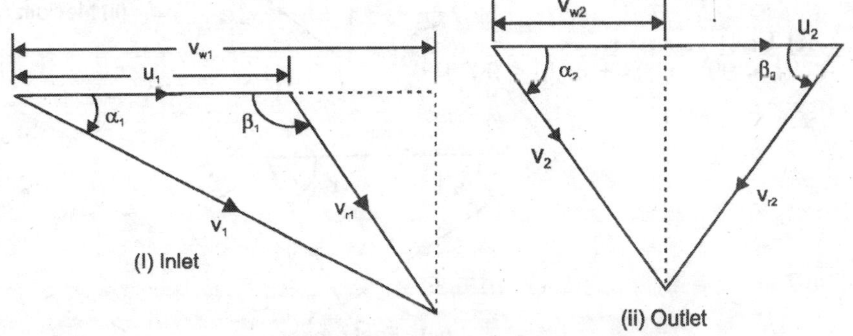


**(a) Actual velocity triangles**

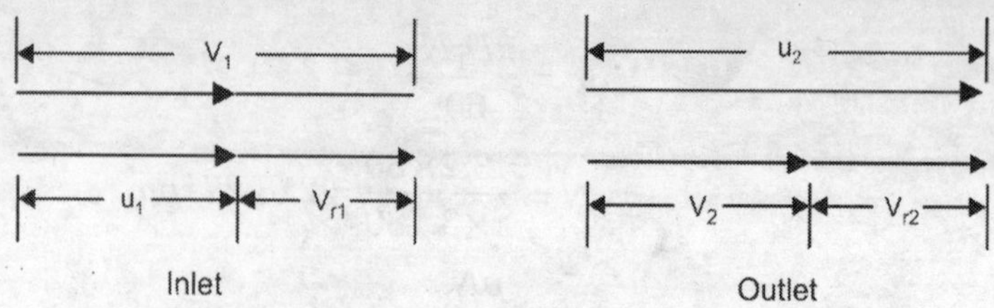


**(b) Assumed velocity kiangles**

**Fig. 5.3 Velocity triangles for Pelton turbine.**

**2. Types of runners:**

The velocity triangles for various types of runners are shown in Fig. 5.4. There are three types of runners.

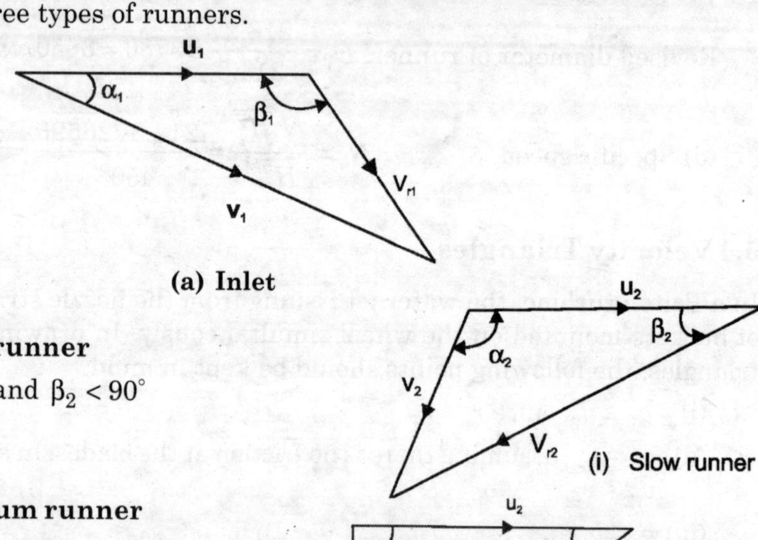


**(i) Slow runner**

$\alpha_2 > 90^\circ$ and $\beta_2 < 90^\circ$

$v_{w2}$ *is* $-ve$

**(ii) Medium runner**

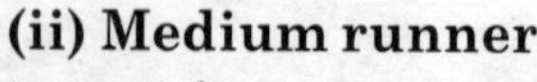

$\alpha_2 = 90^\circ$

$v_{w2} = 0$

$\beta_2 < 90^\circ$

**(iii) Fast runner**

$a_2 < 90^\circ$, $v_{w2}$ *is* $+ve$, $\beta_2 < 90^\circ$

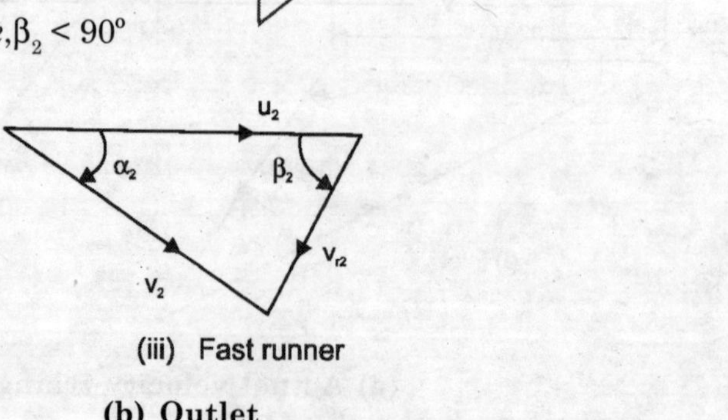


**(b) Outlet**

**Fig. 5.4 Types of runners**

## 5.5 Workdone and efficiency of a Pelton Turbine:

The velocity triangles for the split type buckets of a Pelton turbine are shown in Fig. 5.5

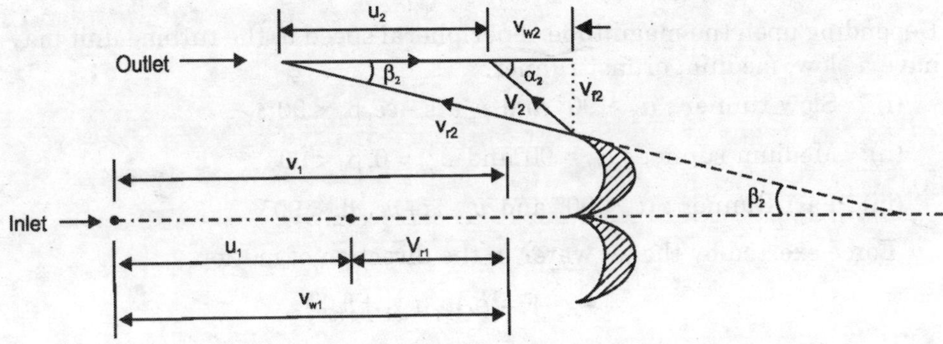


**Fig. 5.5 Velocity triangles for Pelton turbine.**

Let $N$ = rpm of wheel

$D$ = diameter of wheel, m

$d$ = diameter of jet, m

$u$ = mean peripheral velocity of runner $(u_1 = u_2 = u)$

$= \dfrac{\pi DN}{60}\ m/s$

$v_1$ = absolute velocity of water at inlet

$v_{r1}$ = relative velocity of jet at inlet to bucket

$\alpha_1$ = angle between the direction of jet and direction of motion of the bucket (called the guide angle)

$\beta_1$ = angle made by $v_{r1}$ with the direction of motion of bucket at inlet (called vane angle at inlet)

$v_{w1}, v_{f1}$ = components of $v_1$ in the direction of motion of bucket whirl and perpendicular (flow) to it respectively

$v_2$ = absolute velocity of jet leaving the bucket

$v_{r2}$ = relative velocity of jet leaving the bucket

$\alpha_2$ = angle made by $v_2$ with $u_2$

$\beta_2$ = angle made by $v_{r2}$ with $u_2$ (vane angle at outlet)

$v_{w2}, v_{f2}$ = components of $v_2$ in parallel (whirl) and perpendicular (flow) directions to $u_2$ respectively

**Inlet** For $\alpha_1 = 0$ and $\beta_1 = 0$, the velocity triangle at inlet will be a straight line.

$$\therefore \quad v_{r1} = v_1 - u_1 = v_1 - u$$

$$v_{w1} = v_1$$

**Outlet:** Let $v_{r2} = kv_{r1}$

where $k$ = blade friction coefficient

$\simeq 0.85 - 0.95 = 1$ for smooth blades

$$v_{w2} = v_2 \cos\alpha_2 = v_{r2}\cos\beta_2 - u_2 = v_{r2}\cos\beta_2 - u$$

Depending upon the magnitude of peripheral speed u, the turbine unit may have a slow, medium or fast runner.

(i) Slow runner : $\alpha_2 > 90°$ and $v_{w2}$ is $-ve$, $\beta_2 < 90°$

(ii) Medium runner : $\alpha_2 = 90°$ and $v_{w2} = 0, \beta_2 < 90°$

(iii) Fast Runner : $\alpha_2 < 90°$ and $v_{w2}$ is $+ve$, $\beta_2 < 90°$

Force exerted by the jet water in the direction of motion,

$$F = \rho A v_1 (v_{w1} \pm v_{w2})$$

Where $A = \frac{\pi}{4} d^2$ = area of jet

$\rho$ = density of water

Work done per second per unit weight of striking water

$$= \frac{\rho A v_1 (v_{w1} + v_{w2}) u}{\rho A v_1 g} = \frac{1}{g}(v_{w1} + v_{w2}) u$$

Kinetic energy of jet at inlet $= \frac{1}{2} m v_1^2 = \frac{1}{2}(\rho A v_1) v_1^2$

Hydraulic efficiency,

$$\eta_{hyd} = \frac{\text{Work done by turbine per second}}{\text{K.E. of jet per second}}$$

$$= \frac{\rho A v_1 (v_{\omega 1} \pm v_{\omega 2}) u}{\frac{1}{2}(\rho A v_1) v_1^2}$$

$$= \frac{2(v_{w1} \pm v_{w2}) u}{v_1^2}$$

For the velocity triangles shown in Fig. 5.5, we have

$$v_{w1} = v_1, v_{r1} = v_1 - u_1 = v_1 - u$$

$$v_{w2} = v_{w2} = v_{r2}\cos\beta_2 - u_2 = kv_{r1}\cos\beta_2 - u$$

$$= k(v_1 - u)\cos\beta_2 - u$$

$$\therefore \quad \eta_{hyd} = \frac{+ k(v_1 - u)\cos\beta_2 - u] u}{v_1^2}$$

$$= \frac{2[(v_1 - u)(1 + k\cos\beta_2)] u}{v_1^2}$$

The hydraulic efficiency will be maximum for a given value of $v_1$, when

$$\frac{d\eta_{hyd}}{du} = 0$$

Or $$\frac{d}{du}[(v_1 - v)u] = 0$$

Or $$(v_1 - u)\times 1 + u(-1) = 0$$

$$v_1 - 2u = 0$$

$$u = \frac{v_1}{2}$$

Thus, the hydraulic efficiency of Pelton turbine shall be maximum when the velocity of wheel is half of the velocity of jet of water at inlet.

Maximum hydraulic efficiency becomes,

$$(\eta_{hyd})_{max} = \frac{2\left[\left(v_1 - \frac{v_1}{2}\right)(1 + k\cos\beta_2)\right]\frac{v_1}{2}}{v_1^2}$$

$$= \frac{1 + k\cos\beta_2}{2}$$

For no friction on the blades, $k = 1$, and

$$(\eta_{hyd})_{max} = \frac{1 + \cos\beta_2}{2}$$

## 5.6 Definitions of Heads and Efficiencies

The various heads are shown in Fig. 5.6.

**1. Gross head, $H_g$:** The gross or total head is the difference between the water level at the reservoir (also called the head race) and the water level at the tail race.

**2. Net or effective head, $H$:** The head at the inlet of the turbine is called the net or effective head.

$$H = H_g - h_f - h$$

Where $h_f$ = friction head loss between the head race and entrance of the turbine.

$$= \frac{4fLv^2}{2gD}$$

$L$ = length of penstock.

$D$ = diameter of penstock

$v$ = velocity of flow in penstock

$h$ = height of nozzle above the tail race.

V = velocity of flow in penstock

h = height of nozzle above the tail race.

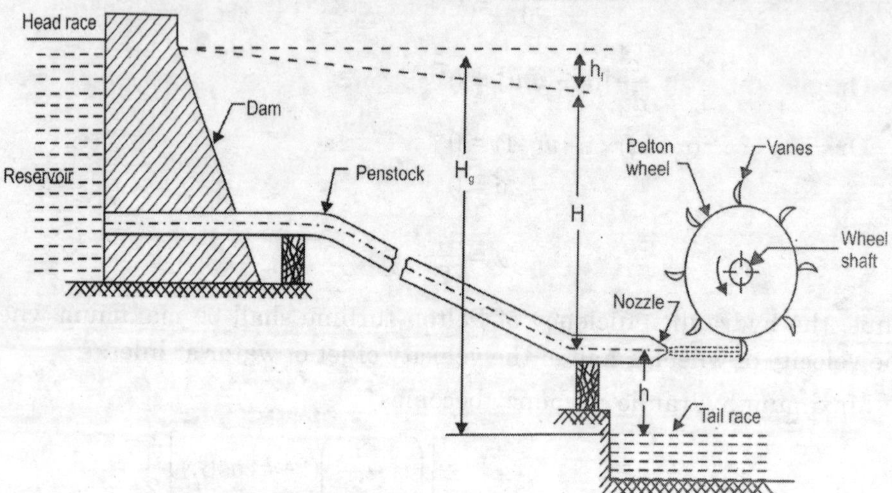


**Fig. 5.6: Various heads acting on Pelton wheel.**

**3. Efficiencies**

The important efficiencies of turbine are:

*(i) Hydraulic efficiency,* $\eta_{hyd}$ : It is defined as the ratio of power developed by the turbine runner to the power supplied by the jet at entrance to the turbine.

$$\eta_{hyd} = \frac{\text{Power developed by the turbine runner}}{\text{Power supplied at the inlet of turbine}}$$

$$= \frac{(v_{w1} \pm v_{w2})u}{gH}$$

$$= \frac{\rho Q_a (v_{\omega 1} \pm v_{\omega 2})u}{\gamma Q_a H} \qquad [\because \gamma = \rho g]$$

$$= \frac{H_r}{H}$$

where $H_r = \frac{1}{g}(v_{\omega 1} \pm v_{\omega 2})u$

= Runner or Euler head

$Q_a$ = actual flow rate to turbine runner

Hydraulic loss within the turbine, $\Delta H = H - H_r$

$v_{\omega 1} - v_{\omega 2}$ = velocities of whirl at inlet and outlet respectively

$u$ = peripheral velocity of vanes

$H$ = net head on the turbine runner

$\gamma$ = specific weight of water

*(ii) Mechanical efficiency, $\eta_{mech}$*: It is defined as the ratio of the power obtained from the shaft of the turbine to the power developed by the runner. These two powers differ by the amount of mechanical losses, viz, bearing friction.

$$\eta_{mech} = \frac{\text{Shaft power}}{\text{Runner (bucket) power}}$$

$$= \frac{P}{\gamma Q_a \left(\frac{v_{\omega 1} \pm v_{\omega 2}}{g}\right) u} = \frac{P}{\gamma Q_a H_r}$$

$$\eta_{mech} = 0.97 \text{ to } 0.99$$

*(iii) Volumetric efficiency, $\eta_{vol}$*: It is the ratio of the volume of water actually striking the runner ($Q_a$) to the volume of water supplied by the jet to the turbine ($Q$).

$$\eta_{vol} = \frac{Q_a}{Q} \simeq 0.97 - 0.99$$

where $Q_a = Q - \Delta Q, \Delta Q$ = quantity of water lost on account of slip.

*(iv) Overall efficiency, $\eta_o$*: It is defined as the ratio of power available at the turbine shaft to the power supplied by the water jet.

$$\eta_o = \frac{\textit{Shaft power}}{\textit{Water power}} = \frac{P}{\gamma Q H}$$

$$\eta_o \simeq 0.85 - 0.90.$$

$$\eta_o = \eta_{hyd} \times \eta_{mech} \times \eta_{vol}$$

$$= \frac{H_r}{H} \times \frac{P}{\gamma Q_a H_r} \times \frac{Q_a}{Q} = \frac{P}{\gamma Q H}$$

If $\eta_g$ = efficiency of generator

Then power output of hydro unit (turbine + electric generator)

$$= (\gamma Q H) \times \eta_o \times \eta_g$$

The product $\eta_o \times \eta_g$ is known as *hydroelectric plant efficiency.*

*(v) Jet efficiency or Head efficiency*

Force exerted by the jet in the direction of motion,

$$F = \rho A v_1 (v_{\omega 1} - v_{\omega 2}) = \rho A v_1 [v_1 \cos\alpha_1 - (u_2 - v_{r2} \cos\beta_2)]$$

$$= \rho A v_1 [v_1 \cos\alpha_1 - (u_2 - v_{r2} \cos\beta_2)]$$

$$= 2\rho A v_1 (v_1 - u) \text{ for } \alpha_1 = 0° \text{ and } \beta_2 = 0°, \text{ k} = 1$$

$$v_1 = K_{v1}\sqrt{2gH}$$

$$u_1 = K_{u1}\sqrt{2gH}$$

$$F = 2\rho A.\, K_{v1}\sqrt{2gH}\,\left[K_{v1}\sqrt{2gH} - K_{u1}\sqrt{2gH}\right]$$

$$= 2\rho A\,(2gH)\left[K_{v1}^2 - K_{v1}.K_{u1}\right]$$

$$= 4\gamma AH\left[K^2_{v1} - K_{v1}.K_{u1}\right]$$

Work done, $W = F \times u_1$

$$= 4\gamma AH\left[K_{u1}.K^2_{v1} - K_{v1}.K^2_{u1}\right] \times \sqrt{2gH}$$

Power developed considering head lost,

$$P_H = \frac{W}{10^3}\,kW$$

Power available from turbine, or water power

$$P_a = \frac{\gamma QH}{10^3} = \frac{\gamma A k_{v1}\sqrt{2gH}.H}{10^3}\,kW$$

By definition, the jet efficiency or head efficiency,

$$\eta_H = \frac{P_H}{P_a} = \frac{4\gamma AH\left[K_{u1}.K^2_{v1} - K_{v1}.K^2_{u1}\right] \times \sqrt{2gH}}{\gamma AH \times K_{v1}\sqrt{2gH}}$$

$$= 4\left[K_{u1}.K_{v1} - K^2_{u1}\right] = 4K_{u1}(K_{v1} - K_{u1})$$

For maximum efficiency, $\frac{d\eta_H}{dK_{u1}} = 0$ for $K_{v1} = const$

$\therefore$ $$K_{v1} - 2K_{v1} = 0$$

or $$K_{u1} = \frac{K_{v1}}{2}$$

or $$u_1 = \frac{v_1}{2}$$

$$(\eta_H)_{max} = 4\left[\frac{K_{v1}^2}{2} - \frac{K_{v1}^2}{4}\right] = K_{v1}^2$$

**Example 5.3** A Pelton wheel is receiving water from a penstock with a gross head of 510m. One third of gross head is lost in friction in the penstock. The rate of flow through the nozzle fitted at the end of penstock is 2.2 m$^3$/s. The angle of deflection of jet is 165°. Calculate:

(a) The power supplied by water jet to runner, and

(b) Hydraulic efficiency of Pelton wheel.

Assume cofficient of velocity, $K_v = 1.0$ and speed ratio, $K_u = 0.45$.

**Solution:**

Given: $H_g = 510m$, $h_f = \dfrac{H_g}{3}$, $Q = 2.2$ m³/s, $\beta_2 = 180° - 165° = 15°$, $K_v = 1.0$, $K_u = 0.45$

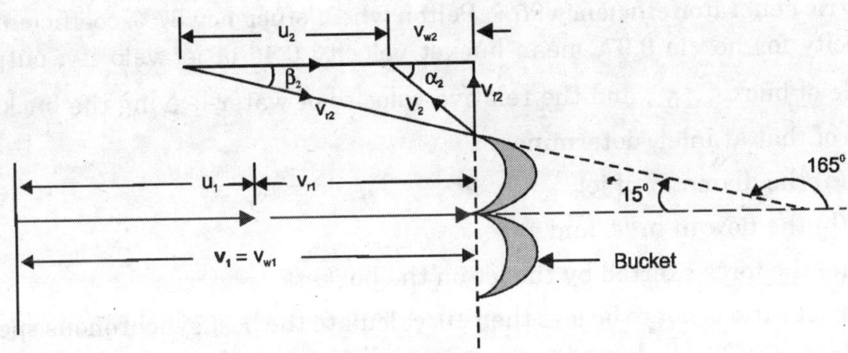


**Fig. 5.7**

Net head,
$$H = H_g - h_f$$
$$= H_g - \frac{H_g}{3} = \frac{2H_g}{3}$$
$$= \frac{2\times510}{3} = 340 \text{ m}$$

(a) Velocity of jet, $v_1 = K_v\sqrt{2gH} = 1.0\sqrt{2\times9.81\times340} = 81.67$ m/s

Wheel velocity, $u = K_u\sqrt{2gH} = 0.45\times81.67 = 36.75$ m/s

$$v_{r1} = v_1 - v = 81.67 - 36.75 = 44.92 \text{ m/s}$$
$$v_{w1} = v_1 = 81.67 \text{ m/s}$$
$$v_{r2} = v_{r1} = 44.92 \text{ m/s}$$
$$v_{w2} = v_{r2}\cos\beta_2 - u$$
$$= 44.92 \cos 15^\circ \quad 36.75 \quad 6.64 \text{ m/s}$$

Work done by the runner per second $= \rho\, Q\,(v_{\omega1} + v_{\omega2})\, u$

$$= 1000 \times 2.2(81.67 + 6.64)\times\frac{36.75}{10^3}$$
$$= 7139.8 \text{ kW}$$

(b) Hydraulic efficiency, $\eta_{hyd} = \dfrac{2(v_{\omega1} + v_{\omega2})\, u}{v_1^2}$

$$= \frac{2(81.67 + 6.64) \times 36.75}{(81.67)^2}$$

$$= 0.973 \text{ or } 97.3\%$$

**Example 5.4** A single jet Pelton turbine is required to drive a generator to develop 10 MW under available head at the nozzle of 760 m. Assuming electric generator efficiency 95%, Pelton wheel efficiency 87%, coefficient of velocity for nozzle 0.97, mean bucket velocity 0.46 of jet velocity, output angle of bucket $15°$, and the relative velocity of water leaving the buckets 0.85 of that at inlet, determine:

(a) the diameter of jet,

(b) the flow in m³/s, and

(c) the force exerted by the jet on the buckets.

If the jet ratio is not to be less then 10, calculate the best synchronous speed for generator at 50 Hz and the corresponding mean diameter of runner.

**Solution.**

Given: $P_g = 10$ MW, $H = 760$ m, $\eta_g = 0.95$

$\eta_t = 0.87$, $K_{v1} = 0.97$, $K_{u1} = 0.46$

$\beta_2 = 15°$, $k = 0.85$, $m > 10$, $f = 50$ Hz

Output of turbine,

$$P_t = \frac{P_g}{\eta_g} = \frac{10{,}000}{0.95} = 10{,}526.3 \text{ kW}$$

Available power,

$$P_a = \frac{P_t}{\eta_t} = \frac{10{,}526.3}{0.87} = 12{,}099.2 \text{ kW}$$

Now
$$P_a = \frac{\gamma Q H}{10^3}$$

$$Q = \frac{12{,}099.2 \times 10^3}{9.81 \times 10^3 \times 760} = 1.6228 \text{ m}^3/\text{s}$$

$$Q = \frac{\pi}{4} d^2 . K_{v1} \sqrt{2gH}$$

$$1.6228 = \frac{\pi}{4} d^2 \times 0.97 \times \sqrt{2 \times 9.81 \times 760}$$

$$d = 0.132 \text{ m or } 132 \text{ mm}$$

For velocity triangles, refer to Fig. 5.8

$$v_1 = K_{v1}\sqrt{2gH} = 0.97 \times \sqrt{2 \times 9.81 \times 760} = 118.45 \text{m/s}$$

$$u_1 = K_{u1}\sqrt{2gH} = 0.46\times\sqrt{2\times9.81\times760} = 56.17\text{m/s}$$

$$v_{r1} = v_1 - u_1 = 118.45 - 56.17 = 62.28\text{m/s}$$

$$v_{r2} = kv_{r1} = 0.85\times62.28 = 52.94\text{m/s}$$

$$v_2 = u_1 = 56.17\text{m/s}$$

$$v_{w1} = v_1 = 118.45\text{m/s}$$

$$v_{w2} = u_2 - v_{r2}\cos\beta_2 = 56.17 - 52.94\cos 15°$$

$$= 5.03 \text{ m/s}$$

Force exerted by jet on buckets,

$$F = \rho Q(v_{w1} - v_{w2})$$

$$= 10^3 \times 1.6228(118.45 \quad 5.03)$$

$$= 1{,}84{,}058 \text{ N}$$

Jet ratio, $m = \dfrac{D}{d} = 10$

$\therefore$ $D = 10 \times 132 = 1320$ mm

$$u_1 = \frac{\pi DN}{60\times10^3}$$

$$56.17 = \frac{\pi\times1320\times N}{60\times10^3}$$

$$N = 812.7 \text{ rpm}$$

$$f = \frac{p.N}{60}$$

Let p = 4, $N_{syn}$ = 750 rpm, which is nearest to 812.7 rpm

Revised value of $D = \dfrac{1320}{750}\times812.7 = 1430$ mm

**Example 5.5** Estimate the time required to stop the wheel by the brake nozzle under the following conditions:

Diameter of brake nozzle = 80 mm

Head of nozzle= 480 mm

Original velocity of buckets = 45 m/s

Corresponding kinetic energy of wheel and all revolving parts = 16 MN.m

Assume that the dynamic pressure of the buckets is equivalent to that on flat radial blades.

**Solution.**

Given: $d_1$ = 80 mm, $H$ = 480m, $u_1$ = 45m/s, $u_2$ = 0

Velocity of nozzle jet, $v_1 = K_{v1}\sqrt{2gH}$

Let $K_{v1} = 0.985$

Then $v_1 = 0.985\sqrt{2\times9.81\times480} = 95.6$ m/s

Cross section area of nozzle,

$$A = \frac{\pi}{4}d_1^2 = \frac{\pi}{4}(80)^2 = 5026.5 \text{ mm}^2$$

Rate of flow through nozzle,

$$Q = Av_1 = 5026.5\times10^{-6}\times95.6 = 0.48 \text{ m}^3/\text{s}$$

Average peripheral velocity of buckets

$$= \frac{1}{2}(45+0) = 22.5m/s$$

Average force acting on buckets,

$$F = \rho Q[v_1 - (-u_1)]$$

$$= 10^3 \times 0.48[95.6+22.5] = 56{,}688 \text{ N}$$

Torque or turning moment of wheel,

$$\text{T} = \frac{FD_1}{2} = 56{,}688\times\frac{D_1}{2} = 28{,}344 \text{ D}_1 \text{ N.m}$$

Let I = moment of inertia of rotating parts

Then $\frac{1}{2}\text{I}\omega_1^2$ = K.E. of all rotating parts = $16 \times 10^6$ N.m

Where $\omega_1 = \frac{2u_1}{D_1} = \frac{2\times45}{D_1} = \frac{90}{D_1}$ rad/s

$$\text{I} = \frac{16\times10^6\times2}{\left(\frac{90}{D_1}\right)^2} = 3950.6 \text{ D}_1^2$$

Now $\quad$ T = I.α = moment of inertia × angular acceleration

$$28{,}344 \text{ D}_1 = 3950.6 \text{ D}_1^2 \times \alpha$$

$$\alpha = \frac{7.175}{D_1}$$

But $\alpha = \frac{\omega}{t}$

or $\quad t = \frac{\omega_1}{\alpha} = \frac{90}{D_1}\times\frac{D_1}{7.175} = 12.5$ s

**Example 5.6** A Pelton turbine is installed 400m below the water surface of reservoir. The jet diameter is 80 mm and velocity coefficient is 0.98. The pipe diameter is 0.6 m, length 4.5 km and friction coefficient is 0.032 in the head loss formula $\frac{fLv^2}{2gD}$. The buckets deflect the jet through $165^\circ$ and run at 0.48 times the jet speed. Bucket friction reduces the outlet velocity by 15 percent of relative velocity at inlet. Mechanical efficiency is 90%. Calculate:

(a) the flow rate, and

(b) the shaft power developed by the turbine.

**Solution.**

Given: $H = 400\text{m}$, $d_1 = 80$ mm, $K_{v1} = 0.98$, $D = 0.6\text{m}$, $L = 4.5$ km, $f = 0.032$, $\beta_2 = 180° - 165° = 15°$, $u_1 = 0.48\ v_1$, $k = 0.85$, $\eta_{mech} = 0.9$

(a) From continuity equation, we have

Flow in pipe = Flow in nozzle jet

$$Av = A_1 v_1$$

$$\frac{\pi}{4} D^2 v = \frac{\pi}{4} d_1^2 v_1$$

$$(0.6)^2 \times v = (80\times10^{-3})^2 \times v_1$$

$$v = 0.01778\ v_1$$

Applying Bernoulli's equation to free surface in reservoir and outlet of the nozzle, we have

Theoretical velocity at nozzle jet = $(v_1)_{th}$

Actual velocity of water jet = $v_1$

$$(v_1)_{th} = \frac{v_1}{K_{v1}}$$

$$\text{Head lost in nozzle} = \frac{(v_1)_{th}^2 - v_1^2}{2g} = \frac{v_1^2}{2g}\left[\frac{1}{K_{v1}^2} - 1\right]$$

$$\text{Head at reservoir} = \frac{v_1^2}{2g} + \frac{fLv^2}{2gD} + \frac{v_1^2}{2g}\left[\frac{1}{K_{v1}^2} - 1\right]$$

$$400 = \frac{fLv^2}{2gD} + \frac{v_1^2}{2g} \times \frac{1}{K_{v1}^2}$$

$$= \frac{0.032\times4.5\times10^3 \times v^2}{2\times9.81\times0.6} + \left(\frac{v}{0.01778}\right)^2 \times \frac{1}{2\times9.81} \times \frac{1}{(0.98)^2}$$

$$= 180.107\ v^2$$

$$v = 1.49 \text{ m/s}$$

$$v_1 = \frac{1.49}{0.01778} = 83.8 \text{ m/s}$$

(b) Flow rate, $Q = A_1 v_1 = \frac{\pi}{4}(0.08)^2 \times 83.8$

$= 0.421 \text{ m}^3/\text{s}$

$v_1 = 0.48\ v_1$

$= 0.48 \times 83.8 = 40.224 \text{ m/s}$

Refer to Fig. 5.8 for velocity triangles.

$$v_{r1} = v_1 - u_1$$
$$= 83.8 - 40.224 = 43.576 \text{ m/s}$$
$$v_{w1} = v_1 = 83.8 \text{ m/s}$$
$$v_{r2} = kv_{r1} = 0.85 \times 43.576$$
$$= 37.04 \text{ m/s}$$
$$v_{\omega 2} = u_2 - v_{r2} \cos \beta_2$$
$$= 40.224 - 37.07 \cos 15° = 4.446 \text{ m/s}$$

Mechanical efficiency, $\eta_{mech} = \frac{\textit{shaft} \text{ power}}{\textit{power} \text{ given to runner}}$

Shaft power, $SP = \eta_{mech} \times \rho Q(v_{w1} - v_{w2})u_1$

$= 0.90 \times 103 \times 0.421\ (83.8 - 4.446) \times 40.224$

$= 1209.42 \text{ kW}$

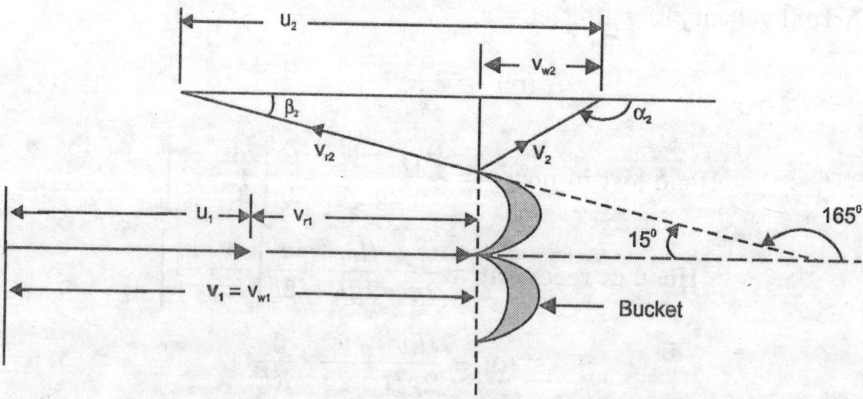


Fig. 5.8

**Example 5.7** The diameter of the nozzle of a Pelton wheel is 40 mm. The head at the nozzle is 100 metres. The velocity coefficient of the nozzle is 0.96 and the wheel diameter is 400 mm. The relative velocity at outlet is 90 percent of that at inlet. The angle through which water is deflected is 170 degree. The blade speed to jet speed is 0.45. Determine the unit speed and unit power. [IES, 1983]

**Solution.**

Given:

$$d_1 = 40 \text{ mm}, H = 100 \text{ m}, K_{v1} = 0.96$$

$$D_1 = 400 \text{ mm}, k = 0.9$$

$$\beta_2 = 180° - 170° = 10°$$

$$K_{u1} = 0.45$$

$$v_1 = K_{v1}\sqrt{2gH} = 0.96\sqrt{2 \times 9.81 \times 100}$$

$$= 42.5 \text{ m/s}$$

$$u_1 = K_{u1}\sqrt{2gH} = 19.9 \text{ m/s}$$

Refer to fig. 5.7

$$v_{w1} = v_1 = 42.5 \text{ m/s}$$

$$v_{r1} = v_1 - u = 42.5 - 19.9 = 22.6 \text{ m/s}$$

$$v_{r2} = kv_{r1} = 0.9 \times 22.6 = 20.34 \text{ m/s}$$

$$v_2 = u_1 = 19.9 \text{ m/s}$$

$$u_{w2} = v_{r2} \cos \beta_2 - u_2$$

$$= 20.34 \cos 10° - 19.9$$

$$= 0.13 \text{m/s } (-ve)$$

$$Q = \frac{\pi}{4} d_1^2 \ v_1$$

$$= \frac{\pi}{4}(40)^2 \times 10^{-6} \times 42.5$$

$$= 0.0534 \text{ m}^3/\text{s}$$

Power developed,

$$P = \rho Q (v_{w1} + v_{w2})u$$

$$= 103 \times 0.0534 \ (42.5 + 0.13) \times 19.9$$

$$= 45.3 \text{ kW}$$

$$u_1 = \frac{\pi D_1 N_1}{60} = \frac{\pi \times 0.4 \times N_1}{60} = 19.9$$

$$N_1 = 950 \ rpm$$

Unit power,

$$P_1 = \frac{P}{H^{3/2}} = \frac{45.3}{100^{3/2}} = 0.0453$$

Unit speed,

$$N_1 = \frac{N}{\sqrt{H}} = \frac{950}{\sqrt{100}} = 95$$

**Example 5.8** A Pelton turbine is to work at the foot of dam whose reservoir level is 220m. The head at the full opening at the turbine nozzle is 200 m and the coefficient of velocity is 0.98. The turbine is to operate at 200 rpm and develop a power of 3.7 MW. Assuming the blade to jet speed ratio as 0.46, estimate the wheel diameter. If the blade outlet angle is 16°, determine the blade and hydraulic efficiencies. Neglect frictional losses. [IES, 2005]

**Solution**

Given: $H$ = 200 m, $K_{v1}$ = 0.98, $N$ = 200 rpm, $P_t$ = 3.7 MW, $K_{u1}$ = 0.46
$\beta_2$ = 16°, $H_g$ = 220 m

$$v_1 = K_{v1}\sqrt{2gH} = 0.98\sqrt{2\times 9.81\times 200}$$
$$= 61.39 \text{ m/s}$$
$$u_1 = K_{u1}\sqrt{2gH} = 0.46\sqrt{2\times 9.81\times 200}$$
$$= 28.81^{\text{m/s}}$$
$$\text{P}_\text{t} = \frac{\gamma QH}{10^3}$$
$$3.7\times 10^3 = \frac{9.81\times 10^3\times Q\times 200}{10^3}$$
$$Q = 1.886 \text{ m}^3\text{/s}$$
$$P_w = \frac{\rho Q\,(v_{\omega 1} + v_{\omega 2})\text{ u}}{10^3}\ kW$$

Refer to Fig. 5.7

$$v_{w1} = v_1 = 61.39 \text{ m/s}$$
$$v_{r1} = v_1 - v_1 = 61.39 - 28.81 = 32.58\text{m/s}$$
$$v_{r2} = v_{r1} \text{ for no friction loss on blades}$$
$$v_2 = u_1 = u$$
$$v_{w2} = v_{r2}\cos\beta_2 - u_2$$
$$= 32.58\cos 16° - 28.81 = 2.50 \text{ m/s}$$
$$P_w = \frac{10^3\times 1.886\ (61.39+2.50)\times 28.81}{10^3}$$
$$= 3471.5 \text{ kW}$$

Power input to buckets,

$$\text{P}_b = \frac{1}{2}\ (\rho Q)\ v_1^2/10^3$$
$$= \frac{1}{2}\left(10^3\times 1.8886\right)\times(61.39)^2/10^3$$
$$= 3553.9 \text{ kW}$$

Blade (or bucket) efficiency, $\eta_b = \dfrac{P_b}{P_t} = \dfrac{3553.9}{3700} = 0.9605$ or 96.05%

Hydraulic efficiency, $\eta_{hyd} = \dfrac{P_w}{P_t} = \dfrac{3471.5}{3700} = 0.9382$ or 93.82%

**Example 5.9** A single jet Pelton wheel runs at 300 rpm under a head of 510 m. The jet diameter is 200 mm. The deflection of jet inside the bucket is 165° and its relative velocity is reduced by 15% due to friction. Determine:

(a) the water power,

(b) the resultant force on the bucket,

(c) the brake power, and

(d) the overall efficiency

Take: Mechanical losses = 3%, coefficient of velocity = 0.98, and speed ratio = 0.46.

**Solution.**

Given: $N = 300$ rpm, $H = 510$m, $d_r = 200$ mm, $\beta_2 = 180° - 165° = 15°$, $K = 0.85$, $K_{v1} = 0.98$, $K_{u1} = 0.46$, $\eta_{mech} = 100 - 3 = 97\%$

(a) Velocity of jet, $v_1 = K_{v1}\sqrt{2gH} = 0.98\sqrt{2\times 9.81\times 510}$

$= 98$ m/s

Discharge through the nozzle, $Q = \frac{\pi}{4}d_1^2 v_1 = \frac{\pi}{4}(0.2)^2 \times 98 = 3.0788$ m³/s

$$\text{Water power} = \frac{\gamma QH}{10^3} = \frac{9.81\times 10^3\times 3.0788\times 510}{10^3}$$

$= 15403$ kW

(b) Peripheral speed of wheel, $u_1 = K_{u1}\sqrt{2gH} = 0.98\sqrt{2\times 9.81\times 510}$

$= 46$ m/s

Refer to Fig.5.8, as $(v_{r2}\cos\beta_2) < u_2$

$v_{r1} = v_1 - u_1 = 97 - 46 = 52$ m/s

$v_{w1} = v_1 = 98$ m/s

$v_{r2} = kv_{r2} = 0.85 \times 52 = 44.2$ m/s

$u = u_2 = u_1 = 46$ m/s

$v_{w2} = u_2 - v_{r1}\cos\beta_2 = 46 - 44.2\cos 15°$

$= 3.31$ m/s

Resultant force on bucket, $F = \rho Q(v_{w1} - v_{w2})$

$= 10^3 \times 3.0788\,(98 - 3.31) = 291.53$ kN

(c) Power developed by the wheel, $P_a = F \times u = 291.53 \times 46 = 13410$ kW

Brake power, $P_t = P_a\,\eta_{mech} - 13410 \times 0.97 - 13008$ kW

(d) Overall efficiency, $\eta_o = \frac{\textit{brake}\text{ power}}{\textit{water}\text{ power}} = \frac{13008}{15403}$

$= 0.8445$ or $84.45\%$

**Example 5.10** A Pelton wheel of 1.1m mean bucket diameter works under a head of 500m. The deflection of jet is 165° and its relative velocity is reduced over the bucket by 15 percent due to friction. If the diameter of jet is 100mm and water is to leave the bucket without any whirl, calculate:

(a) Rotational speed of wheel

(b) Ratio of bucket speed to jet velocity,

(c) Impulsive force and power developed by the wheel

(d) Available power (water power)

(e) Power input to buckets, and

(f) Efficiency of the wheel with power input to bucket as reference input.

Take coefficient of velocity = 0.97.

**Solution.**

Given: $D_1 = 1.1$ m, $H = 500$ m, $\beta_2 = 180° - 165° = 15°$, $k = 0.85$, $d_1 = 100$ mm, $\alpha_2 = 90°$, $K_{v1} = 0.97$

(a) Velocity of jet, $v_1 = K_{v1}\sqrt{2gH} = 0.97\sqrt{2 \times 9.81 \times 500}$

$= 96.07$ m/s

Bucket speed, $u = u_1 = u_2$

$v_{r1} = v_1 - u_1 = 96.07 - u_1$

$v_{r2} = kv_{r1} = 0.85\ (96.07 - u_1)$

For $\alpha_2 = 90°$, $v_{r2} \cos \beta_2 = u_2$

$0.85\ (96.07 - u) \cos 15° = u$

$78.88 - 0.82\ u = u$

$u = 43.34$ m/s

$$u_1 = \frac{\pi D_1 N}{60}$$

$$N = \frac{60 \times 43.34}{\pi \times 1.1} = 752.5 \text{ rpm}$$

(b) $$\frac{u}{v_1} = \frac{43.34}{96.07} = 0.45$$

(c) $v_{w1} = v_1 = 96.07$ m/s

$v_{w2} = 0$

Impulsive force, $F = \rho Q(v_{w1} - v_{w2})$

$$Q = \frac{\pi}{4} d_1^2 v_1 = \frac{\pi}{4}(0.1)^2 \times 96.07 = 0.7545 \text{ m}^3/\text{s}$$

$F = 10^3 \times 0.7545 \times 96.70 = 72484.8$ N

Power developed by wheel, $$P_w = \frac{F \times u}{10^3} = \frac{72484.8 \times 43.34}{10^3}$$

$= 3141.5$ kW

(d) Available power, $$P_a = \frac{\gamma QH}{10^3} = \frac{9.81 \times 10^3 \times 0.7545 \times 500}{10^3}$$

$= 3700.8$ kW

(e) Power input to buckets, $P_b = \frac{1}{2}(\rho Q)\ v_1^2\ /10^3$

$$= \frac{1}{2} \times 1000 \times 0.7545 \times (96.07)^2 / 10^3$$

$$= 3481.808 \text{ kW}$$

(f) Efficiency of wheel, $\eta_{wheel} = \frac{P_w}{P_b} = \frac{3141.5}{3481.808}$

$$= 0.9023 \text{ or } 90.23\%$$

**Example 5.11** The design particulars of a Pelton turbine are given below: Head at distributor = 630m; discharge = 13.5 m³/s; brake power = 65 MW; speed of rotation = 500 rpm; runner diameter = 2m; number of jets = 4; jet diameter = 200mm; angle through which jet is deflected by the bucket = 150°; and mechanical efficiency of turbine = 96%. Determine the hydraulic power losses in the distribution nozzle assembly and the buckets.

**Solution.**

Given: $H = 630$ m, $Q = 13.5$ m³/s, $P_t = 65$ MW, $N = 500$ rpm, $D_1 = 2$ m, $n = 4$
$d_1 = 200$ mm, $\beta_2 = 180° - 150° = 30°$, $\eta_{mech} = 0.96$

Actual power developed, $P_a = \frac{P_t}{\eta_{mech}} = \frac{65}{0.96} = 67.708 \text{ MW}$

Power developed per jet $= \frac{P_a}{n} = \frac{67.708}{4} = 16.927 \text{ MW}$

Discharge per jet, $q = \frac{Q}{n} = \frac{13.5}{4} = 3.375 = 3.375 \text{ m}^3/\text{s}$

Now $q = A_1 v_1$

$$3.375 = \frac{\pi}{4}(0.2)^2 \times v_1$$

$$v_1 = 107.43 \text{ m/s}$$

Peripheral speed of runner, $u = \frac{\pi D_1 N}{60} = \frac{\pi \times 2 \times 500}{60} = 52.36 \text{ m/s}$

Power developed per jet $= \rho q(v_1 - u)(1 + k\cos\beta_2)u$

$\therefore \quad 10^3 \times 3.375(107.43 - 52.36)(1 + k\cos 30°) \times 52.36 = 16.927 \times 10^6$

$$k = 0.85$$

$$v_{r1} = v_1 - u_1 = 107.43 - 52.36 = 55.07 \text{ m/s}$$

$$v_{r2} = k v_{r1} = 0.85 \times 55.07 = 46.81 \text{ m/s}$$

Theoretical velocity of jet, $(v_1)_{th} = \sqrt{2gH} = \sqrt{2 \times 9.81 \times 630} = 111.18 \text{ m/s}$

Actual velocity of jet, $v_1 = K_{v1}\sqrt{2gH}$

$\therefore \quad K_{v1} = \frac{107.43}{111.18} = 0.966$

Head lost in nozzle $= \left(1 - K_{v1}^2\right) H = [1 - (0.966)^2] \times 630 = 42.11 \text{ m}$

$$\text{Head lost in nozzle} = \frac{V_{r1}^2 - V_{r2}^2}{2g} = \frac{V_{r1}^2}{2g}\left(1-k^2\right)$$

$$= \frac{(55.07)^2}{2\times 9.81}\left[1-(0.85)^2\right] = 42.89 \text{ m}$$

Head lost in nozzle and buckets, $H_L = 42.11 + 42.89 = 85$ m

$$\text{Power loss} = \frac{\gamma Q H_L}{10^3} = \frac{9.81\times 10^3 \times 13.5\times 85}{10^3}$$

$$= 11256.975 \text{ kW}$$

**Example 5.12** In a hydroelectric plant, a number of Pelton wheels are to be used under the following conditions:

Total output = 30 MW; gross head = 245 m; speed = 375 rpm; number of jets per wheel = 2; velocity coefficient for nozzle = 0.97; overall efficiency = 81.5%; dimensionless specific speed not to exceed per jet = 0.22 rev; frictional head loss in pipe line = 12 m; speed ratio = 0.46.

Calculate:

(a) the number of wheels required,

(b) the diameter of jets and wheels,

(c) the hydraulic efficiency, if the blade deflects the water jet through 165° and reduces its relative velocity by 15%, and

(d) the percentage of input power lost as kinetic energy at exit.

**Solution.**

Given: $P = 30$ MW, $H_g = 245$ m, $N = 375$ rpm, $n = 2$, $K_{v1} = 0.97$, $\eta_o = 0.815$, $Ns = 0.022$, $h_f = 12$ m, $K_{u1} = 0.46$, $k = 0.85$, $\beta_2 = 15°$

Net head, $H = H_g - h_f = 245 - 12 = 233$ m

(a) Dimensionless specific speed, $N_s = \dfrac{N\sqrt{P_t}}{\sqrt{\rho}\,(gH)^{5/4}}$

$$0.022 = \frac{\frac{375}{60}\times\sqrt{P_t}}{\sqrt{1000}\,(9.81\times 233)^{5/4}}$$

$$P_t = 3.0949 \text{ MW/Jet}$$

$$\text{Number of jets} = \frac{P}{P_t} = \frac{30}{3.0949} = 9.7 \simeq 10$$

$$\text{Number of wheels} = \frac{10}{n} = \frac{10}{2} = 5$$

(b)

$$Q = \frac{P}{\gamma H \eta_o} = \frac{3.0949\times 10^6}{9.81\times 10^3 \times 233\times 0.815}$$

$$= 1.66 \text{ m}^3\text{/s}$$

$$\text{v}_1 = K_{v1}\sqrt{2gH} = 0.97\sqrt{2\times 9.81\times 233}$$

$$= 65.58 \text{ m/s}$$

Dia of jet, $$d_1 = \sqrt{\frac{4Q}{\pi \text{v}_1}} = \sqrt{\frac{4\times 1.66}{\pi\times 65.58}}$$

$$= 0.1795 \text{ m or } 179.5 \text{ mm}$$

Speed of wheel, $$u = K_{u1}\times v_1 = 0.46\times 65.58 = 30.167 \text{ m/s}$$

$$= \frac{\pi D_1 N}{60} = \frac{\pi D_1\times 375}{60}$$

$$D_1 = 1.54 \text{ m}$$

(c) $$\text{v}_1 = \text{v}_{w1} = 65.58 \text{ m/s}$$

$$\text{v}_{r1} = \text{v}_1 - u_1 = 65.58 - 30.167 = 35.413 \text{ m/s}$$

$$\text{v}_{r2} = k\text{v}_{r1} = 0.85\times 35.413 = 30.10 \text{ m/s}$$

$$\text{v}_{w2} = u_2 - \text{v}_{r2}\cos\beta_2 = 30.167 - 30.10\cos 15^\circ$$

$$= 1.09 \text{m/s}$$

Shaft power, $$\text{P}_s = \frac{\rho Q(\text{v}_{w1}-\text{v}_{w2})u}{10^3}$$

$$= \frac{10^3\times 1.66(65.58-1.09)\times 30.167}{10^3}$$

$$= 3229.5 \text{ kW}$$

$$\eta_{hyd} = \frac{P_s}{\gamma QH} = \frac{3229.5}{9.81\times 1.66\times 233}$$

$$= 0.851 \text{ or } 85.1\%$$

(d) $$v_2 = \sqrt{v_{w2}^2 + v_{f2}^2}$$

$$= \sqrt{(1.09)^2 + \left(30.10 \sin 15^\circ\right)^2}$$

$$= 7.866 \text{ m/s}$$

$$\text{K.E. lost per } kg = \frac{v_2^2}{2}$$

$$\% \text{ lost K.E.} = \frac{v_2^2}{2gH} = \frac{(7.866)^2\times 100}{2\times 9.81\times 233} = 1.353\ \%$$

**Example 5.13** A single jet Pelton turbine runs at 320 rpm under a head of 600m. The jet diameter is 180 mm. The deflection of jet inside the bucket is 160°. The relative velocity is reduced by 15% due to friction. Calculate:

(a) Water power,

(b) Resultant force on the bucket, and

(c) Overall efficiency.

Take mechanical losses = 3%, coefficient of velocity = 0.98 and speed ratio = 0.45.

**Solution.**

**Given:** $N = 320$ rpm, $H = 600$ m, $d = 180$ mm, $k = 0.85$, $K_v = 0.98$, $K_u = 0.45$

(a) Jet velocity,
$$v_1 = K_v\sqrt{2gH}$$
$$= 0.98\sqrt{2\times9.81\times600} = 106.3 \text{ m/s}$$
$$Q = \text{area of jet} \times \text{jet velocity}$$
$$= \frac{\pi}{4}(0.18)^2\times106.3 = 2.706 \text{ m}^3/\text{s}$$
$$\text{Water power} = \frac{\gamma QH}{10^3} = \frac{9.81\times10^3\times2.706\times600}{10^6}$$
$$= 15.926 \text{ MW}$$

(b)
$$u = K_u\sqrt{2gH} = 0.45\sqrt{2\times9.81\times600}$$
$$= 48.8 \text{ m/s}$$

Referring to Fig. 5.8, we have
$$v_{w1} = v_1 = 106.3 \text{ m/s}$$
$$v_{r1} = v_1 - u_1 = 106.3 - 48.8 = 57.5$$
$$\beta_2 = 180° - 160° = 20°$$
$$v_{r2} = kv_{r1} = 0.85\, v_{r1} = 0.85 \times 57.5$$
$$= 48.87 \text{ m/s}$$
$$v_{w2} = u_2 - v_{r2}\cos 20° = 2.88 \text{ m/s}$$

Resultant force on bucket,
$$F = \rho Q\,(v_{w1} - v_{w2})$$
$$= 10^3 \times 2.706\,(106.3 - 2.88) = 279854 \text{ N}$$

(c)
$$\text{Brake power} = \frac{F\times u}{10^3} = \frac{279854\times48.8}{10^3} = 13657\, kW$$
$$\text{Overall efficiency} = \frac{\textit{Brake power}}{\textit{Water power}}$$
$$= \frac{13657\times100}{15926} = 85.75\,\%$$

**Example 5.14** A Pelton wheel of 1 m mean bucket diameter works under a head of 500 m. The deflection of jet is 165° and its relative velocity is reduced over the bucket by 15 percent due to friction. The diameter of jet is 100 mm and water leaves the bucket without any whirl. Calculate: (a) wheel speed, (b) jet velocity, (c) power developed by wheel, (d) water power, (e) power input to bucket, and (f) efficiency of wheel.

Take coefficient of velocity = 0.97

**Solution.**

Given: $D = 1$ m, $d = 0.1$ m, $H = 500$ m, $\beta_2 = 180° - 165° = 15°$, $K_v = 0.97$, $k = 1 - 0.15 = 0.85$, $\alpha_2 = 90°$

(a) Jet velocity, $v_1 = K_v\sqrt{2gH} = 0.97\sqrt{2 \times 9.81 \times 500}$

$= 96.07$ m/s

Now $u_1 = u_2 = u$

$$v_{r1} = v_1 - u_1 = 96.07 - u$$

$$v_{r2} = kv_{r1} = 0.85 \times (96.07 - u)$$

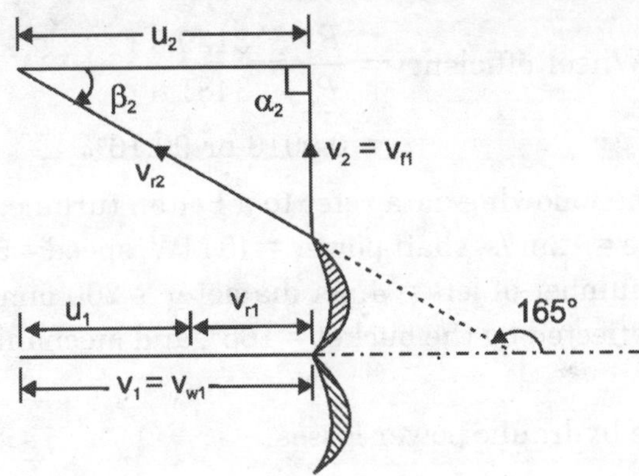


**Fig. 5.9**

The velocity triangles are shown in Fig. 5.9

$$u = u_{r2} \cos \beta_2 = 0.85\ (96.07 - u) \cos 15°$$

$$= 78.877 - 0.821\ u$$

$$u = \frac{78.877}{1.821} = 43.31 \text{ m/s}$$

Now $u = \dfrac{\pi DN}{60}$

$$N = \frac{43.31 \times 60}{\pi \times 1} = 827.2 \text{ rpm}$$

(b) Jet velocity, $v_1 = 96.07$ m/s

(c) $Q = \dfrac{\pi}{4} d^2 v_1 = \dfrac{\pi}{4}(0.1)^2 \times 96.07$

$= 0.7545$ m³/s

Impulsive force, $F = \rho Q\ (v_{w1} \pm v_{w2})$

Here $v_{w2} = 0$ and $v_{w1} = v_1$

$\therefore$ $F = 10^3 \times 0.7545 \times 96.07 = 75485$ N

Power developed by the wheel, $P_w = F \times u$

$$= 72485 \times 43.31 = 31.39.3 \text{ kW}$$

(d) Water power $= \gamma QH$

$$= 9.81 \times 10^3 \times 0.7545 \times 500$$

$$= 3700.8 \text{ kW}$$

(e) Power input to buckets, $P_b = \frac{1}{2} m v_1^2 = \frac{1}{2} \rho Q v_1^2$

$$= \frac{1}{2} \times 10^3 \times 0.7545 \times (96.07)^2$$

$$= 3481.8 \text{ kW}$$

(f) Wheel efficiency $= \frac{P_w}{P_b} = \frac{3139.3}{3481.8}$

$$= 0.9016 \text{ or } 90.16\%$$

**Example 5.15** The following data refer to a Pelton turbine: Available head = 500 m, discharge = 12m³/s, shaft power = 45 MW, speed = 500 rpm, runner diameter = 2 m, number of jets = 4, jet diameter = 200 mm, angle through which the jet is deflected by the bucket = 165°, and mechanical efficiency of turbine = 96%.

Determine the hydraulic power losses.

**Solution.**

Given: $H = 500$ m, $Q = 12$ m³/s, $P_s = 45$ MW, $N = 500$ rpm; $D = 2$ m, $n = 4$, $d = 200$ mm, $\beta_2 = 180° - 165° = 15°$, $\eta_{mech} = 0.96$

$$\eta_{mech} = \frac{\textit{Shaft power}, P_s}{\textit{Power developed by the runner}, P_r}$$

$$P_r = \frac{45 \times 10^3}{0.96} = 46875 \, kW$$

Power developed per jet $= \frac{P_r}{n} = \frac{46875}{4} = 11718.75 \, kW$

Discharge per jet, $q = \frac{Q}{n} = \frac{12}{4} = 3\text{m}^3/\text{s}$

Now, $q$ = area of jet × velocity of jet

$$3 = \frac{\pi}{4}(0.2)^2 \times v_1$$

$$v_1 = 95.5 \text{ m/s}$$

$$u = \frac{\pi DN}{60} = \frac{\pi \times 2 \times 500}{60} = 52.36 \text{ m/s}$$

Work developed per jet $= \rho q(v_1 - u)(1 + k \cos \beta_2)u$

$$= 10^3 \times 3(95.5 - 52.36)\ (1 + k \cos 15°) \times 52.36$$

$$= 6776.43\ (1 + k \cos 15°)\text{kW} = 11718.75$$

$$k = \frac{0.7293}{\cos 15^\circ} = 0.755$$

$$v_{r2} = kv_{r1} = k(v_1 - u_1) = k(v_1 - u)$$

$$\text{Actual jet velocity} = K_v\sqrt{2gH}$$

$$\text{Head of loss in nozzle} = \frac{v_1^2 - K_v^2 \times 2gH}{2g} = \frac{2gH(1-K_v^2)}{2g}$$

$$= (1-K_v^2)H$$

Now

$$K_v = \frac{v_1}{\sqrt{2gH}} = \frac{95.5}{\sqrt{2\times 9.81\times 500}} = 0.964$$

$$\text{Head loss in nozzle} = [1-(0.964)^2]\times 500 = 35.35 \text{ m}$$

$$\text{Head loss in buckets} = \frac{v_{r1}^2 - v_{r2}^2}{2g} = \frac{v_{r1}^2 - k^2 v_{r2}^2}{2g}$$

$$= \frac{v_{r1}^2}{2g}(1-k^2) = \frac{(v_1 - u)^2}{2g}(1-k^2)$$

$$= \frac{(95.5-52.36)^2}{2\times 9.81}[1-(0.755)^2]$$

$$= 40.78 \text{ m}$$

Total head loss in nozzle and buckets,

$$H_l = 35.35 + 40.78 = 76.13 \text{ m}$$

$$\text{Power loss} = \gamma Q H_l = 9.81 \times 12 \times 76.13 = 8962 \text{ kW}$$

**Example 5.16:** A Pelton turbine is to work at the foot of a dam whose reservoir level is 200 m. The head at the full opening at the turbine nozzle of diameter 200 mm and the coefficient of velocity is 0.98. The turbine is to operate at 200 rpm and develops a power of 3.7 MW. Assuming the blade to jet speed ratio as 0.46, estimate the wheel diameter. If the blade outlet angle is 16°, determine the blade and hydraulic efficiencies. Neglect friction losses.

**Solution.**

Given: $H = 200$ m, $K_v = 0.98$, $N = 200$ rpm, $d = 200$ mm, $P_t = 3.7$ MW, $K_u = 0.46$, $\beta_2 = 16°$, $k = 1.0$.

$$v_1 = K_v\sqrt{2gH} = 0.98\sqrt{2\times 9.81\times 200}$$

$$= 61.389 \text{ m/s}$$

$$Q = A \times v_1 = \frac{\pi}{4}(0.2)^2 \times 61.389 = 1.9286 \text{ m}^3/\text{s}$$

$$P_t = \gamma Q H \eta_b$$

Blade efficiency, $$\eta_b = \frac{3.7\times10^6}{9.81\times10^3\times1.9286\times200}$$

$$= 0.9778 \text{ or } 97.78\%$$

$$K_u = \frac{u}{v_1}$$

$$u = 0.46 \times 61.389 = 28.239 \text{ m/s}$$

Also $$u = \frac{\pi DN}{60}$$

$$D = \frac{28.239\times60}{\pi\times200} = 2.697 \text{ m}$$

Hydraulic efficiency, $$\eta_{hyd} = 2\left(K_u - K_u^2\right)(1 + k\cos\beta_2)$$

$$= 2[0.46-(0.46)^2]\left(1+1\times\cos16^\circ\right)$$

$$= 0.9743 \text{ or } 97.43\%$$

## Summary

1. Work done by the jet on runner per second $= \rho A v_1 (v_{w1} \pm v_{w2})u$
2. Work done per second per unit weight of striking water

$$= \frac{1}{g}(v_{w1} \pm v_{w2})u$$

3. Hydraulic efficiency, $\eta_{hyd} = \dfrac{2(v_{w1} \pm v_{w2})u}{v_1^2}$

   $\eta_{hyd}$ is maximum when $u = \dfrac{v_1}{2}$

   $$(\eta_{hyd})_{max} = \frac{1 + k\cos\beta_2}{2}$$

4. Mechanical efficiency, $\eta_{mech} = \dfrac{P}{\gamma Q_a H_r} = \dfrac{\textit{Shaft power}}{\textit{Bucket power}}$

   Where $H_r = \dfrac{1}{g}(v_{w1} \pm v_{w2})u =$ Euler or runner head

5. Volumetric efficiency, $\eta_{vol} = \dfrac{Q_a}{Q} = \dfrac{\text{Actual discharge}}{\text{Theoretical discharge}}$

6. Overall efficiency, $\eta_o = \dfrac{P}{\gamma QH} = \dfrac{\text{Shaft power}}{\text{Water power}}$

   Also, $\eta_o = \eta_{hyd} \times \eta_{mech} \times \eta_{vol}$

7. Jet velocity, $v_1 = K_{v1}\sqrt{2gH}, K_{v1} = 0.98 - 0.99$

8. Wheel velocity, $u = K_{u1}\sqrt{2gH} = \dfrac{\pi DN}{60}, K_{u1} = 0.43 - 0.45$

9. Jet ratio, $m = \dfrac{D_1}{d_1} = 11 - 16$

10. Number of buckets, $z = 15 + 0.5$ m

11. Specific speed = $N_s = \dfrac{N\sqrt{nP_t}}{H^{5/4}}$, $n$ = number of jets

12. Jet or Head efficiency, $\eta_H = 4K_{u1}(K_{v1} - K_{u1})$

    For maximum head efficiency, $K_{u1} = \dfrac{K_{v1}}{2}$

    and $(\eta_H)_{max} = K_{v1}$

13. Net or effective head, $H = H_g - h_f - h$

    where $h_f = \dfrac{fLv^2}{2gD}$, $H_g$ = gross head, $h$ = height of nozzle above tail race

| Type of runner | $\alpha_2$ | $\beta_2$ | $v_{\omega 2}$ |
|---|---|---|---|
| 1. Slow | $> 90°$ | $< 90°$ | $-ve$ |
| 2. Medium | $\simeq 90°$ | $< 90°$ | Zero |
| 3. Fast | $< 90°$ | $< 90°$ | $+ve$ |

## Multiple Choice Questions

1. The flow in impulse turbine is
   (a) tangential (b) radial
   (c) axial (d) mixed flow

2. The jet ratio of a Pelton wheel is defined as
   (a) Wheel dia (*D*)/jet dia (*d*) (b) $\alpha/D$
   (c) $(D/d)^2$ (d) $(d/D)^2$

3. The number of bucket in a Pelton turbine are given by
   (a) 0.5 m (b) 10 + 0.5 m
   (c) 15 + 0.5 m (d) 20 + 0.5 m
   Where m = jet ratio

4. The maximum efficiency of a series of curved vanes mounted on a wheel is
   (a) $\frac{1+\cos\beta_2}{2}$ (b) $\frac{1-\cos\beta_2}{2}$
   (c) $\frac{1+\sin\beta_2}{2}$ (d) $\frac{1-\sin\beta_2}{2}$

5. Impulse turbines are used for
   (a) high head and low discharge
   (b) low head and high discharge
   (c) low head and low discharge
   (d) high head and high discharge

6. Specific speed of a turbine is defined as
   (a) $\frac{N\sqrt{P_t}}{H^{5/4}}$ (b) $\frac{N\sqrt{P_t}}{H^{3/4}}$
   (c) $\frac{N\sqrt{P_t}}{H^{3/2}}$ (d) $\frac{N\sqrt{P_t}}{H^{5/2}}$

7. The ratio of actual work available at the turbine to energy imparted to wheel is called
   (a) mechanical efficiency (b) hydraulic efficiency
   (c) turbine efficiency (d) overall efficiency

8. The ratio of work done on the wheel to the energy actually supplied to the turbine is called
   (a) mechanical efficiency (b) hydraulic efficiency
   (c) turbine efficiency (d) overall efficiency

**Answers:**

1. (a) 2. (b) 3. (c) 4.(a) 5. (a) 6.(a) 7. (d) 8. (b)

## Review Questions

1. Describe with the help of a simple sketch the main components of a Pelton turbine.
2. Define coefficient of velocity and speed ratio.
3. What is vena– contracta?
4. What is the usual relationship between minimum diameter of water jet of a nozzle and nozzle diameter?
5. What is the function of spear and deflector in a Pelton turbine?
6. How brake is applied to a Pelton wheel to stop it?
7. Sketch the arrangements of single overhung and double overhung Pelton wheels.
8. On what factors does the number of jets depend in case of Pelton turbines?
9. Define jet ratio and explain its importance.
10. On what factors do the selections of speed of Pelton turbine depend?
11. How would you determine the minimum number of buckets for a Pelton turbine?
12. Define the following: Head efficiency, volumetric efficiency, hydraulic efficiency, mechanical efficiency, and overall efficiency of a Pelton turbine.
13. Derive an expression giving the relationship between the jet speed and bucket speed for maximum efficiency of a Pelton wheel.
14. Define gross head and net head on a Pelton turbine.
15. What is the relationship between the various efficiencies of a Pelton wheel?
16. How do you determine the head loss in the penstock?
17. Define synchronous speed of a turbine.
18. Write the expression for maximum hydraulic efficiency with blade friction of a Pelton turbine.
19. Define specific speed of Pelton turbine with multiple jets.

## Numerical Problems

1. A Pelton wheel has a mean bucket speed o f 10 m/s with a jet of water flowing at the rate of 0.7 $m^3/s$ under a head of 30m. The buckets deflect the jet through an angle of 160°. Calculate the power developed and the efficiency of the turbine. Assume coefficient of velocity as 0.98.

   **[Ans.** 186.9 kW, 94.54%]

2. A jet of water coming out of nozzle strikes the buckets of a Pelton wheel which when stationary would deflect the jet through 165°. The velocity of water at exit is 0.9 times at the inlet and the bucket speed is 0.45 times the jet speed. If the speed of the Pelton wheel is 300 rpm and the effective head is 150 m, determine:

   (a) Hydraulic efficiency, and

   (b) Diameter of the Pelton wheel.

   Take coefficient of velocity, $K_v = 0.98$

   **[Ans.** 92.5%, 1.55m]

3. A Pelton turbine is to be designed to develop 750 kW at 400 rpm. It is to be supplied with water from a reservoir whose level is 250m above the wheel through a pipe 900m long. The pipeline losses are to be 5% of gross head. The coefficient of friction is 0.005. The bucket speed is to be 0.46 of jet speed and efficiency of wheel is 85%. Calculate:

   (a) The pipe line diameter

   (b) The jet diameter, and

   (c) The wheel diameter.

4. The following data is available for a Pelton turbine:

   Brake power = 130 kW, Head = 300m, speed = 600 rpm, $\eta_t = 0.75$, $K_{v1} = 0.98$, $K_{u1} = 0.45$. Calculate:

   (a) the discharge

   (b) the least jet diameter

   (c) the mean runner diameter

   (d) the jet ratio, and

   (e) the number of buckets.

5. The buckets of a Pelton wheel deflect the jet impinging on them through 160°. Pressure at the nozzle is 4 MPa. The centre line of the buckets is on a circle of diameter 2.5m.

Calculate:

(a) the best speed of the wheel,

(b) the theoretical efficiency

(c) the theoretical power developed at a nozzle flow of 3m$^3$/s, and

(d) the specific speed.

6. A Pelton wheel has a mean bucket speed of 12 m/s and is supplied with water at the rate of 0.75 m$^3$/s under a head of 30m. If the buckets deflect the jet through an angle of 160°, find the power and the efficiency of the turbine.

7. A Pelton turbine is required to develop 2700 kW under a head of 300 m. Its efficiency is 80%. Specific speed is 16 and speed ratio is 0.47. Assume $K_{v1} = 0.97$ for the nozzle. Calculate the jet diameter and the rpm of the wheel.

8. A jet of water impinges on a series of curved vanes at an angle of 30° to the direction of motion of the vanes while entering and leaves the vanes horizontally. The head under which the jet issues from the nozzle is 30 m, the coefficient of velocity for the nozzle is 0.9 and the diameter of jet after leaving the nozzle is 50mm. The speed of the vanes is 10 m/s and the relative velocity of water at outlet is 0.8 times the relative velocity at inlet.

   Calculate:

   (a) the angle of vane tips at inlet,

   (b) the power developed by the jet, and

   (c) the efficiency of the system.

9. A Pelton turbine is required to develop 3500 kW at 300 rpm under a head of 350 m. Determine the least diameter of the jet impinging on the buckets of the turbine wheel if the coefficient of velocity is 0.985. Assume overall efficiency of turbine as 90%. If speed ratio is 0.46, what would be the mean diameter of wheel? Calculate the size of the nozzle used if its diameter is 25 % larger than the least diameter of jet. How many buckets will be used for the wheel for the best efficiency?

10. A twin jet Pelton wheel has a mean bucket circle diameter of 1,680 mm and runs at 500 rpm. When the jets are 150 mm diameter, the available head at the nozzle is 480 m. Assuming co-efficient of velocity for the nozzle as 0.98, outlet angle of bucket 15°, relative velocity for water leaving the bucket 0.88 of that of inlet; and windage and mechanical losses 3% of the water power supplied, find:

   (a) the water power supplied and be are power

(b) the force of one jet on the buchet; and

(c) the overall efficiency.

11. A Pelton turbine has a mean bucket speed of 40 m/s, and is supplied with water at the rate of 0.55 $m^3/s$, the head of water behind the nozzle being 400 m. If the jet is deflected by the buckets through 170°, find power developed and the efficiency of the wheel, Assume $\beta_1 = 0°$.

12. A Pelton wheel is to be designed to the following specifications:-

    Power = 12,000 kW

    Head = 380 m

    Speed = 750 rpm

    Overall Efficiency = 86 percent

    Jet diameter not to exceed $\frac{1}{8}$ times the wheel diameter

    Determine:

    (a) the wheel diameter;

    (b) number of jets required;

    (c) the diameter of the jet.

13. A double overhung Pelton wheel unit is to operate a 30,000 kW generator under the effective head of 305 m at the base of the nozzle. Find the size of jet, mean diameter of runner, synchronous speed and specific speed of each wheel. Assume generator efficiency 93%, Pelton wheel efficiency 85%, co-efficient of nozzle velocity 0.97, speed ratio 0.46 and jet ratio 12.

14. A Pelton wheel has a mean speed of 12 m/s and is supplied with water at the rate of 1,370 litres per second under a head of 30.5m. If the buckets deflect the jet through and angle of 160°, find the power and efficiency of the wheel.

# CHAPTER 6 Francis Turbine

## 6.1 Introduction

Francis turbine is an inward mixed flow reaction turbine. The water under pressure enters the runner from the guide vanes towards the centre in radial direction and discharges out of the runner axially. The Francis turbine operates under

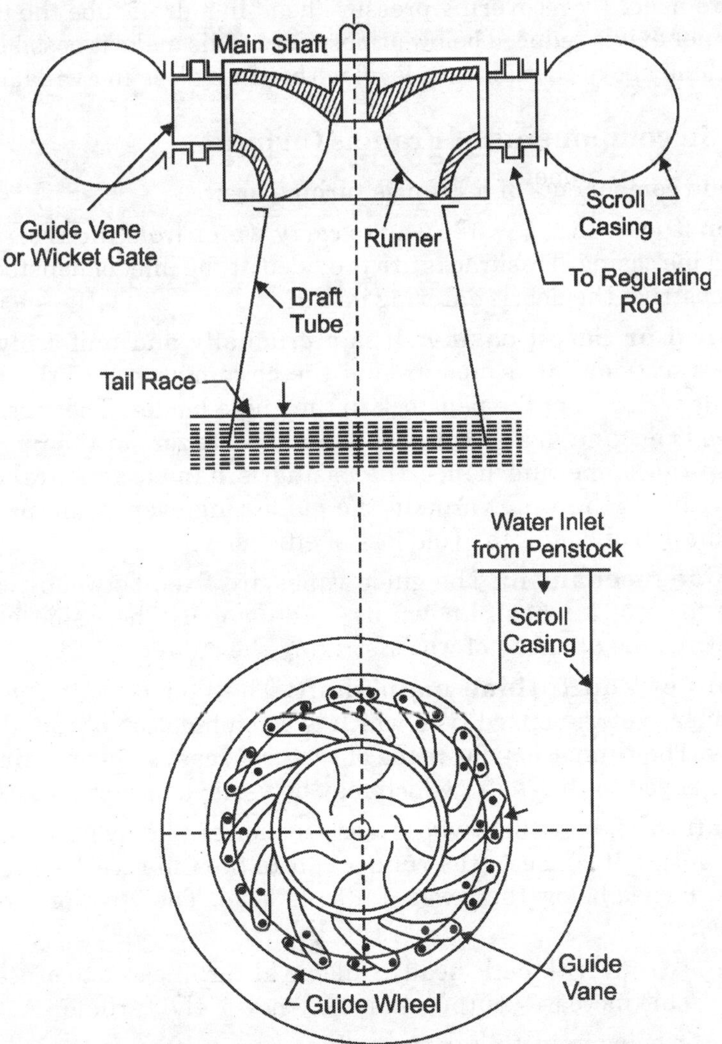


Fig. 6.1 Outlines of Francis Turbine, Vertical Closed Type

medium heads and requires medium quantity of water. Water is brought down to the turbine through the penstock and directed to a number of stationary orifices fixed all around the circumference of the runner as shown in Fig.6.1.These stationary orifices are commonly termed as guide vanes or wicket gates.

A part of the head acting on the turbine is transformed into kinetic energy and the rest remains as pressure head. There is a difference of pressure between the guide vanes and the runner which is called the reaction pressure, and is responsible for the motion of the runner .The pressure at the inlet to the turbine is more than that at the outlet. Therefore, the water in the turbine must flow in a closed conduit. After doing its work the water is discharged to the tail race though a closed tube of gradually enlarging section known as the draft tube .The draft tube converts kinetic head to pressure head. By recovering pressure head in a draft tube the pressure at the runner exit is reduced below atmosphere. This make it possible to install the turbine above the tail race level without any loss in available head.

## 6.2 Main components of Francis turbine

The main components of a Francis turbine are:

1. **Penstock :** It is a water way to carry water from the reservoir to the turbine casing. Trashracks are provided at the inlet of penstock in order to obstruct the debris entering into it.
2. **Spiral or Scroll casing:** It is a gradually and uniformly reducing cross-sectional area pipe around the circumference of the turbine .It leads water from the penstock to the guide blades. The cross-sectional area is maximum at the entrance and nearly zero at the tip. This gives a spiral shape and hence the casing is named as spiral casing. It distributes the water around the guide ring evenly and prevents the formation of eddies to avoid loss of efficiency.
3. **Guide mechanism:** The guide vanes are fixed between two rings in the form of a wheel, known as guide wheel. The guide mechanism controls the quantity of water striking the runner.
4. **Runner and turbine main shaft:** The flow of water through the runner is of the mixed type which is a combination of radial and axial flow. The runner may be made of cast iron, cast steel or stainless steel. It is keyed to the shaft., which may be vertical or horizontal.
5. **Draft tube:** It is a pipe whose cross-section gradually increases towards the outlet. It connects the runner exit to the tail race. The tube should be drowned below the lowest tail race level. The functions of the draft tube are;
   - (i) To convert kinetic head available at turbine exit to pressure head. This increases output and efficiency of the turbine.
   - (ii) It helps in installing the turbine above the tail race level.

(iii) It prevents the splashing of water coming out of the runner and guides the water to the tail race.

## 6.3 Comparison between Reaction and Impulse Turbines

Table 6.1 Gives the comparison between reaction and impulse turbines.

**Table 6.1 Comparison between reaction and impulse turbines.**

| Parameters | Reaction turbine | Impulse turbine |
|---|---|---|
| 1. Conversion of fluid energy. | Only a fraction of the available fluid energy is converted into kinetic energy before the fluid enters the runner. | All the available fluid energy is converted into kinetic energy by a nozzle. It is the fluid jet which strikes the runner blades. |
| 2. Change in pressure and velocity. | Both pressure and velocity change as water passes through the runner. Pressure at inlet is much higher than at the outlet. | Only the velocity of jet changes. The pressure throughout remains atmospheric. |
| 3. Admittance of water over the wheel. | Water is admitted over the entire circumference of the wheel. | Water may be allowed to enter a part or whole of the wheel circumference with the help of nozzles. |
| 4. Casing | The runner must be enclosed within a water tight casing (scroll casing). | Water tight casing is not necessary. Casing has no hydraulic function to perform. It only serves to prevent splashing and guide water to the tail race. |
| 5. Extent to which water fills the wheel/turbine. | Water completely fills all the passages between the blades and while flowing between inlet and outlet sections does work on the blades. | The turbine does not run full and air has free access to the buckets |
| 6. Installation of unit. | The unit may be installed above or below the tail race level. The turbine is connected to the tail race through a draft tube. | The turbine is always installed above the tail race level and no draft tube is used. |
| 7. Flow regulation. | The flow regulation is carried out by means of a guide-vane assembly | The flow regulation is done by means of a needle valve fitted into the nozzle. |
| 8. Relative velocity of water. | Increases due to continuous drop in pressure. | Reduces due to blade friction. |

## 6.4 Advantages and Disadvantages of Francis Turbine over Pelton Turbine

### Advantages

1. The variation in operating head can be more easily controlled.
2. The ratio of maximum to minimum heads can be even two.
3. The mechanical efficiency decreases slowly with wear.
4. The size of runner, generator and power house required is smaller.

### Disadvantages

1. Dirty water can cause very rapid wear.
2. The over haul and inspection is much more difficult.
3. Cavitation is more.
4. Working of turbine below 50% head for a long period decreases its efficiency.

## 6.5 Comparison of Inward Flow and Outward Flow Reaction Turbines

The comparison of inward and outward flow reaction turbine is given in Table 6.2.

**Table 6.2 Comparison between inward and outward flow reaction turbine**

| Parameters | Inward flow | Outward flow |
|---|---|---|
| 1. Entry of water | Water enters at the outer periphery, flows inwards and towards the centre of the turbine and discharges at the inner periphery. | Water enters at the inner periphery, flows outwards and discharges at the outer periphery. |
| 2. Centrifugal head imparted | Negative centrifugal head reduces the relative velocity of water at the outlet. | Positive centrifugal head in ceases the relative velocity of water at the outlet. |
| 3. Discharge | Remains constant | increases |
| 4. Speed control | Easy | Very difficult |
| 5. Speed adjustment | The turbine adjusts the speed by itself. | The turbine cannot adjust the speed by itself. |
| 6. Suitability | Suitable for medium to high heads, large outputs and units. | Suitable for low or medium heads. |
| 7. Applications | For power projects. | Practically obsolete. |

## 6.6 Power Developed

Fig. 6.2. shows general layout of a hydroelectric power plant using a reaction turbine.

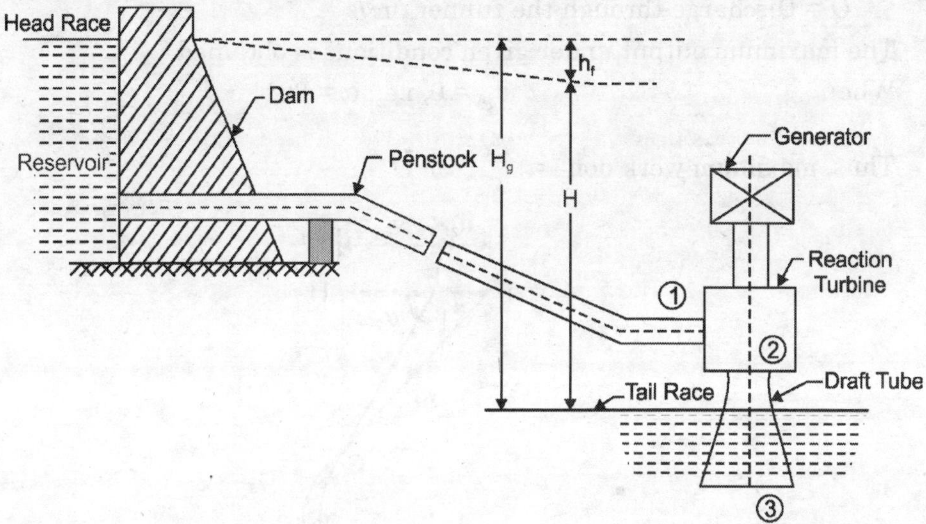


**Fig. 6.2 General layout of a hydroelectric power plant using reaction turbine.**

Let $H_g$ = gross head = difference between head race and tail race

$h_f$ = frictional head loss in penstock

$H$ = net head = $H_g - h_f$ = available, working or operation head

Also, $H$ = Total energy available at exit from penstock – Total energy available at exit from the draft tube.

$$= \left[\frac{p_1}{\gamma} + \frac{v_1^2}{2g} + z_1\right] - \left[\frac{p_3}{\gamma} + \frac{v_3^2}{2g} + z_3\right]$$

**Workdone by the runner**

Fig 6.3 shows the velocity triangles for an inward flow reaction turbine. The resultant dynamic face exerted by water on the runner vanes in the direction of rotation,

$$F = \rho Q\ (v_{w1} \pm v_{w2}) \qquad ...(6.1)$$

$$\text{Torque, } T = \rho Q(v_{w1}R_1 \pm v_{w2}R_2) \qquad ...(6.2)$$

$$\text{Work done} = \rho Q(v_{w1}u_1 \pm v_{w2}u_2) = T \times w$$

$$= \frac{\gamma Q}{g}(v_{w1}u_1 \pm v_{w2}u_2) \qquad ...(6.3)$$

Where $u_1 = \omega R_1$ and $u_2 = \omega R_2$

$$\omega = \frac{2\pi N}{60} \text{ rad/s, } N = \text{rpm of runner}$$

$R_1, R_2$ = Radii of runner at inlet and outlet respectively.

$Q$ = Discharge through the runner, m³/s

The maximum output under given conditions is obtained

When $v_{w2} = 0$, i.e., $\alpha = 90°$

Thus, maximum work done = $\frac{\gamma Q}{g}(v_{w1}u_1)$

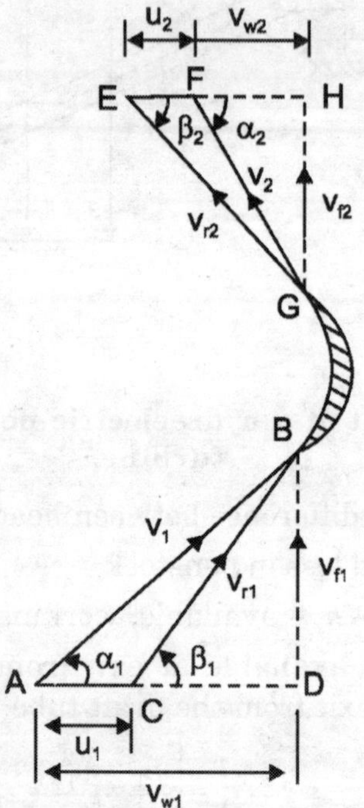


**Fig. 6.3 Velocity triangles for an inward flow Francis turbine.**

Power developed by the turbine,

$$P_H = \rho Q(v_{w1}u - v_{w2}u_2)/10^3 \text{ kW} \quad ...(6.4)$$

## 6.7 Efficiencies of Francis turbine

**1. Head efficiency, $\eta_H$:** It is the ratio of net head available to the turbine after subtracting head lost and the operating head.

Let $H$ = oprating head on the turbine

$\Delta H$= total head loss in turbine

**Then head efficiency,** $$\eta_H = \frac{H - \Delta H}{H} = 1 - \frac{\Delta H}{H} \quad ...(6.5)$$



Head power, $P_H = P_a.\eta_H = \gamma QH.\eta_H$

$P_a$ = Available or water power

$$\text{or} \quad \frac{\gamma Q}{g}(v_{w1}u_1 - v_{w2}u_2) = \gamma QH.\eta_H \qquad ...(6.6)$$

$$\therefore \quad \eta_H = \frac{v_{w1}\,.\,u_1 - v_{w2}\,.\,u_2}{gH} = \frac{2(v_{w1}.u_1 - v_{w2}.u_2)}{2gH}$$

Now $\quad v_{w1} = Kv_{w1}.\sqrt{2gH}$

and $\quad v_{w2} = K_{v_{w2}}.\sqrt{2gH}$

$\quad u_1 = k_{u1}\sqrt{2gH}$

and $\quad u_2 = k_{u2}\sqrt{2gH}$

$$\therefore \quad \eta_H = 2\left(K_{v_{w1}}.K_{u1} - K_{v_{w2}}.K_{u2}\right)$$

But $\quad u_2 = \frac{R_2}{R_1}.u_1 \qquad \left[\therefore \omega = \frac{u_1}{R_1} = \frac{u_2}{R_2}\right]$

$$\therefore \quad K_{u2} = \frac{R_2}{R_1}.K_{u1}$$

$$\therefore \quad \eta_H = 2K_{u1}\left[K_{v_{w1}} - K_{v_{w2}}.\frac{K_{u2}}{K_{u1}}\right]$$

$$= 2K_{u1}\left[K_{v_{w1}} - K_{v_{w2}}.\frac{R_2}{R_1}\right] \qquad ...(6.7)$$

For radial discharge, $\alpha_2 = 90°$ and $K_{u2} = 0$

$$\therefore \quad \eta_H = 2K_{u_1}.K_{v_{w1}} \qquad ...(6.7a)$$

**2. Volumetric Efficiency, $\eta_{vol}$ :** As the fluid passes though the turbine, a certain fraction of it escapes through the gap between the guide apparatus and the runner without doing work on the runner.

Let $\Delta Q$ = part of the fluid which goes waste without doing work on the runner

$Q$ = total discharge entering the turbine.

Then, volumetric efficiency, $\eta_{vol} = \frac{Q - \Delta Q}{Q} = 1 - \frac{\Delta Q}{Q}$ ...(6.8)

**3. Hydraulic Efficiency, $Q_{hyd}$ :** It is defined as the ratio of hydraulic power utilized or developed by the runner and the net power supplied by water at entrance to the turbine.

$$\eta_{hyd} = \frac{\rho Q(v_{w1}u_1 - v_{w2}u_2)}{\gamma QH}$$

$$= \frac{v_{w1}u_1 - v_{w2}u_2}{gH} \quad ...(6.9)$$

Total hydraulic loss= total head loss + volumetric loss

Actual hydraulic power of turbine, $P_{hyd} = \gamma(Q - \Delta Q)(H - \Delta H)$

Available or water power, $P_a = \gamma QH$

$$\eta_{hyd} = \frac{P_{hyd}}{P_a} = \left(1 - \frac{\Delta Q}{Q}\right)\left(1 - \frac{\Delta H}{H}\right) = \eta_{vol}.\eta$$

$$= \eta_{vol}.\eta_H \quad ...(6.9b)$$

4. **Mechanical Efficiency, $\eta_{mech}$ :** The power developed by the runner is decreased by the mechanical losses caused by friction between the rotating parts like the shaft and runner, and stationary parts like guide, through bearing and sealings. On account of these losses, the power available at the turbine shaft is less than the power delivered to it by the runner. Thus

$$\eta_{mech} = \frac{\text{power available at the turbine shaft (shaft power)}}{\text{power developed by the runner}} \quad ...(6.10a)$$

Brake power, $P_t = P_{hyd} \times \eta_{mech}$ ...(6.10b)

5. **Overall efficiency, $\eta_o$ :** It is defined as the ratio of power available to the turbine shaft to the power input to the turbine.

$$\eta_o \text{ or } \eta_t = \frac{\textit{shaft power}}{\textit{water}\text{ power}} = \frac{P_t}{P_a} = \frac{P_{hyd} \times \eta_{mech}}{P_a} = \frac{P_a \eta_{vol}.\eta_H.\eta_{mech}}{P_a}$$

$$= \eta_{vol}.\eta_H.\eta_{mech} \quad ...(6.11)$$

## 6.7 Discharge Through Francis Turbine

Quantity of water flowing per second,

$Q$ = area across flow, $A_f \times$ velocity of flow$_{f1}$ ...(6.12)

Where $A_f = (\pi D_1 - zt_1)B_1 = K_1 \pi D_1 B_1$

$D_1$ = Intel diameter of runner

$z_1$ = number of blades in runner

$t_1$ = thickness of runner blades at inlet

$B_1$ = height of runner blades $\simeq$ height of guide blades

$K_1$ = blade thickness factor = 0.90 – 0.95

$v_{f1}$ = velocity of flow at inlet

$= v_1 \sin \alpha_1$

Let $$n = \frac{B_1}{D_1} = 0.1 - 0.45$$

Then $$Q = K_1 \pi n D_1^2 . v_{f1} \quad ...(6.13)$$

Flow ratio, $$K_{f1} = \frac{v_{f1}}{\sqrt{2gH}} = 0.15 - 0.30 \quad ...(6.14)$$

Speed ratio, $$K_{u1} = \frac{u_1}{\sqrt{2gH}} = 0.6 - 0.9 \quad ...(6.15)$$

Coefficient of reaction, $$K_r = \frac{p_1 - p_2}{\gamma H} \quad ...(6.16)$$

Degree of reaction, $$R = 1 - \frac{1}{2}\left[\frac{\cot\alpha_1}{\cot\alpha_1 - \cot\beta_1}\right] \quad ...(6.17)$$

## 6.8 Design parameters of Francis turbine runner

The runner of a Francis turbine is required to be designed to develop a known power P, when running at a known speed N rpm under a known head H. The design of the runner involves the determination of its size and the vane angles.

The design is carried out as follows:

1. Assume suitable values of $\eta_o$, $\eta_{hyd}$, $n$, $K_f$ and $K$.
2. Determine the required discharge $Q$ from the relation
$$P = \gamma Q H \times \eta_o \quad ...(6.18)$$
3. Calculate the velocity of flow from the discharge and flow area. Effective area at the runner inlet,
$$A_1 = (\pi D_1 - zt_1) B_1 = K_1 \pi D_1 B_1$$

Where $K_1$ = vane thickness factor at inlet $\simeq 0.95$

Discharge, $$Q = K_1 \pi D_1 B_1 v_{f1} \quad ...(6.19)$$

Flow velocity, $$v_{f1} = \frac{Q}{K_1 \pi D_1 B_1} \quad (\because B_1 = nD_1)$$

$$= \frac{Q}{K_1 \pi n D_1^2} \quad ...(6.20)$$

Also $$v_{f1} = K_f \sqrt{2gH}$$

$$\therefore \quad K_f \sqrt{2gH} = \frac{Q}{K_1 \pi n D_1^2}$$

Or $$D_1 = \left[\frac{Q}{K_f\sqrt{2gH}.K_1\pi n}\right]^{1/2} \quad ...(6.21)$$

Then $B_1 = nD_1$

4. Calculate the peripheral velocity $u_1$ at inlet, from the relation

$$u_1 = \frac{\pi D_1 N}{60} \; m/s$$

5. Calculate the velocity of whirl at inlet $v_{w1}$, from the relation

$$\eta_{hyd} = \frac{v_{w1}u_1}{gH}$$

Or $$v_{w1} = \frac{\eta_{hyd}gH}{u_1} \quad ...(6.22)$$

6. Calculate the guide vane angle $\alpha_1$ and runner vane angle $\beta_1$, from the inlet velocity triangle.

$$\tan\alpha_1 = \frac{v_{f1}}{v_{w1}} \quad ...(6.23)$$

and $$\tan\beta_1 = \frac{v_{f1}}{v_{w1} - u_1} \quad ...(6.24)$$

7. Runner diameter at outlet, $D_2 \simeq \frac{D_1}{2}$ and $u_2 = \frac{u_1}{2}$.

8. Calculate velocity of flow at exit. For continuity of flow, we have

$$Q = K_1\pi D_1 B_1 v_{f1} = K_2\pi D_2 B_2 v_{f2}$$

$$\therefore \quad \frac{v_{f1}}{v_{f2}} = \frac{K_2}{K_1} \times \frac{D_2}{D_1} \times \frac{B_2}{B_1}$$

for $v_{f1} = v_{f2}$ and $K_1 = K_2$, we get

$$B_2 = 2B_1$$

9. Calculate the runner vane angle $\beta_2$ at exit from the outlet velocity triangle.

For radial discharge, $\alpha_2 = 90°$. Thus

$$\tan\beta_2 = \frac{v_{f2}}{u_2} \quad ...(6.25)$$

10. Number of vanes on the runner = Number of guide vanes ± 1

$\simeq$ 16-24

**Example 6.1** A Francis turbine works at 450 rpm under a head of 120m. Its diameter at inlet is 1.2m and the flow area is 0.4m². The angles made by absolute and relative velocities at inlet are 20° and 60° respectively with the tangential velocity. Determine:

(a) The volume flow rate

(b) The power developed, and

(c) The hydraulic efficiency.

**Solution**

Given: $N = 450$ rpm, $H = 120$m, $D_1 = 1.2$m, $\pi D_1 B_1 = 0.4$ m², $\alpha_1 = 20°$, $\beta_1 = 60°$

The velocity triangles are shown in Fig. 6.4.

(a) $$u_1 = \frac{\pi D_1 N}{60} = \frac{\pi \times 1.2 \times 450}{60} = 28.27\, m/s$$

Form inlet velocity triangle,

$$\tan \alpha_1 = \frac{v_{f1}}{v_{w1}}$$

$$v_{f1} = v_{w1} \tan 20° = 0.364 v_{w1}$$

$$\tan \alpha_1 = \frac{v_{f1}}{v_{w1} - u_1}$$

$$\tan 60° = \frac{0.364 v_{w1}}{v_{w1} - 28.27} = 1.732$$

$$v_{w1} = 35.79\, m/s$$

$$v_{f1} = 0.364 \times 35.79 = 13.027\, m/s$$

Volume flow rate, $$Q = Av_{f1} = 0.4 \times 13.027 = 5.211 \text{ m}^3/\text{s}$$

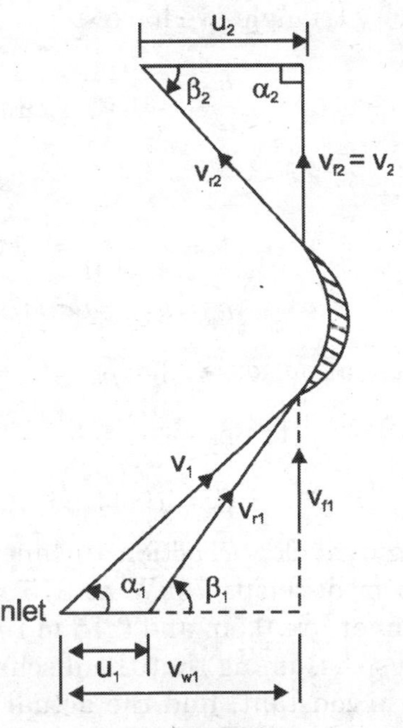


**Fig. 6.4**

(b) Work done per second = $\rho Q v_{w1} u_1$

$$= 1000 \times 5.211 \times 35.79 \times 28.27 \times 10^{-3}$$

$$= 5272.4 \text{ kW}$$

(c) Hydraulic efficiency, $\eta_{hyd} = \frac{v_{w1}u}{gH} = \frac{35.79 \times 28.27}{9.81 \times 120}$

$$= 0.8595 \text{ or } 85.95\%$$

**Example 6.2** An inward flow reactive turbine has outer and inner diameters of the wheel as 1m and 0.5m respectively. The vanes are radial at inlet and the discharge is radial at outlet. The water enters the vanes at an angle of 10. Assuming the velocity of flow as constant and equal to 3 m/s, find the speed of the wheel and vane angle at outlet.

**Solution.**

Given: $D_1 = 1\text{m}$, $D_2 = 0.5\text{m}$, $\alpha_1 = 10°$, $\beta_1 = 90°$, $\alpha_2 = 90°$, $v_{f1} = v_{f2} = 3\text{m/s}$

The velocity triangles are shown in Fig.6.5.

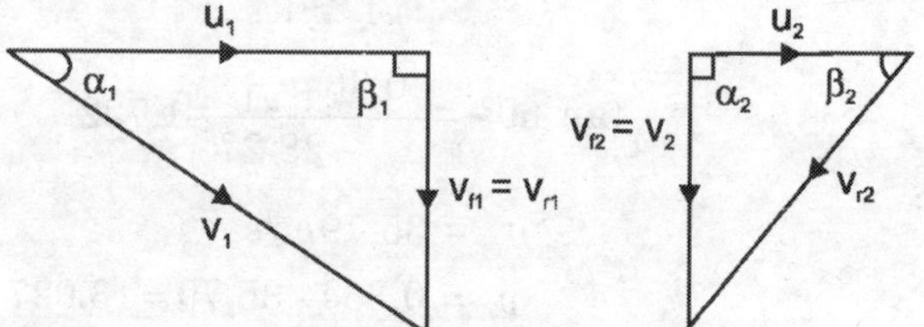


**Fig. 6.5 Velocity triangles.**

From the inlet velocity triangle, we have

$$u_1 = \frac{v_{f1}}{\tan\alpha_1} = \frac{3}{\tan 10^\circ} = 17 m/s$$

Also $u_1 = \frac{\pi D_1 N}{60}$

$$N = \frac{17 \times 60}{\pi \times 1.0} = 324.7 rpm$$

$$u_2 = u_1/_2 = 8.5 m/s$$

From velocity triangle at outlet, we have

$$\tan\beta_2 = \frac{v_{f2}}{u_2} = \frac{3}{8.5} = 0.35294$$

$$\beta_2 = 19.44° \text{ or } 19° - 27'$$

**Example 6.3** An inward flow reaction turbine with a supply of 550lit/s under a head of 15 m develops 75kW at 375 rpm. The inner and outer diameters of the runner are 0.5m and 0.75 m respectively. The velocity of water at exit is 3 m/s. Assuming that the discharge is radial and that the width of the wheel is constant, find the actual and theoretical hydraulic efficiencies of the turbine and the inlet angles of the guide and wheel vanes.

**Solution.**

Given: $Q = 0.55\ m^3/s$, $D_1 = 0.75$ m, $D_2 = 0.5$ m, $H = 15$ m, $P_t = 75$ kW, $N = 375$ rpm, $v_{f2} = 3$m/s, $\alpha_2 = 90°$, $B_1 = B_2$

Peripheral velocity of wheel at inlet,

$$u_1 = \frac{\pi D_1 N}{60} = \frac{\pi \times 0.75 \times 375}{60} = 14.73\,m/s$$

For $\alpha_2 = 90°$, $v_{f2} = v_2 = 3$m/s

Work done/s by the turbine per kg of water $= \dfrac{v_{w1} u_1}{g}$

= Head utilized by the turbine

$$= H - \frac{v_2^2}{2g}$$

$$= \frac{v_{w1} \times 14.73}{9.81} = 15 - \frac{3^2}{2 \times 9.81}$$

$$v_{w1} = 9.68 \text{ m/s}$$

Power developed by the turbine wheel,

$$P_H = \frac{\rho Q . v_{\omega 1} u_1}{10^3} = \frac{10^3 \times 0.55 \times 9.68 \times 14.73}{10^3}$$

$$= 78.42 \text{ kW}$$

Available or water power, $P_a = \dfrac{\gamma Q H}{10^3} = \dfrac{9.81 \times 10^3 \times 0.55 \times 15}{10^3} = 80.93$ kW

Brake power, $P_t = 75$ kW

Over all efficiency of turbine, $\eta_t = \dfrac{P_t}{P_a} = \dfrac{75}{80.93} = 0.9267$ or 92.67%

= Actual hydraulic efficiency

Theoretical hydraulic efficiency, $\eta_{hyd} = \dfrac{P_H}{P_a} = \dfrac{78.42}{80.93} = 0.9689$ or 96.89%

Now $Q = \pi D_1 B_1 v_{f1} = \pi D_2 B_2 v_{f2}$

Or
$$v_{f1} = \frac{D_2}{D_1} v_{f1} = \frac{0.5}{0.75} \times 3 = 2\,m/s$$

From inlet velocity triangle, we have

$$\tan \beta_1 = \frac{v_{f1}}{u_1 - v_{w1}} = \frac{2}{14.73 - 9.68} = 0.39604$$

$$\beta_1 = 21.6° \text{ or } 21° - 36'$$

Now $v_1 = \sqrt{v_{w1}^2 + v_{f1}^2} = \sqrt{(9.68)^2 + 2^2} = 9.88\,m/s$

and $$v_{w1} = v_1 \cos \alpha_1$$

$$\therefore \quad \cos \alpha_1 = \frac{9.68}{9.88} = 0.97975$$

$$\alpha_1 = 11.548° \text{ or } 11° - 33'$$

**Example 6.4** An inward flow reaction turbine rotates at 370 rpm. The wheel vanes are radial at inlet and inner diameter of the wheel is half of the outer diameter. The constant velocity of flow in the wheel is 2 m/s. Water enters the wheel at 10° to the tangent to the wheel at inlet. The breadth of the wheel at inlet is 75 mm and the area of flow blocked by the vanes is 5% of the gross area of flow at inlet. Calculate:

(a) the net available head at the wheel,

(b) the vane angle of wheel at outlet,

(c) the inner and outer diameter of wheel, and

(d) the theoretical water power developed by the wheel.

**Solution.**

Given: $N = 370$ rpm, $\beta_1 = 90°$, $\alpha_1 = 10°$, $D_2 = \frac{1}{2}D_1$, $v_{f1} = v_{f2} = 2$m/s, $B_1 = 75$ mm, $K_1 = 1 - 0.05 = 0.95$.

Now
$$v_{f1} = v_1 \sin \alpha_1$$
$$2 = v_1 \sin 10°$$
$$v_1 = 11.52 \text{ m/s}$$
$$v_{w1} = v \cos \alpha_1 = 11.52 \cos 10°$$
$$= 11.34 \text{ m/s}$$

For $\beta_1 = 90°$, $v_{w1} = u_1$

$$\therefore \quad 11.34 = \frac{\pi D_1 \times 370}{60}$$

$$D_1 = 0.585\text{m}$$

$$D_2 = 0.5 \times 0.585 = 0.293 \text{ m}$$

(a) Assuming radial discharge at outlet, net available head at the wheel,

$$H = \frac{v_{w1}u_1}{g}$$

$$= \frac{11.34 \times 11.34}{9.81} = 13.1m$$

(b) $$u_2 = \frac{\pi \times 0.293 \times 370}{60} = 5.67\, m/s$$

$$\tan \beta_2 = \frac{v_{f2}}{u_2} = \frac{2}{5.67} = 0.35273$$

$$\beta_2 = 19.43° \text{ or } 19° - 26'$$

(c)
$$Q = K_1 \pi D_1 B_1 v_{f1}$$
$$= 0.95 \times \pi \times 0.585 \times 0.075 \times 2$$
$$= 0.262 \text{ m}^3\text{/s}$$

Theoretical water power developed by the wheel,

$$P_\alpha = \frac{\gamma QH}{10^3} = \frac{9.81 \times 10^3 \times 0.262 \times 13.1}{10^3}$$
$$= 33.67 \text{ kW}$$

**Example 6.5** An inward flow reaction turbine is required to develop 3750 kW when operating under a net head of 30 m. The specific speed is to be nearly 270.

Assuming guide vane angle at full gate opening 30°, hydraulic efficient 90%, overall efficiency 87%, radial velocity of flow at inlet $0.3\sqrt{2gH}$, blade thickness coefficient 5%, determine:

(a) the nearest synchronous speed to drive an alternator to give a frequency of 50 Hz,

(b) the diameter and the width of runner at inlet, and

(c) the theoretical inlet angle of the runner vanes.

**Solution.**

Given: $P_t = 370$ kW, $H = 30$m, $N_S = 270$, $\alpha_1 = 30°$, $\eta_{hyd} = 0.90$, $\eta_0 = 0.87$, $v_{f1} = 0.3\sqrt{2gH}$, $K_1 = 0.95$, $F = 50$ Hz

(a) Specific speed of turbine, $N_S = \dfrac{N\sqrt{P_t}}{H^{5/4}}$

$$270 = \frac{N\sqrt{3750}}{30^{5/4}}$$
$$N = 309.6 \text{ rpm}$$

Nearest synchronous speed for $p = 10$, $N_{syn} = \dfrac{3000}{p} = \dfrac{3000}{10} = 300\ rpm$

$\therefore$
$$N_{syn} = 300 \text{ rpm}$$

(b)
$$v_{f1} = 0.3\sqrt{2 \times 9.81 \times 30} = 7.28\,m/s$$

Now
$$v_{f1} = v_1 \sin \alpha_1$$
$$v_1 = \frac{7.28}{\sin 30^\circ} = 14.53\,m/s$$
$$v_{w1} = v_1 \cos \alpha_1 = 14.56 \cos 30° = 12.61 \text{ m/s}$$

Work done by the turbine runner for radial discharge at outlet per kg of water flow

$$= \frac{v_{w1}u_1}{g} = \eta_{hyd}.H$$

$$\therefore \qquad u_1 = \frac{0.9\times9.81\times30}{12.61} = 21 m/s$$

Also
$$u_1 = \frac{\pi D_1 N}{60}$$

$$D_1 = \frac{21\times60}{\pi\times300} = 1.337m$$

Now
$$P_t = \frac{\gamma QH}{10^3}\times\eta_o$$

$$Q = \frac{3750\times10^3}{9.81\times10^3\times30\times0.87} = 14.646 m^3/s$$

Also
$$Q = K_1\pi D_1 B_1 v_{f1}$$

$$B_1 = \frac{14.646}{0.95\times\pi\times1.337\times7.28} = 0.504m$$

(c) Now $\tan\beta_1 = \dfrac{v_{f1}}{u_1 - v_{w1}} = \dfrac{7.28}{21-12.61} = 0.8677$

$$\beta_1 = 40.95° \text{ or } 40° - 57'$$

**Example 6.6** In an inward flow reaction turbine the peripheral velocity of the wheel at inlet is given by, $u_1 = \phi\sqrt{2gH}$, and the radial velocity of flow, $v_{f1} = \Psi\sqrt{2gH}$ The breadth of the wheel at inlet is '$n$' times the diameter of the wheel at inlet. If the turbine efficiency is 85% and the area taken by the vanes at inlet is 5 percent of the peripheul area at inlet ,prove that the specific speed of the turbine is equal to $888.2\phi\sqrt{\Psi.n}$.

A runner of the above type having a specific speed of 220rpm is required to develop 6700 kW under a head of 85 m. Taking $\Psi$ = 0.18 and $n$ = 0.2, calculate the rpm and the diameter of the runner.

**Solution.**

Given: $K_{u1} = \phi,\ K_{v_{f1}} = \Psi,\ B_1 = nD_1,\ \eta_t = 0.85,\ K_1 = 0.95$,

$N_s = 220rpm,\ P_t = 6700KW,\ H = 85m,\ \Psi = 0.18,\ n = 0.2$

(a)
$$Q = K_1\pi D_1 B_1 v_{f1} = 0.95\times\pi\times nD_1^2\times\Psi\sqrt{2gH}$$

Or $$Q_{11} = \frac{Q}{D_1^2\sqrt{H}} = 0.95\times\pi\times n\times\Psi\sqrt{2\times 9.81}$$

$$= 13.2197n\psi$$

Now $u_1 = \phi\sqrt{2gH} = \frac{\pi D_1 N}{60}$

$$N = \frac{60u_1}{\pi D_1} = \frac{60\phi\sqrt{2gH}}{\pi D_1}$$

Specific speed, $$N_s = \frac{N\sqrt{P_t}}{H^{5/4}} = \frac{60\phi\sqrt{2gH}}{\pi D_1}\times\sqrt{\frac{\gamma QH\eta_t}{10^3}}\times\frac{1}{H^{5/4}}$$

$$= \frac{60\times\phi\times\sqrt{2\times 9.81\times H}}{\pi D_1}$$

$$\times\sqrt{\frac{9.81\times 10^3\times 13.2197\times n\times\psi\times D_1^2\sqrt{H}\times H\times 0.85}{10^3}}\times\frac{1}{H^{5/4}}$$

$$= \frac{84.5962\times\phi\sqrt{H}}{D_1}\times 10.5\times D_1\times H^{3/4}\times\sqrt{n\psi}\times\frac{1}{H^{5/4}}$$

$$= 888.2\phi\sqrt{n\psi}$$

Now $$N_S = \frac{N\sqrt{P_t}}{H^{5/4}}$$

$$220 = \frac{N\sqrt{6700}}{85^{5/4}}$$

$$N = 693.7 \text{ rpm}$$

$$Q_{11} = 13.2197\times 0.2\times 0.18 = 0.475$$

$$Q = \frac{10^3\times P_t}{\gamma H\eta_t} = \frac{10^3\times 6700}{9.81\times 10^3\times 85\times 0.85}$$

$$= 9.453 \text{ m}^3\text{/s}$$

$$= Q_{11}D_1^2\sqrt{H} = 0.475\, D_1^2\times\sqrt{85}$$

$$D_1 = 1.47 \text{ m}$$

**Example 6.7** In an inward flow reaction turbine, the head on the turbine is 30m. The external and internal diameters are 1.5 m and 0.75 m respectively. The velocity of flow through the runner is constant and equal to 3m/s. The guide blade angle is 12° and the runner vanes are radial at inlet. If the discharge at outlet is radial, determine:

(a) The speed of the turbine

(b) The vane angle at outlet of the runner, and

(c) Hydraulic efficiency.

**Solution.**

Given: $H = 30$m, $D_1 = 1.5$m, $D_2 = 0.75$m, $v_{f1} = v_{f2} = 3$ m/s, $\alpha_1 = 12°$, $\beta_1 = 90°$, $\alpha_2 = 90°$, *i.e.*, $v_2 = v_{f2}$

The velocity triangles are shown in Fig.6.6.

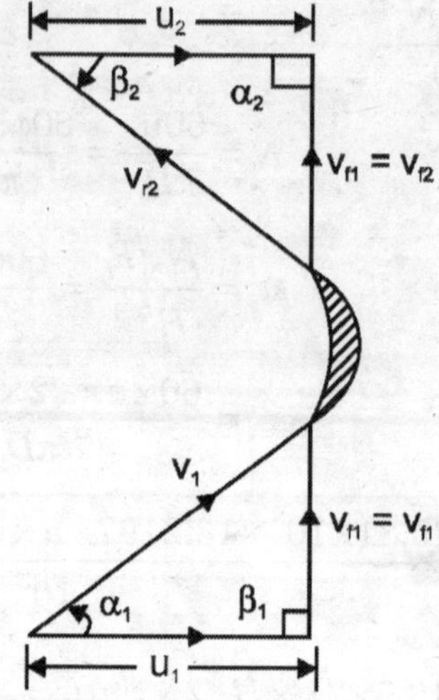


**Fig. 6.6 Velocity triangles.**

(a) From inlet velocity triangle, we have

$$\tan \alpha_1 = \frac{v_{f1}}{u_1}$$

$$\tan 12° = \frac{3}{u_1}$$

$$u_1 = 14.1 \text{ m/s}$$

Also $$u_1 = \frac{\pi D_1 N}{60}$$

$$N = \frac{14.1 \times 60}{\pi \times 1.5} = 179.7 rpm$$

(b) $$u_2 = \frac{\pi D_2 N}{60} = \frac{\pi \times 0.75 \times 179.7}{60} = 7.057 \text{ m/s}$$

From outlet velocity triangle, we have

$$\tan \beta_2 = \frac{v_{f2}}{u_2} = \frac{3}{7.057} = 0.4251$$

$$\beta_2 = 23.03°$$

(c) $$v_{w1} = u_1$$

$$\eta_{hyd} = \frac{u_1 v_{w1}}{gH} = \frac{u_1^2}{gH} = \frac{(14.1)^2}{9.81 \times 30}$$

$$= 0.6755 \text{ or } 67.55\%$$

**Example 6.8** An inward flow reaction turbine running at 420rpm requires a discharge of 15 m³/s. Its overall efficiency is 90 percent. The velocity at inlet of the spiral casing is 8 m/s and pressure head at this point is 225 m The centre line of the spiral casing inlet is 2.5 m above the tail race. The runner diameter at inlet is 2m and width is 0.25 m. The hydraulic efficiency is 94 percent and the flow is radial at the outlet from the runner. Determine: (a) power developed by the turbine, (b) specific speed, (c) guide vane angle, (d) runner blade angle at inlet, and (e) percentage of net head which is kinetic at entry to the runner. Neglect thickness of blades.

**Solution.**

Given: $N = 420$ rpm, $Q = 15$ m³/s, $\eta_o = 0.9$, $v_c = 8$ m/s, $H_c = 225$ m, $h = 2.5$ m, $D_1 = 2$m, $B_1 = 0.25$ m, $\eta_{hyd} = 0.94$, $\alpha_2 = 90°$

(a) Head at inlet to spiral casing,

$$H = H_c + \frac{v_c^2}{2g} + h = 225 + \frac{8^2}{2\times 9.81} + 2.5$$

$$= 230.76 \text{ m}$$

$$\text{Overall efficiency, } \eta_o = \frac{\text{power developed by the turbine}}{\gamma Q H}$$

$$0.9 = \frac{P_t}{9.81\times 10^3 \times 15 \times 230.76}$$

$$P_t = 30560.7 \text{ kW}$$

(b) Specific speed, $N_S = \dfrac{N\sqrt{P_t}}{H^{5/4}} = \dfrac{420\sqrt{30560.7}}{(230.76)^{5/4}} = 81.63$

(c) The velocity triangles are shown in Fig 6.7.

$$\text{Discharge, } Q = \pi D_1 B_1 v_{f1}$$

$$15 = \pi \times 2 \times 0.25 \times v_{f1}$$

$$v_{f1} = 9.55 \text{ m/s}$$

$$u_1 = \frac{\pi D_1 N}{60} = \frac{\pi \times 2 \times 420}{60} = 44\, m/s$$

$$v_{w2} = 0 \text{ as } \alpha_2 = 90°$$

$$\eta_{hyd} = \frac{u_1 v_{w1}}{gH}$$

$$0.94 = \frac{44 \times v_{w1}}{9.81 \times 230.76}$$

$$v_{w1} = 48.36 \text{ m/s}$$

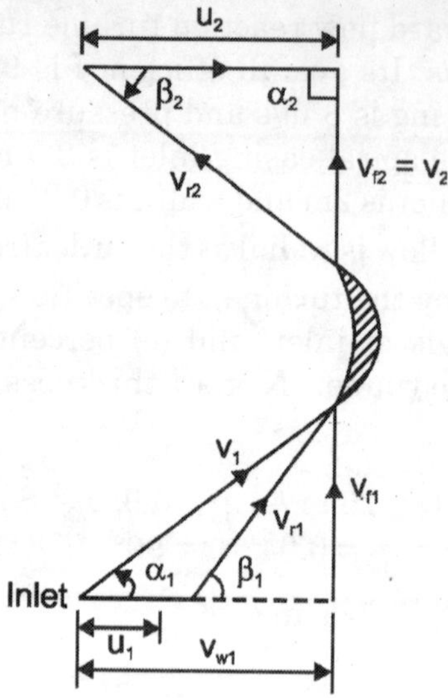


**Fig. 6.7**

From inlet velocity triangle, we have

$$\tan \alpha_1 = \frac{v_{f1}}{v_{w1}} = \frac{9.55}{48.36} = 0.19747$$

$$\alpha_1 = 11.17°$$

(d)

$$\tan \beta_1 = \frac{v_{f1}}{v_{w1} - u_1} = \frac{9.55}{48.36 - 44}$$

$$= \frac{9.55}{4.36} = 2.19$$

$$\beta_1 = 65.46°$$

(e)

$$\cos \alpha_1 = \frac{v_{w1}}{v_1}$$

$$v_1 = \frac{48.36}{\cos 11.17^\circ} = 49.3 m/s$$

$$\text{Kinetic head at entry, } H_k = \frac{v_1^2}{2g} = \frac{(49.3)^2}{2 \times 9.81} = 123.8\ m$$

$$\text{Percentage of net head} = \frac{H_k}{H} = \frac{123.8}{230.76}$$

$$= 0.5367 \text{ or } 53.67\%$$

**Example 6.9** A reaction turbine has to be designed to develop 370 kWunder a head of 70 m while running at a speed equal to 750 rpm.

Ratio of width of runner to outer diameter of runner = 0.1

Ratio of inner diameter to outer diameter of runner = 0.5

Flow ratio = 0.5

Hydraulic efficiency = 0.95

Mechanical efficiency = 0.84

Circumferential area occupied by thickness of vanes = 5%. Assuming velocity of flow to be constant, calculate guide vane angle at inlet and runner vane angles at inlet and outlet. The flow is radial. [IES,2003]

**Solution.**

Given: $P = 370$ kW, $H = 70$ m, $N = 750$ rpm, $\frac{B_1}{D_1} = 0.1$, $\frac{D_2}{D_1} = 0.5$, $K_f = 0.15$,

$\eta_{hyd} = 0.95$, $\eta_{mech} = 0.84$, $K_1 = 1 - 0.05 = 0.95$, $v_{f1} = v_{f2} = v_f$

$$\text{Flow velocity, } v_{f1} = k_f\sqrt{2gH} = 0.15\sqrt{2\times9.81\times70}$$

$$= 5.56 \text{ m/s}$$

$$v_{f2} = v_{f1} = 5.56 \text{ m/s}$$

Overall efficiency of turbine, $\eta_0 = \eta_{hyd} \times \eta_{mech} = 0.95 \times 0.84 = 0.798$

Shaft power available from the turbine,

$$P = \gamma QH\eta_o$$

$$370 = \frac{9.81\times10^3\times Q\times70\times0.798}{10^3}$$

$$Q = 0.675 \text{ m}^3\text{/s}$$

Also

$$Q = \pi D_1 B_1 v_{f1} \times K_1$$

$$0.675 = \pi D_1 \times 0.1 D_1 \times 5.56 \times 0.95$$

$$D_1 = 0.638\text{m}$$

$$B_1 = 0.1\ D_1 = 0.1 \times 0.638 = 0.0638 \text{ m}$$

$$D_2 = 0.5\ D_1 = 0.5 \times 0.638 = 0.319 \text{ m}$$

Now

$$\pi D_1 B_1 v_{f1} = \pi D_2 B_2 v_{f2}$$

and

$$v_{f1} = v_{f2}$$

∴

$$B_2 = \frac{D_1 B_1}{D_2} = 2 \times 0.0638 = 0.1276 \text{ m}$$

The velocity diagrams are shown in Fig .6.8.

$$u_1 = \frac{\pi D_1 N_1}{60} = \frac{\pi\times0.638\times750}{60} = 250.5 \text{ m/s}$$

$$\eta_{hyd} = \frac{u_1 v_{w1}}{gH} \text{ as } v_{w2} = 0$$

$$0.95 = \frac{25.05\times v_{w1}}{9.81\times70}$$

$$v_{w1} = 26.04 \text{ m/s}$$

$$\tan \alpha_1 = \frac{v_{f1}}{v_{w1}} = \frac{5.56}{26.04} = 0.2135$$

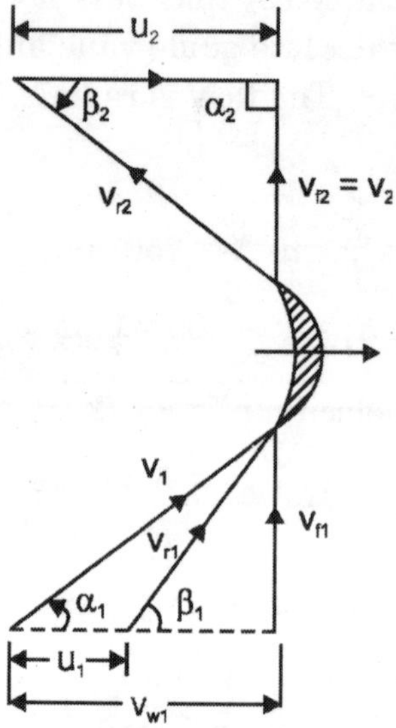

Fig.6.8

$$\alpha_1 = 12.05°$$

$$\tan \beta_1 = \frac{v_{f1}}{v_{w1} - u_1} = \frac{5.56}{26.04 - 25.05} = 5.616$$

$$\beta_1 = 79.9°$$

Now
$$u_2 = \frac{1}{2} u_1 = 12.525 \text{m/s}$$

$$\tan \beta_2 = \frac{v_{f2}}{u_2} = \frac{5.56}{12.525} = 0.444$$

$$\beta_2 = 23.9°$$

**Example 6.10** (a) Show that the hydraulic efficiency of a Francis turbine having velocity of flow through the runner as constant, is given by:

$$\eta_{hyd} = \frac{1}{1 + \dfrac{\tan^2 \alpha_1}{2\left(1 - \dfrac{\tan \alpha_1}{\tan \beta_1}\right)}}$$

where $\alpha_1$ = guide blade angle, $\beta_1$ = runner vane angle at inlet.

The turbine has radial discharge at outlet.

(b) If the vanes are radial at inlet, then show that

$$\eta_{hyd} = \frac{2}{2 + \tan^2 \alpha_1}$$

**Solution.**

Given: $v_{f1} = v_{f2}$, $\alpha_2 = 90°$

(a) The velocity triangles are shown in Fig.6.9.

From inlet velocity triangle, we have

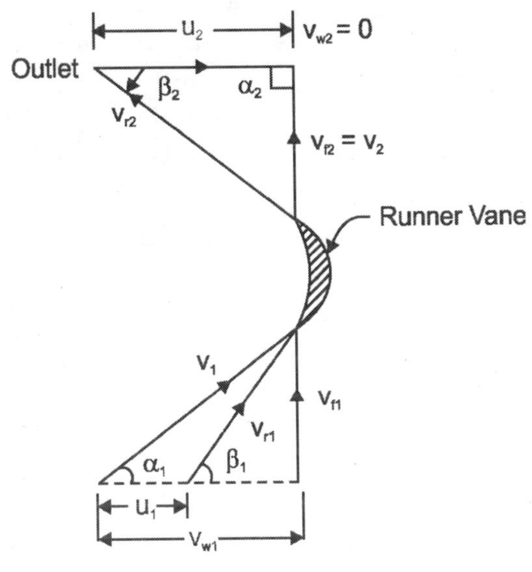


**Fig. 6.9**

$$\tan \alpha_1 = \frac{v_{f1}}{v_{w1}}$$

$$\tan \beta_1 = \frac{v_{f1}}{v_{w1} - u_1}$$

or $$v_{w1} - u_1 = \frac{v_{f1}}{\tan\beta_1} = \frac{v_{w1}\tan\alpha_1}{\tan\beta_1}$$

$$u_1 = v_{w1}\left[1 - \frac{\tan\alpha_1}{\tan\beta_1}\right]$$

$$\text{Head on turbine, } H = \frac{v_{w1}u_1}{g} + \frac{v_2^2}{2g} = \frac{v_{w1}u_1}{g} + \frac{v_{f1}^2}{2g}$$

or $$H = \frac{v_{w1}}{g} \times v_{w1}\left[1 - \frac{\tan\alpha_1}{\tan\beta_1}\right] + \frac{[v_{w1}\tan\alpha_1]^2}{2g}$$

$$= \frac{v_{w1}^2}{g}\left[1 - \frac{\tan\alpha_1}{\tan\beta_1} + \frac{\tan^2\alpha_1}{2}\right]$$

$$\text{Hydraulic efficiency, } \eta_{hyd} = \frac{v_{w1}u_1}{gH}$$

$$= \frac{v_{w1}v_{w1}\left[1-\frac{\tan\alpha_1}{\tan\beta_1}\right]}{v^2_{w1}\left[1-\frac{\tan\alpha_1}{\tan\beta_1}+\frac{\tan^2\alpha_1}{2}\right]}$$

$$= \left[1-\frac{\tan\alpha_1}{\tan\beta_1}+\frac{\tan^2\alpha_1}{2}\right]$$

Dividing numerator and denominator by $\left[1-\frac{\tan\alpha_1}{\tan\beta_1}\right]$, we get

$$\eta_{hyd} = \frac{1}{1+\frac{\tan^2\alpha_1}{2\left[1-\frac{\tan\alpha_1}{\tan\beta_1}\right]}}$$

Hence proved.

(b) For $\beta_1 = 90°$, we have $\eta_{hyd} = \frac{1}{1+\frac{\tan^2\alpha_1}{2[1-0]}} = \frac{2}{2+\tan^2\alpha_1}$

**Example 6.11** In a francis turbine, the velocity of flow inlet to exit of the runner remains constant. If the turbine discharge radially, show that the degree of reaction R can be expressed as:

$$R = 1 - \frac{1}{2}\left[\frac{\cot\alpha}{\cot\alpha-\cot\beta}\right]$$

where $\alpha$ and $\beta$ are the guide and runner vanes angles respectively.

The degree of reactive R is equal to the radio of pressure drop to the hydraulic work done in the runner, assuming that the losses in the runner are negligible.

**Solution.**

For radial discharge, $v_{w2} = 0$

Applying Bernoulli's equation between inlet and exit of runner, we have

$$\frac{p_1}{\gamma}+\frac{v_1^2}{2g} = \frac{p_2}{\gamma}+\frac{v_2^2}{2g}+\frac{v_{w1}u_1}{g}$$

$$\therefore \quad \frac{p_1-p_2}{\gamma} = \frac{v_2^2-v_1^2}{2g}+\frac{v_{w1}u_1}{g}$$

$$\text{Degree of reaction, } R = \frac{\frac{1}{\gamma}(p_1 - p_2)}{\frac{v_{w1}u_1}{g}} = \frac{\frac{v_2^2 - v_1^2}{2g} + \frac{v_{w1}u_1}{g}}{\frac{v_{w1}u_1}{g}}$$

$$= 1 + \frac{1}{2}\left[\frac{v_2^2 - v_1^2}{v_{\omega 1}u_1}\right]$$

From Fig.6.10,we have

$$v_2 = v_{f2} = v_{f1}$$

$$v_{w1} = v_1 \cos\alpha$$

$$v_{f1} = v_1 \sin\alpha$$

$$u_1 = v_{w1} - v_{f1}\cot\beta$$

$$= v_1 \cos\alpha - v_1 \sin\alpha \cot\beta$$

$$= v_1 (\cos\alpha - v_1 \sin\alpha \cot\beta)$$

$$v_2 = v_1 \sin\alpha$$

$$R = 1 + \frac{1}{2}\left[\frac{\upsilon_1^2 \sin^2\alpha - \upsilon_1^2}{\upsilon_1 \cos\alpha(\cos\alpha - \sin\alpha\cot\beta)\upsilon_1}\right]$$

$$= 1 + \frac{1}{2}\left[\frac{\sin^2\alpha - 1}{\cos\alpha(\cos\alpha - \sin\alpha\cot\beta)}\right]$$

$$= 1 + \frac{1}{2}\left[\frac{-\cos^2\alpha - 1}{\cos^2\alpha(1 - \tan\alpha\cot\beta)}\right]$$

$$= 1 + \frac{1}{2}\left[\frac{1}{1 - \tan\alpha\cot\beta}\right]$$

$$= 1 + \frac{1}{2}\left[\frac{\cot\alpha}{\cot\alpha - \cot\beta}\right]$$

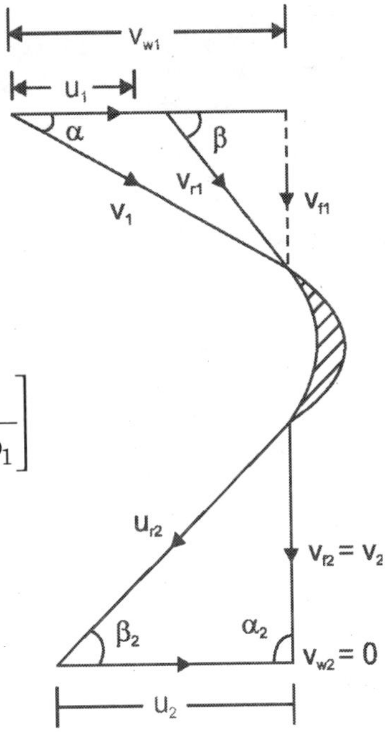

**Fig. 6.10**

**Example 6.12** An inward flow reaction turbine has a degree of reaction of 0.6. The peripheral velocity of the vanes at entry is 15 m/s. The radial velocity of flow is constant at 3.2 m/s . The diameter of the rotor at entry is twice that at the exit. Determine the blade angles at entry and exit assuming that the absolute velocity of the water leaving the runner is in the radial direction. [IES,1985]

**Solution.**

Given: $R = 0.6$, $u_1 = 15$ m/s, $v_{f1} = v_{f2} = 3.2$ m/s, $D_1 = 2D_1$, $\alpha_2 = 90°$

As derived in Example 6.11, we have

$$R = 1 - \frac{1}{2}\left[\frac{\cot\alpha_1}{\cot\alpha_1 - \cot\beta_1}\right]$$

$$0.6 = 1 - \frac{1}{2}\left[\frac{\cot\alpha_1}{\cot\alpha_1 - \cot\beta_1}\right]$$

$$\frac{\cot\alpha_1}{\cot\alpha_1 - \cot\beta_1} = 0.8$$

or

$$\cot\alpha_1 = 0.8 \cot\alpha_1 - 0.8 \cot\beta_1$$

$$\cot\alpha_1 + 4\cot\beta_1 = 0$$

$$u_2 = \frac{u_1 D_2}{D_1} = 15 \times \frac{1}{2} = 7.5 \text{ m/s}$$

$$\tan\beta_2 = \frac{v_{f2}}{u_2} = \frac{3.2}{7.5} = 0.4267$$

$$\beta_2 = 23.1$$

$$v_{f1} = v_2 = 3.2 \text{ m/s}$$

$$\tan\beta_1 = \frac{v_{f1}}{v_{w1} - u_1}$$

$$v_{w1} = v_1 \cos\alpha_1$$

$$\sin\alpha_1 = \frac{v_{f1}}{v_1}$$

$$v_1 = \frac{v_2}{\sin\alpha_1} = \frac{3.2}{\sin\alpha_1}$$

$$= 3.2(-4\cot\beta_1) - 12.8\cot\beta_1 \text{ (from eq.(1)}$$

$$\cot\beta_1 \cot\beta_1 = \frac{-12.8\cot\beta_1 - 15}{3.2}$$

$$3.2\cot\beta_1 = -12.8\cot\beta_1 - 15$$

$$16\cot\beta_1 = -15$$

$$\cot\beta_1 = -0.9375$$

$$\beta_1 = -46.85°$$

**Example 6.13** An inward flow reaction turbine works under a head of 30 m and 10 m$^3$/s discharge. The speed of the runner is 300 rpm. The peripheral velocity of wheel at inlet is $0.9\sqrt{2gH}$ and radial velocity of flow is $0.3\sqrt{2gH}$, where H is the head on the turbine. If the overall efficiency and hydraulic efficiency of the turbine are 80% and 90% respectively, determine: (a) the power developed, (b) diameter and width of runner at inlet, (c) guide blade at inlet, (d) inlet angle at runner vane, and (e) diameter of runner at outlet. Assume that the discharge at outlet is radial.

**Solution.**

Given: $H = 30\text{m}$, $Q = 10\ \text{m}^3/\text{s}$, $N = 300$ rpm, $u_1 = 0.9\sqrt{2gH}$, $v_{f1} = 0.3\sqrt{2gH}$, $\eta_{hyd} = 0.9$, $\eta_o = 0.8$, $\alpha_2 = 90°$
The velocity triangles are shown in Fig. 6.11.

(a) Overall efficiency, $\eta_o = \dfrac{P}{\gamma QH}$

$$P = \frac{\eta_o \gamma QH}{10^3} = \frac{0.8 \times 9.81 \times 10^3 \times 10 \times 30}{10^3}$$

$$= 2354.4 \text{ kW}$$

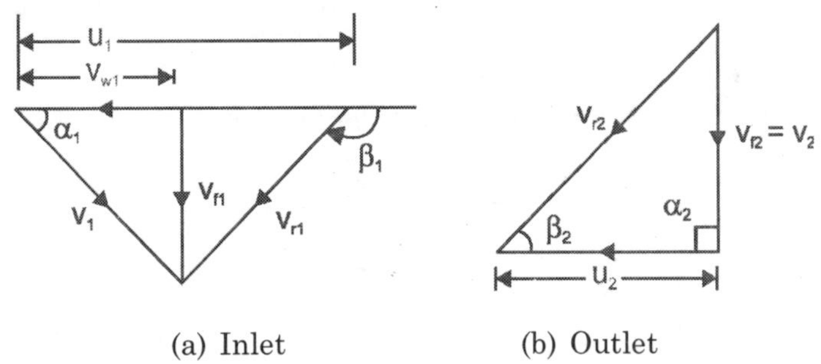


(a) Inlet (b) Outlet

**Fig. 6.11 Velocity triangles.**

(b)
$$u_1 = \frac{\pi D_1 N}{60}$$

$$D_1 = \frac{60\ u_1}{\pi N} = \frac{60 \times 0.9\sqrt{2 \times 9.81 \times 30}}{10^3} = 1.38 \text{ m}$$

$$Q = \pi D_1\ B_1\ v_{f1}$$

$$B_1 = \frac{10}{\pi \times 1.39 \times 0.3\sqrt{2 \times 9.81 \times 30}} = 0.315 \text{ m}$$

(c)
$$\tan \alpha_1 = \frac{v_{f1}}{v_{w1}}$$

$$\eta_{hyd} = \frac{u_1 v_{w1}}{gH}$$

$$v_{w1} = \frac{0.9 \times 9.81 \times 30}{0.9\sqrt{2 \times 9.81 \times 30}}$$

$$= 12.13 \text{ m/s}$$

$$\tan \alpha_1 = \frac{0.3\sqrt{2 \times 9.81 \times 30}}{12.13} = 0.6$$

$$\alpha_1 = 30.96°$$

(d)
$$\tan \beta_1 = \frac{v_{f1}}{v_{w1} - u_1}$$

$$= \frac{0.3\sqrt{2\times 9.81\times 30}}{12.13 - 0.9\sqrt{2\times 9.81\times 30}} = -0.75$$

$$\beta_1 = 180° - 36.88° = 143.12°$$

(e)
$$\sin(180° - \beta_1) = \frac{v_{f1}}{v_{r1}}$$

$$v_{r1} = \frac{0.3\sqrt{2\times 9.81\times 30}}{\sin 143.12°} = 12.13 \text{ m/s}$$

Assuming $v_{f1} = v_{f2}; v_2 = v_{f2} = 0.3\sqrt{2\times 9.81\times 30} = 7.278$ m/s

For no blade friction, $v_{r2} = v_{r1} = 12.13$ m/s

$$\sin \beta_2 = \frac{v_2}{v_{r2}} = \frac{7.278}{12.13} = 0.6$$

$$\cos \beta_2 = \frac{u_2}{v_{r2}}$$

$$u_2 = 12.13 \cos 36.87° = 9.7 \text{ m/s}$$

$$u_2 = \frac{\pi D_2 N}{60}$$

$$D_2 = \frac{60\times 9.7}{\pi\times 300} = 0.618 \text{ m}$$

## Summary

1. Francis turbine is an inward flow reaction turbine.

2. Degree of reaction, $R = \dfrac{\text{pressure drop in the runner}}{\text{hydraulic work done on the runner}}$

$$= \frac{[(p_1 - p_2)/\gamma]}{[v_{w1}u_1/g]}$$

$$= 1 - \frac{1}{2}\left[\frac{\cot \alpha_1}{\cot \alpha_1 - \cot \beta_1}\right]$$

3. Power developed by the turbine, $P_H = \dfrac{\rho Q(v_{w1}u_1 - v_{w2}u_2)}{10^3}$ kW

Available or water power developed by the turbine,

$$P_a \text{ or } P_a \text{ or } P_w = \frac{\gamma Q H}{10^3} \text{ kW}$$

4. For radial discharge, $\alpha_2 = 90°$ and $v_{w2} = 0$, $v_{f2} = v_2$

5. Maximum work done on the runner $= \rho Q\, v_{w1}\, u_1$

6. Efficiencies of Francis turbine:

(i) Head efficiency, $\eta_H = \dfrac{\textit{Net head}}{\textit{Operating head}} = 1 - \dfrac{\Delta H}{H}$

$$= 2K_{u1}\left[K_{vw1} - K_{vw2} \cdot \frac{R_2}{R_1}\right]$$

$$= 2K_{u1} \cdot K_{vw1} \text{ for radial discharge}$$

(ii) Volumetric efficiency, $\eta_{vol} = 1 - \dfrac{\Delta Q}{Q}$

(iii) Hydraulic efficiency, $\eta_{hyd} = \dfrac{v_{w1}u_1 - v_{w2}u_2}{gH}$

$$= \left(1 - \frac{\Delta Q}{Q}\right)\left(1 - \frac{\Delta H}{H}\right)$$

$$= \eta_{vol} \cdot \eta_H$$

Actual hydraulic power developed by the turbine,

$$P_{hyd} = \frac{\gamma(Q - \Delta Q)(H - \Delta H)}{10^3}\ kW$$

$$\eta_{hyd} = \frac{P_{hyd}}{P_a} = \frac{Hydraulic\ power\ (actual)}{Water\ power}$$

(iv) Mechanical efficiency, $\eta_{mech} = \dfrac{Shaft\ Power}{Power\ Developed\ by\ the\ Runner}$

Brake power, $P_t = P_{hyd} \cdot \eta_{mech}$

(v) Overall efficiency, $\eta_o$ or $\eta_t = \dfrac{Shaft\ power}{Water\ power} = \dfrac{P_t}{P_a} = \dfrac{P_{hyd} \times \eta_{mech}}{P_a}$

$$= \frac{P_a\ \eta_{vol}\ \eta_H\ \eta_{mech}}{P_a}$$

$$= \eta_{vol} \cdot \eta_H \cdot \eta_{mech}$$

7. Area across flow, $A_{f1} = (\pi D_1 - z t_1)\, B_1 = K_1 \pi D_1 B_1 = K_1 \pi n\, D_1^2$
Discharge, $Q = A_{f1} \cdot v_{f1} = A_{f2}\, v_{f2}$, $A_{f2} = K_2\ \pi D_2\ B_2 = K_2\ \pi n\ D_2^2$
$K_1 = K_2$ blade thickness factor / coefficient
$= 0.90 - 0.95$

$$n = \frac{B_1}{D_1} = 0.1 - 0.45$$

Flow ratio, $K_{f1} = \dfrac{v_{f1}}{\sqrt{2gH}} = 0.15 - 0.30$

Speed ratio, $K_{u1} = \dfrac{u_1}{\sqrt{2gH}} = 0.6 - 0.9$

8. Number of vanes on the runner = Number of guide vanes $\pm$ 1 $\simeq 16 - 24$

## Multi–Choice Questions

1. Reaction turbines are used for
   (a) high head and low discharge
   (b) medium head and high discharge
   (c) low head and low discharge
   (d) high head and high discharge
2. The specific speed of a Francis turbine ranges from :
   (a) 5 –10 (b) 10 – 40
   (c) 50 – 340 (d) 350 – 800
3. The runner blades in a Francis turbine ranges from :
   (a) 2 – 4 (b) 5 – 8
   (c) 9 – 15 (d) 16 – 24
4. The (number of runner blades) – (number of guide blades) are kept as
   (a) $\pm 1$ (b) $\pm 2$
   (c) $\pm 3$ (d) Zero
5. The ratio of actual work available from the turbine to energy imparted to the wheel is called
   (a) mechanical efficiency (b) hydraulic efficiency
   (c) turbine efficiency (d) overall efficiency
6. The ratio of work done on the wheel to the energy actually supplied to the turbine is called
   (a) mechanical efficiency (b) hydraulic efficiency
   (c) turbine efficiency (d) overall efficiency
7. The degree of reaction of a turbine is defined as
   (a) $\dfrac{\text{static pressure drop}}{\text{total energy transfer}}$
   (b) $\dfrac{\text{total energy transfer}}{\text{static pressure drop}}$
   (c) $\dfrac{\text{change of velocity energy across the turbine}}{\text{total energy transfer}}$
   (d) $\dfrac{\text{static pressure drop}}{\text{hydraulic work done}}$

### Answers

1. (b) 2. (c) 3. (d) 4. (a) 5. (a) 6. (b) 7. (d)

## Review Questions

1. Define a reaction turbine.
2. Differentiate between an impulse and reaction turbine.
3. Name main components of a Francis turbine.
4. What is the function of scroll casing?
5. What is a draft tube? What are its functions?
6. Why outward flow reaction turbines are not used these days?
7. Define degree of reaction.
8. Define gross head and net head.
9. Define head efficiency, volumetric efficiency, and hydraulic efficiency.
10. Define mechanical efficiency and overall efficiency.
11. What is the relationship between overall, volumetric, head, and mechanical efficiencies?
12. Differential between available power and head power developed by the turbine.
13. Define water power and shaft power.
14. What is blade thickness factor?
15. Define flow ratio and speed ratio.
16. Draw velocity triangles for a Francis turbine with radial discharge.

## Numerical Problems

1. An inward flow reaction turbine of inlet diameter 1.2 m operates under a head of 150 m and requires a discharge of 6 $m^3/s$ at a rotational speed of 400 rpm. The guide vane angle is 20° and the water leaves the runner blade radially.

   If the runner is 0.1m wide at the inlet, calculate:

   (a) The torque and power supplied to the shaft, and

   (b) The efficiency of the turbine.

   [**Ans.** 157.42 k.N.m, 6593 kW; 74.7%]

2. A Francis turbine develops 2 MW when running at 250 rpm under a head of 24 m. Calculate:

   (a) specific speed of the turbine, and

   (b) the inlet diameter of the runner if the speed ratio is 0.8.

   [**Ans.** 210.47, 1.326 m]

3. An inward flow reaction turbine, having an overall efficiency of 75% is required to generate 130 kW under a head of 6 m when running at 250 rpm. The velocity of the periphery of wheel is $0.8\sqrt{2gH}$, and the radial velocity of flow is $0.5\sqrt{2gH}$. The hydraulic losses in the turbine are 20% of available energy. Determine:
   (a) the angle of the guide blade at inlet,
   (b) the wheel vane angle at inlet,
   (c) the diameter of the wheel, and
   (d) the width of the wheel at inlet
   Assume radial discharge.
4. The following data is available for an inward flow reaction turbine:
   Power developed = 150 kW
   Head = 10 m
   Speed = 175 rpm
   Ratio of outlet to inlet diameter = 0.65
   Peripheral speed at inlet = $0.75\sqrt{2gH}$
   Constant flow velocity = $0.16\sqrt{2gH}$
   Hydraulic efficiency = 86%
   Overall efficiency = 80%
   Assuming radial discharge, calculate:
   (a) the diameter of runner at inlet and outlet,
   (b) the discharge,
   (c) the guide vane angle, and
   (d) the runner vane angles at inlet and outlet.
5. A Francis turbine gave the following performance:
   $H = 5$ m, $Q = 1.5$ m$^3$/s, Brake power = 65 kW, speed = 200 rpm, Gate opening = 90%, and inlet diameter = 0.75 m. Assume radial discharge:
   Calculate:
   (a) the efficiency, and
   (b) the speed ratio.
6. An inward flow reaction turbine has external and internal diameters as 0.9 m and 0.45 m respectively. The turbine is running at 200 r.p.m. and width of turbine at inlet is 0.2 m. The velocity of flow through the runner is constant and is aqual to 1.8 m/s. The guide blades make an angle of 10° to the tangent of the wheel and discharge at the outlet of turbine is radial. Draw the inlet and outlet velocity triangles and determine:
   (a) Relative velocity at inlet
   (b) The runner blade angles

(c) Width of the runner at outlet

(d) Head at the inlet of the turbine

(e) Power developed; and

(f) Hydraulic efficiency of the turbine.

[**Ans.** (a) 1.963 m/s; (b) 66.48°; 20.9°; (c) 0.4 m; (d) 9.97 m; (e) 97.8 kW; (f) 98.34% ]

7. An inward flow reaction turbine of inlet diameter 1.2 m operates under a head of 150 m and requires a discharge of 6 $m^3/s$ at a rotational speed of 400 r.p.m. The guide vane angle is 20° and the water leaves the runner blade axially. If the runner is 0.1 m wide at the inlet, Calculate:

   (a) The torque and power supplied to the shaft, and

   (b) The efficiency of the turbine.

   [**Ans.** (a) 156 kNm; 6533 kW; (b) 74%]

8. The inward flow reaction turbine develops 735 kW at 750 r.p.m. under a net head of 100 m. The guide vane makes an angle of 15° with the tangent at inlet. The axial length of the blade at inlet is 0.1 times the outer diameter. The radial velocity of flow through the wheel is constant and the discharge from the wheel is constant and the discharge from the wheel is radial. The blade thickness blocks 5 percent of the area of flow at inlet. The hydraulic efficiency is 84%.

   (a) The wheel diameter,

   (b) The wheel width, and

   (c) The blade angle at inlet.

   [**Ans.** (a) 0.574 m; (b) 0.0574 m; (c) 38.7°]

9. A Francis turbine has to be designed to develop 367.5 kW under a head of $H = 70$ m while running at $N = 750$ r.p.m. Ratio of width of runner to diameter of runner, $n = 0.1$, inner diameter is half the outer diameter. Flow ratio = 0.15, hydraulic efficiency = 95%, mechanical efficiency = 84%. Four percent of the circumferential area of runner to be occupied by the thickness of vanes, velocity of flow is constant and the discharge is radial at exit. Calculate:

   (a) The diameter of the wheel

   (b) The quantity of water supplied, and

   (c) The guide vane angle at inlet and runner vane angles at inlet and exit.

   [**Ans.** (a) 0.633 m; (b) 0.67$m^3/s$ (c) 11°58′, 76°50′, 24°5′]

10. A Francis turbine with an overall efficiency of 75 percent is required to produce 149.26 kW. It is working under a head of 7.62 m. The peripheral velocity = $0.26\sqrt{2gH}$ and the radial velocity of flow at inlet is $0.96\sqrt{2gH}$. The wheel runs at 150 r.p.m. and the hydraulic losses in the turbine

are 22 percent of the available energy. Assuming radial discharge, determine:

(a) The guide blade angle,

(b) The wheel vane angle at inlet,

(c) Diameter of the wheel at inlet, and

(d) Width of the wheel at inlet.

[**Ans.** (a) 32.6°; (b) 37.7°; (c) 0.404 m; (d) 0.17 m]

11. Design a Francis turbine runner with the following data:

Net head ... 68 m

Speed of the runner ... 750 r.p.m.

Output ... 330.9 kW.

Hydraulic efficiency ... 94%

Overall efficiency ... 85%

Flow ratio ... 0.15

Breadth ratio ... 0.1

Inner diameter of runner $= \frac{1}{2} \times$ outer diameter

Also assume 5 percent of circumferential area of the runner to be occupied by the thickness of the vanes.

Velocity of flow remains constant throughout and flow is radial at exit.

[**Ans.** $D_1 = 0.6$ m; $D_2 = 0.3$ m; $\beta_1 = 0.06$ m; 11°38′, 60°40′, 24°57′]

12. A Francis turbine has a wheel diameter of 1 m at the entrance and 0.5 m at the exit. The vane angle at the entrance is 90° and the guide vane angle is 15°. The water at the exit leaves the vanes without any tangential velocity. The head is 30 m and the radial component of flow is constant. What would be the speed of the wheel in r.p.m. and vane angle at exit? State whether the speed calculated is synchronous one or not. If not, what speed would you recommend to couple the turbine with an alternator of 50 Hz?

[**Ans.** 324 r.p.m; 28°12′ ; 300 r.p.m.]

13. The following data pertain to an inward flow reaction turbine:

Overall efficiency ... 75%

Power given by the turbine ... 128.7 kW

Head ... 6 m

The velocity of the peripheral of wheel $= 0.5\sqrt{2gH}$

Radial velocity of flow $= 0.5\sqrt{2gH}$

Speed of wheel = 250 r.p.pm.

Hydraulic losses in the turbine = 22 percent of the available energy.

Determine the following:

(a) Guide blade angle at inlet,

(b) The wheel vane angle at inlet,

(c) The diameter of the wheel, and

(d) The width of wheel at inlet.

Assume the discharge to be radial.

**[Ans.** (a) 34°48′, (b) 49°36′, (c) 662 mm; (d) 371 mm]

14. The following data pertain to an inward flow reaction turbine:

| | |
|---|---|
| Power to be developed | ... 625 kW |
| Speed of the runner | ... 1000 r.p.m. |
| Head | ... 100 m |
| Guide vane angle | ... 16° |

Internal diameter = 0.6 times the external diameter

| | |
|---|---|
| Hydraulic efficiency | ... 88% |
| Overall efficiency | ... 86% |
| Allowance of blade thickness | ... 5% |

Axial length of blade inlet = 0.1 times the outer diameter

| | |
|---|---|
| Radial velocity of flow | ... constant |

Find the leading dimensions of the runner.

**[Ans.** $D_1$ = 0.527 m; $D_2$ = 0.316 m; $B_1$ = 52.7 mm; 67.5°; 31°37′]

15. An inward flow turbine wheel works under a head of 20 m and makes 380 rpm. The diameter of the outer circumference of the wheel is 60 cm and of the inner circumference 30 cm. The velocity of water entering the wheel is 13 m/s and the angle it makes with the tangent to the wheel is 10°. Assuming the radial velocity of flow through the wheel to be constant, determine the direction of the tangent to the vane of the vane of the wheel at the inlet and outlet. Determine also the hydraulic efficiency of the turbine.

**[Ans.** 110° – 30′ ; 21° tangent to the periphery; 77.5%]

16. Find the leading dimensions of the runner of an inward flow reaction water turbine to develop 825.4 kW at 1,000 rpm under a head of 100 m assuming a guide vane angle of 16° : axial length of blade at inlet 0.1 times the outer diameter, inner radius 0.6 time the outer radius; radial velocity of flow constant; final discharge radial; hydraulic efficiency 0.88; overall efficiency 0.86; allowance for blade thickness 5%.

**[Ans.** $D_1$ = 0.527 m; $D_2$ = 0.316 m; $B_1$ = 5.27 cm; $112\frac{1}{2}$°; 31° – 37′]

17. The following data refers to an inward flow reaction turbine:
    Supply = 1 $m^3$/s at 25 m head,
    Wheel diameters = 75 cm at outlet and 50 cm at inlet,
    Radial exit velocity = 2.5 m/s
    Inlet vane angle = 35°
    Calculate the power and rpm of the turbine. Assume the width of the wheel as constant, and turbine efficiency 80%.

    [**Ans.** 196.2 kW; 326 rpm]

18. An Inward flow turbine works under a total head of 30 m. The velocity of wheel periphery at inlet is 15 m/s. The outlet pipe of the turbine is 30 cm in diameter and the turbine is supplied with 250 litres of water per second. The radial velocity of flow through the wheel is the same as the velocity in the outlet pipe. Neglecting friction, determine –
    (a) the vane angle at inlet,
    (b) the guide blade angle,
    (c) HP of the turbine.

    [**Ans.** 10° – 37′; 40° – 12′; 73.6 kW]

19. Discuss the general theory of the inward flow reaction turbine and deduce an expression for the maximum hydraulic efficiency. State the reasons that lead to the adoption of comparatively smaller inlet angles of runner blades for higher heads and larger ones for lower heads. Show also that when the inlet angle is 90° for the medium heads, the velocity of flow being constant, the hydraulic efficiency is expressed by $\frac{2}{2+\tan\alpha}$, where α is the guide blade angle.

20. The following information pertains to an inward flow reaction turbine: Output = 370 kW; Speed = 752 r.p.m.; Head = 70 m. Axial length of blade at inlet = 0.1 × wheel diameter at inlet; ratio of outer and inner diameters = 0.5; flow ratio = 0.17 ; hydraulic efficiency = 92%; overall efficiency = 84%.

    Assume radial discharge, constant velocity of flow throughout and area and area blocked by blade thickness as 5% of area of flow. Determine:
    (a) diameters at inlet and outlet
    (b) widths of wheel at inlet and outlet
    (c) guide blade angle
    (d) vane angles at inlet and outlet. [IES, 1988]

C H A P T E R 7

# Kaplan Turbinne

## 7.1 Introduction

If the specific speed of a turbine rises the flow tends to be axial, i.e. parallel to the axis of rotation of the turbine axis. Such a turbine is called axial flow turbine. The head at the inlet to the turbine is the sum of pressure head and kinetic head. Such a turbine is called as reaction turbine.

The shaft of an axial flow reaction turbine is vertical. There are two types of axial flow reaction turbines: 1. Propeller turbine, and 2. Kaplan turbine.

All parts of an axial flow reaction turbine are similar to those of Francis turbine. But the water enters the runner in an axial direction, travels axially across the blades passage and leaves axially. The pressure at the inlet of the blades is larger than the pressure at the exit of the blades. The change in the magnitude of the relative velocity across the blades is responsible for the energy transfer or reaction effect.

The runner blades on a propeller turbine are fixed as in case of a Francis turbine. In the case of a Kaplan turbine the runner blades are made adjustable. Both these turbines are used where comparatively low head and large quantity of water are available. The propeller turbine is used economically when it has to work constantly under full load, otherwise Kaplan turbine is preferred.

In the axial flow reaction turbines, the lower part of the shaft is made larger, which is known as 'hub' or 'boss'. The vanes are fixed on the hub and it acts as a runner. The runner blade actuating mechanism is accommodated in the boss and the main turbine shaft, which is made hollow.

## 7.2 Differences Between Francis and Kaplan Turbine.

The differences between Francis and Kaplan turbines are given in Table 7.1.

**Table 7.1 Differences Between Francis and Kaplan Turbine**

| Aspects | Francis turbine | Kaplan turbine |
|---|---|---|
| 1. Type of turbine | Radially inwards Or mixed flow | Axial flow |
| 2. Disposition of shaft | Horizontal or vertical | Only vertical |

| | | |
|---|---|---|
| 3. Adjustability of runner blades | Not adjustable | Adjustable |
| 4. Number of vanes | Large, 16 to 24 | Small, 3 to 8 |
| 5. Resistance overcome | Large | Small |
| 6. Head | Medium, 60 to 250 m. | Low, upto 30m |
| 7. Flow rate | Medium | Large |
| 8. Specific speed | 50-250 | 250-850 |
| 9. Type of governor | Ordinary | Heavy duty |

## 7.3 Kaplan Turbine

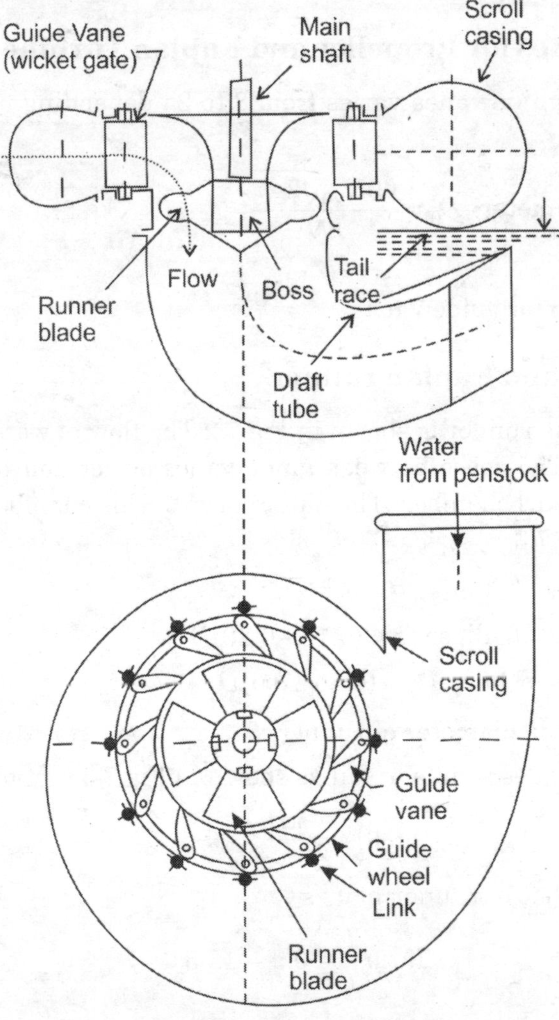


**Fig. 7.1 Outlines of a Kaplan Turbine**

The Kaplan turbine, as shown in Fig. 7.1, is an axial flow reaction turbine. It operates entirely in a closed conduit from inlet to tail race. The runner blades are made adjustable. The turbine operates where comparative low head and large quantity of water are available. The main components of a Kaplan turbine are similar to those of Frances turbine. It is generally installed on a canal. The enlargement of width of canal near the turbine is called a forebay.

There is a considerable space between the ends of guide vanes and the leading edge of runner. The direction of flow changes from radial to axial in this space. Simultaneously in this space, the flow pattern changes approximately to free spiral vortex. The guide vanes impart the whirl component to the flow and the blades remove the wheel making the discharge purely axial and transforming the whirl into useful power.

## 7.4 Guide Vanes for Propeller and Kaplan Turbine

The number of guide vanes varies from 8 to 24 depending upon the guide wheel diameter.

Guide wheel diameter, $D_0 = \dfrac{60K_{u0}\sqrt{2gh}}{\pi N}$

Where $N$ = rpm

$K_{u0}$ = speed ratio for guide vanes

## 7.5 Propeller and Kaplan runner

A Kaplan turbine runner is shown in Fig.7.2.The flow of water at inlet and outlet is axial. The space between guide vanes outlet and runner inlet is known as the whirl chamber .The outer curve of this chamber is an ellipse as shown in Fig.7.3.

Semi –major axis of ellipse, $a = 0.13D$

Semi –minor axis of ellipse, $b = (0.16 - 0.20)D$

Diameter of boss or hub, $D_b = (0.35 - 0.6)D$

Where $D$ = outside diameter of runner.

Section of the blades is an aerofoil as shown in Fig.7.3 section $x - x$.

$$\text{Pitch, } t = \frac{\pi D}{z}$$

Wheel $z$ = number of runner blades

$$\text{Ratio } \frac{l}{t} = 0.9 - 1.05$$

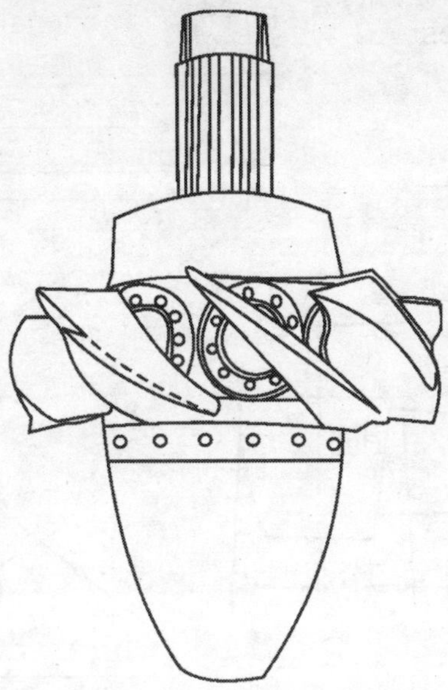

Fig. 7.2 Kaplan turbine runner

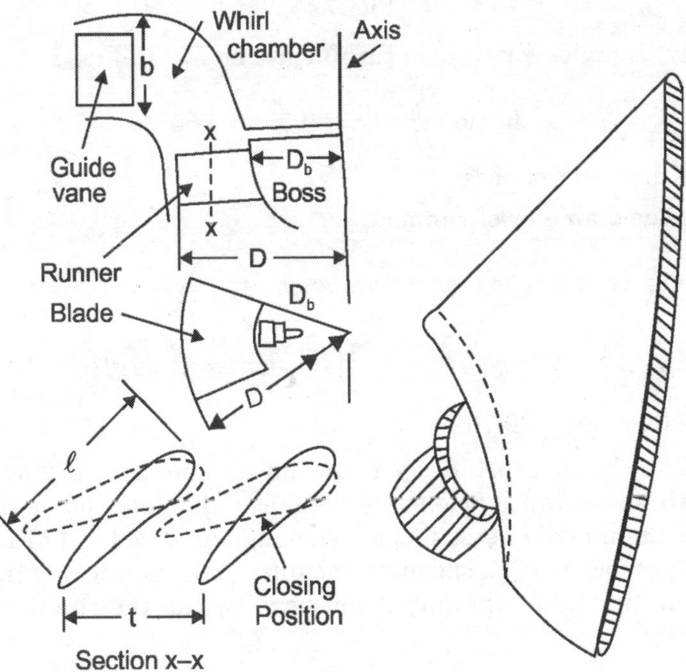


Fig. 7.3 Outlines of Kalpan or propeller runner with one blade.

## 7.6 Velocity triangles

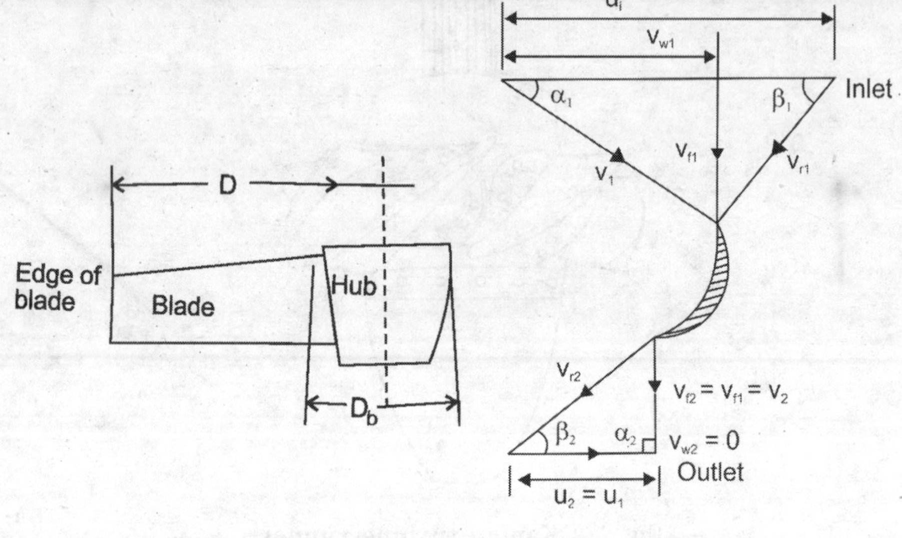


(a) Hub and blade (b) Velocity triangles

**Fig. 7.4**

The velocity triangles for Kaplan turbine are shown in Fig. 7.4.

$$\text{Ratio } \frac{D_b}{D} = n = 0.35 \text{ to } 0.60$$

Where $D_b$ = diameter of boss

$D$ = outside diameter of runner.

$$\text{Discharge } Q = \text{area across flow} \times \text{flow velocity} = \frac{\pi}{4}\left(D^2 - D_b^2\right) \times v_f$$

$$= \frac{\pi}{4}\left(D^2 - D_b^2\right) \times K_f\sqrt{2gH}$$

Where $K_f$ = flow ratio $\simeq$ 0.70

**Example 7.1** A Kaplan turbine operating under a net head of 20m develops 16 MW with an overall efficiency of 80%. The diameter of the runner is 4.2m, while the hub diameter is 2 m. The angular speed is 17 rad/s. if the hydraulic efficiency is 90%, calculate the inlet and exit angle of the runner blades at the tip and at the hub if the flow leaving the runner is purely axial.

**Solution.**

Given: $H$ = 20m, $P$ = 16 MW, $\eta_o$ = 0.8, $D$ = 4.2m, $D_b$ = 2m, $\omega$ = 17 rad/s, $\eta_{hyd}$ = 0.9

$$P = \rho g Q \eta_o$$

$$1000 \times 9.81 \times Q \times 20 \times 0.8 = 16 \times 10^6$$

$$Q = 102.04 \text{ m}^3/\text{s}$$

$$Q = \frac{\pi}{4}\left(D^2 - D_b^2\right)v_f$$

$$102.04 = \frac{\pi}{4}\left((4.2)^2 - 2^2\right)v_f$$

$$v_f = 9.53 \text{ m/s}$$

$$\eta_{hyd} = \frac{v_{w1}u}{gH}$$

$$0.90 = \frac{v_{w1}u}{9.81\times 20}$$

$$v_{w1}\,u = 176.4$$

Mean speed of runner, $u = \omega r_m = 17\times\frac{1}{2}\left(\frac{4.2+2}{2}\right)$

$$= 26.35 \text{ m/s}$$

From exit of velocity triangle, $\tan\beta_2 = \frac{v_f}{u} = \frac{9.53}{26.35} = 0.3617$

$$\beta_2 = 19.9°$$

$$v_{w1} = \frac{176.4}{26.35} = 6.7 \text{ m/s}$$

From inlet velocity triangle, $\tan\beta_1 = \frac{v_{f1}}{u - v_{w1}}$

$$= \frac{9.53}{26.35 - 6.7} = 0.485$$

$$\beta_1 = 25.9°$$

**Example 7.2** Calculate the diameter and speed of the runner of a Kaplan turbine developing 6 MW under an effective head of 5 m. Overall efficiency of the turbine is 90 percent. The diameter of the boss in 0.4 times the external diameter of the runner. The turbine speed ratio is 2.0 and flow ratio 0.5, What is the specific speed of the turbine?

**Solution.**

Given: $P_t = 6$ MW, $H = 5$ m, $\eta_o = 0.9$, $D_b = 0.4$ D, $K_u = 2.0$ $K_f = 0.6$

$$K_u = \frac{u_1}{\sqrt{2gH}}$$

$$2.0 = \frac{u_1}{\sqrt{2\times 9.81\times 5}}$$

$$u_1 = 19.81 \text{ m/s}$$

$$K_f = \frac{v_{f1}}{\sqrt{2gH}}$$

$$0.6 = \frac{v_{f1}}{\sqrt{2\times 9.81\times 5}}$$

$$v_{f1} = 5.94 \text{ m/s}$$

$$\eta_o = \frac{P_t}{\gamma QH}$$

$$0.9 = \frac{6\times 10^6}{9.81\times 10^3 \times Q\times 5}$$

$$Q = 135.9 \text{ m/s}$$

Now $$Q = \frac{\pi}{4}\left(D^2 - D_b^2\right)\times v_{f1}$$

$$135.9 = \frac{\pi}{4}\left[1-(0.4)^2\right]D^2 \times 5.94$$

$$\text{D} = 5.89 \text{ m}$$

$$u_1 = \frac{\pi D_0 N}{60} = \frac{\pi\times 5.89\times N}{60}$$

$$N = 64.23 \text{ rpm}$$

Specific speed, $$N_s = \frac{N\sqrt{P_t}}{H^{5/4}}$$

$$= \frac{64.23\sqrt{6\times 10^3}}{5^{5/4}} = 665.4$$

**Example 7.3** A propeller reaction turbine of runner diameter 5 m is running at 50 rpm. The guide blade angle at inlet is 150° and the runner blade angle at outlet is 20° to the direction of vane. The axial flow area of water through the runner is 30 m², If the runner blade angle at inlet is radial, calculate: (a) hydraulic efficiency of the turbine, (b) discharge through the turbine, and (c) power developed by the runner.

**Solution.**

**Given:** $D = 5$ m, $N = 50$ rpm, $\alpha_1 = 150°$, $\beta_2 = 20°$, $A_f = 30$ m², $\beta_1 = 90°$

For $A_{f1} = A_{f2} = A_f, v_{f1} = v_{f2} = v_f$

$$u_1 = \frac{\pi DN}{60}$$

$$= \frac{\pi\times 5\times 50}{60} = 13.9 \text{ m/s}$$

Also $u_1 = u_2 = u = 13.09$ m/s

The velocity triangles are shown in Fig.7.5.

From inlet velocity triangle, we have

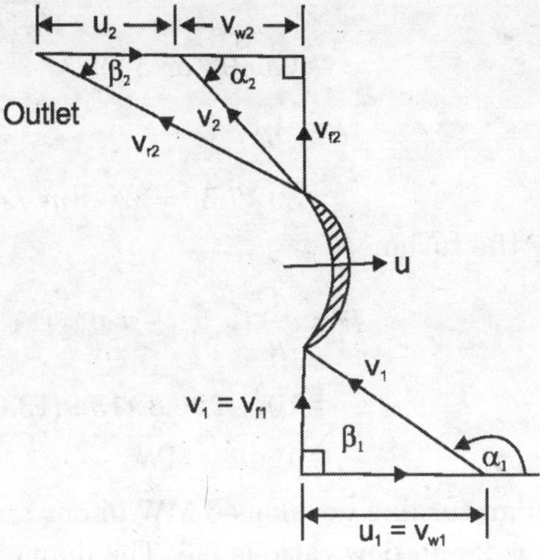


**Fig. 7.5**

$$\tan(180° - \alpha_1) = \frac{v_{f1}}{u_1}$$

$$\tan(180° - 150°) = \frac{v_{f1}}{13.09}$$

$$v_{f1} = 13.09 \tan\ 30°$$

$$= 7.56 \text{ m/s}$$

$$v_{w1} = u_1 = 13.09 \text{ m/s}$$

From outlet velocity triangle, we have

$$\tan \beta_2 = \frac{v_{f1}}{u_2 + v_{\omega 2}}$$

$$\tan\ 20° = \frac{7.56}{13.09 + v_{w2}}$$

$$0.36397 = \frac{7.56}{13.09 + v_{w2}}$$

$$v_{w2} = 7.68 \text{ m/s}$$

$$v_2 = \sqrt{v_{f2}^2 + v_{w2}^2} = \sqrt{(7.56)^2 + (7.68)^2}$$

$$= 10.78 \text{ m/s}$$

$$H = \frac{1}{g}(u_1 v_{w1} - u_2 v_{w2}) + \frac{v_2^2}{2g}$$

$$= \frac{1}{9.81}[13.09(13.09 - 7.68)] + \frac{(10.78)^2}{2 \times 9.81}$$

$$= 7.219 + 5.923 = 13.142 \text{ m}$$

$$h_{hyd} = \frac{u_1 v_{w1} - u_2 v_{w2}}{gH} = \frac{7.219}{13.142}$$

$$= 0.5493 \text{ or } 54.93\%$$

$$Q = A_f v_f$$

$$= 30 \times 7.56 = 226.8 m^3/s$$

Power developed by the turbine,

$$P_t = \frac{\gamma Q}{g}(u_1 v_{w1} - u_2 v_{w2})$$

$$= 10^3 \times 226.8 \times 13.9(13.09 - 7.68)$$

$$= 16.061 \text{ MW}$$

**Example 7.4** A kaplan turbine develops 6 MW under an effective head of 5 m. its speed ratio is 2 and flow ratio is 0.6. The diameter of boss is 0.35 times the external diameter of runner. Mechanical efficiency of the turbine is 90 %. Calculate:

(a) The diameter of runner, and

(b) The speed of runner and its specific speed.

**Solution.**

Given: $P_t = 6$ MW, $H = 5$m, $K_{u1} = 2$, $K_{f1} = 0.6$, $\eta_{mech} = \eta_t = 0.9$, $D_b = 0.35\ D$

(a)

$$P_t = \frac{\gamma Q H}{10^3} \times \eta_t$$

$$6000 = \frac{9.81 \times 10^3 \times Q \times 5 \times 0.9}{10^3}$$

$$Q = 135.9 \text{m}^3/\text{s}$$

$$v_{f1} = v_{f2} = K_{f1}\sqrt{2gH} = 0.6\sqrt{2 \times 9.81 \times 5}$$

$$= 5.94 \text{ m/s}$$

$$u_1 = u_2 = K_{u1}\sqrt{2gH} = 2\sqrt{2 \times 9.81 \times 5}$$

$$= 19.81 \text{ m/s}$$

$$Q = A_f v_{f1}$$

$$A_f = \frac{135.9}{5.94} = 22.88 \text{ m}^2$$

$$= \frac{\pi}{4}\left(D^2 - D_b^2\right)$$

$$= \frac{\pi}{4}D^2\left[1 - (0.35)^2\right]$$

$$D = 5.762 \text{ m or } 5762 \text{ mm}$$

(b)
$$u = \frac{\pi DN}{60}$$

$$N = \frac{19.81 \times 60}{\pi \times 5.762} = 65.66 \text{ rpm}$$

$$N_s = \frac{N\sqrt{P_t}}{H^{5/4}} = \frac{65.66\sqrt{6000}}{5^{5/4}} = 680.2$$

**Example 7.5** A propeller turbine runner has an outer diameter of 4.5 m and inner diameter of 2 m. It developed 21 MW when working under a head of 20 m. The turbine is directly coupled to an alternator having 22 pairs of poles. The hydraulic efficiency is 94 % and overall efficiency is 88 %. Find the discharge through the turbine. Runner vale angles at inlet and outlet, at the hub and at the edge of the blade.

**Solution.**

Given: $P_t$ = 21 MW, $D$ = 4.5 m, $D_b$ = 2 m, $H$ = 20 m, $p$ = 22, $h_{hyd}$ = 0.94, $\eta_t$ = 0.88

Assuming $f$ = 50 Hz

$$\text{Speed of turbine, } N = \frac{60f}{p} = \frac{60 \times 50}{22} = 126.4 \text{ rpm}$$

Generally in a propeller turbine, discharge is radial, i.e., $v_{w2} = 0$. (see Fig.7.6)

$$u_1 = u_2 = \frac{\pi DN}{60} = \frac{\pi \times 4.5 \times 136.4}{60}$$

$$= 32.14 \text{ m/s}$$

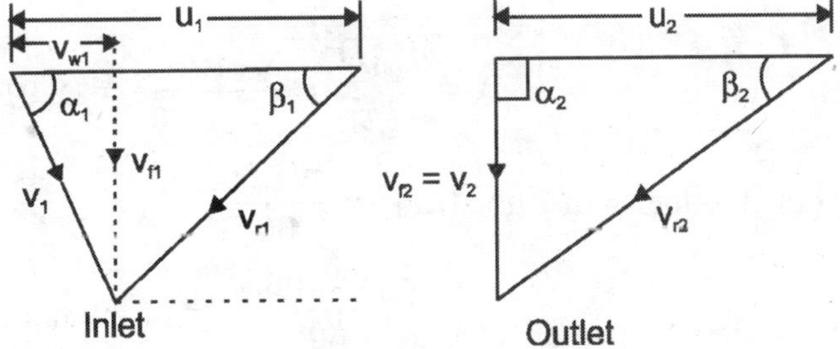


**Fig. 7.6 Velocity triangles.**

$$P_t = \frac{\gamma QH}{10^3} \times \eta$$

$$21{,}00 = \frac{9.81 \times 10^3 \times Q \times 20}{10^3} \times 0.88$$

$$Q = 121.63 \text{m}^3/\text{s}$$

$$\text{Hydraulic efficiency, } \eta_{hyd} = \frac{v_{w1}u_1}{gH}$$

$$0.94 = \frac{v_{w1} \times 32.14}{9.81 \times 20}$$

$$v_{w1} = 5.74 \text{ m/s}$$

Velocity of flow,
$$v_{f1} = v_{f2} = \frac{Q}{A_f}$$

$$= \frac{121.63}{\frac{\pi}{4}\left[(4.5)^2 - 2^2\right]} = 9.53 \text{ m/s}$$

$$\tan \alpha_1 = \frac{v_{f1}}{v_{w2}} = \frac{9.53}{5.74} = 1.66028$$

$$\alpha_1 = 58.94°$$

$$\tan \beta_1 = \frac{v_{f1}}{u_1 - v_{w1}} = \frac{9.53}{32.14 - 5.74} = 0.36098$$

$$\beta_1 = 13.85°$$

$$\tan \beta_2 = \frac{v_{f2}}{u_2} = \frac{9.53}{32.14} = 0.29651$$

$$\beta_2 = 16.51°$$

At the hub: $D_b = 2\text{m}$

For free vortex distribution along the inlet edge of the blade, we have

$$v_{w1}\frac{D}{2} = (v_{w1})_b \frac{D_b}{2}$$

$$\therefore \quad (v_{w1})_b = \frac{v_{w1}D}{D_b} = \frac{5.74 \times 4.5}{2} = 12.915 \text{ m/s}$$

$$\text{Blade peripheral velocity at hub, } (u_1)_b = \frac{\pi D_b N}{60}$$

$$= \frac{\pi \times 2 \times 136.4}{60} = 14.28 \text{ m/s}$$

For
$$v_{f1} = v_{f2} = (v_{f1})_b = 9.53 \text{ m/s}$$

Velocity triangles at the hub are shown in Fig. 7.7

$$\tan(\alpha_1)_b = \frac{(v_{f1})_b}{(v_{w1})_b} = \frac{9.53}{12.915} = 0.73790$$

$$(\alpha_1)_b = 36.42°$$

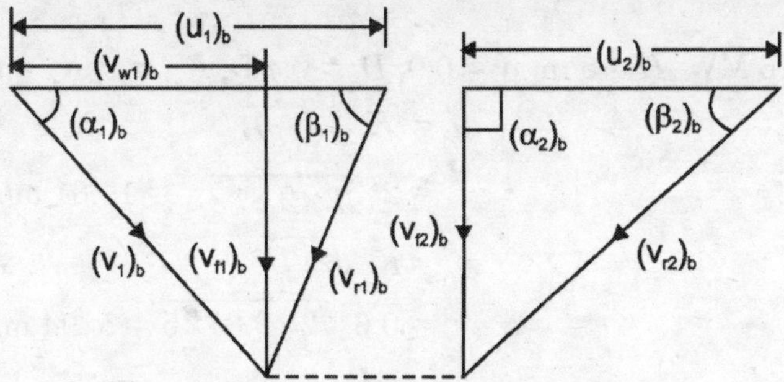


**Fig. 7.7 Velocity triangles at boss.**

**Vane angle at the hub of runner,**

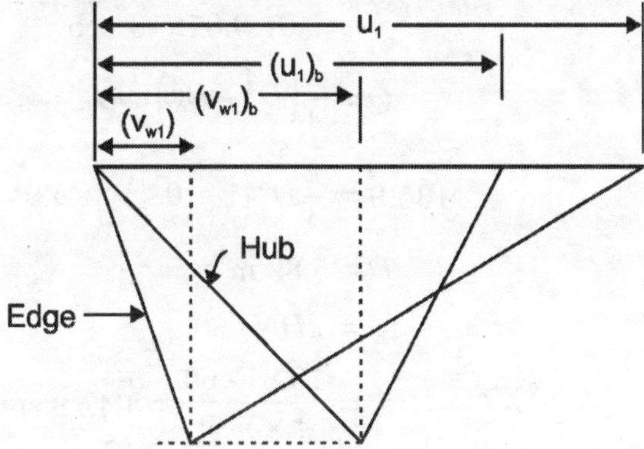


**Fig 7.8 Combined inlet velocity triangles at the edge of blade and that at the hub.**

$$\tan(\beta_1)_b = \frac{(v_{f1})_b}{(u_1)_b - (v_{w1})_b}$$

$$= \frac{9.53}{14.28 - 12.915} = 6.98168$$

$$(\beta_1)_b = 81.85°$$

Inlet velocity triangles at the edge of blade and hub are shown in Fig.7.8.

$$\tan(\beta_2)_b = \frac{(v_{f1})_b}{(u_2)_b} = \frac{9.53}{14.28} = 0.66737$$

$$(\beta_2)_b = 33.72°$$

**Example 7.6** Calculate the diameter and speed of the runner of a Kaplan turbine developing 6 MW under an effective head of 5 m. Overall efficiency of the turbine is 90 %. The diameter of boss is 0.4 times the external diameter of the runner. The turbine speed ratio is 2 and flow ratio 0.6 What is the specific speed of the turbine?

**Solution.**

Given: $P_t = 6$ MW, $H = 5$ m, $\eta_t = 0.9$, $D_b = 0.4$ D, $K_{u1} = 2$, $K_{f1} = 0.6$

$$u_1 = K_{u1}\sqrt{2gH}$$

$$= 2\sqrt{2\times9.81\times5} = 19.81 \text{ m/s}$$

$$v_{f1} = K_{f1}\sqrt{2gH}$$

$$= 0.6\sqrt{2\times9.81\times5} = 5.94 \text{ m/s}$$

$$\eta_t = \frac{P_t}{\gamma QH}$$

$$Q = \frac{6\times10^6}{0.9\times9.81\times10^3\times5} = 135.9 \text{ m}^3/\text{s}$$

$$Q = \frac{\pi}{4}\left(D^2 - D_b^2\right)v_{f1}$$

$$135.9 = \frac{\pi}{4}D^2\left(1-(0.4)^2\right)\times5.94$$

$$D = 5.89 \text{ m}$$

$$u_1 = \pi DN/60$$

$$N = \frac{19.81\times60}{\pi\times5.89} = 64.23 \text{ rpm}$$

$$N_s = \frac{N\sqrt{P_t}}{H^{5/4}} = \frac{64.23\sqrt{6000}}{5^{5/4}} = 665.5$$

## 7.7 Deriaz Turbine

The Deriaz turbine is a variable pitch mixed flow runner suitable for medium heads approximately up to 200 m. The runner is surrounded by a conventional guide and gate mechanism of reaction turbine.

Fig.7.9 shows the schematic diagram of a Deriaz turbine. The flow in the turbine is intermediate between the mixed flow and the axial flow of water through the runner. It is also known as diagonal turbine as the water passes through the runner at $45^0$ to axis.

The principal features of Deriaz runner are:

1. The axis obliquity allows easier location of operating gear without excessive increase in the hub-diameter ratio $\left(\frac{d}{D}\right)$. The width – diameter ratio $\left(\frac{B}{D}\right)$ is also more favourable to good efficiencies.

2. The oblique position of the blade trunnions gives a significant reduction in the loading of the outer trunnion journal. This happens because of the centrifugal force on the blade, acting partly against hydraulic loading. The runner blades are pivoted to the hub.
3. The spherical surfaces of the hub and of the skirt permit rotation of the vanes while maintaining close clearances.
4. Longer levers give securing advantages and less crowded construction in the boss.
5. Simpler runner blades than those of Kaplan runner. If desirable, the blades can be made to recover the whole cross – sectional area.
6. Guide vanes are provided ahead of the blades to regulate and direct the flow.
7. The runner has no outer rim connecting all the blades as these these blades are movable.

The runner of the Deriaz turbine is so shaped that it can be used both as a turbine as well as a pump. This turbine is very suitable for pumped storage hydropower plants.

### 7.7.1 Advantages of Deriaz Turbine

The main advantages of Deriaz turbine are:

1. It gives improved part load efficiency.
2. It can be conveniently used as a turbine – pump unit.
3. The starting torque under water can be reduced by adjusting the runner to shut position.
4. The flow area is completely closed at shut position.
5. The loading on the outer trunnion journal bearing is reduced due to oblique location of blades.
6. There is greater convenience to attach the arrangement for varying the blade angle.

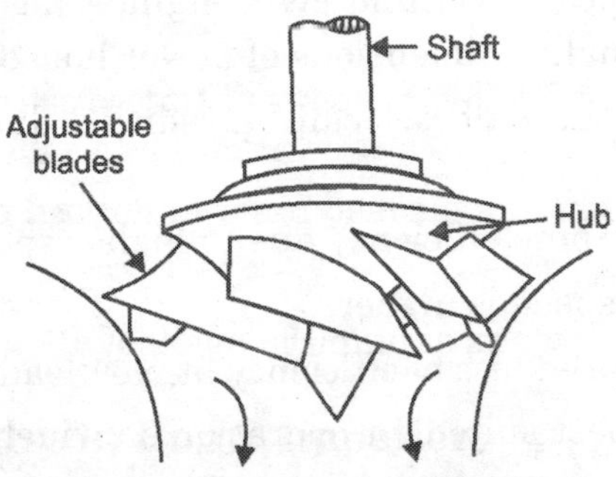


Fig. 7.9 Deriaz Turbine.

### 7.8 Tubular or Bulb Turbine

In a tubular or bulb turbine, the turbo – generator set has an outer casing having the shape of a bulb. The arrangement for such a turbine is shown in Fig.7.10. The bulb unit is a water tight assembly of turbine and generator with horizontal axis, submerged in a stream of water. The axis of rotation coincides with the axis of the passage of water. These turbines are highly suitable for tidal power stations.

The salient features of bulb turbines are:

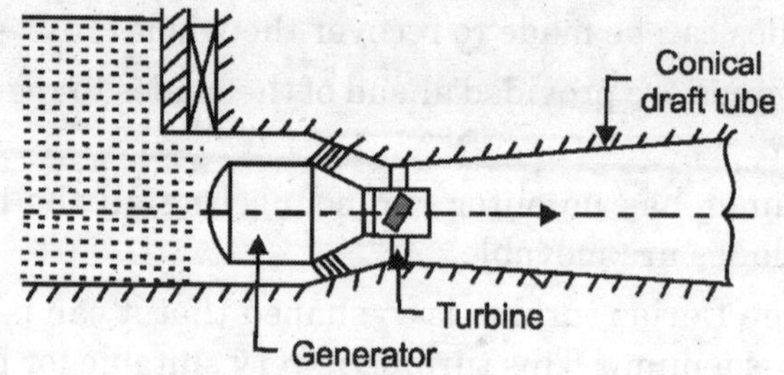


**Fig. 7.10 Bulb turbine.**

1. It is an axial flow turbine with either adjustable or non – adjustable runner vanes similar to Kaplan or propeller turbine.
2. The scroll casing is not provided in these turbines, instead the runner is placed in a tube extending from head water to the tail water.
3. It requires low heads of water varying from 3 m to 15 m.
4. The disposition of shaft may be vertical, inclined or horizontal.

#### 7.8.1 Advantages and disadvantages of bulb Sets

(a) *Advantages of bulb sets over Kaplan turbine*

1. It can be employed at very low head sites.
2. The width of the plant is less because of the absence of spiral casing.
3. Saving in excavation and civil engineering works owing to the relatively smaller dimensions of power house.
4. The size of the bulb set can be reduced due to higher speed of rotation.
5. Increased turbine efficiency due to higher speed of rotation.
6. The specific speed is higher.
7. There is reduced loss of efficiency at part loads.
8. The unit is best suited for operation on widely varying heads.

(b) *Disadvantages*

1. Due to complete submergence under pressure, water leakage into the generator chamber and condensation are source of trouble. High humidity inside the chamber results in gradual deterioration of electrical insulation
2. The erection techniques are costly.

## Summary

1. Propeller and Kaplan turbines are axial flow reaction turbines.
2. These turbines are used under low head and high discharge.
3. In the case of propeller turbine, the blades are fixed on the shaft hub whereas in Kaplan turbine the blades are adjustable.
4. The ratio of boss diameter to outer diameter of runner varies from 0.35 to 0.6.
5. Discharge, $Q = \frac{\pi}{4}\left(D^2 - D_b^2\right) v_{f1}$

   Where $v_{f1} = K_{f1}\sqrt{2gH}$

   $K_{f1}$ = flow ratio = 0.6 – 0.7
6. Peripheral speed of runner, $u_1 = \frac{\pi DN}{60}$

   Where $u_1 = K_{u1}\sqrt{2gH}$

   $K_{u1}$ = speed ratio $\simeq$ 2.0

## Multi-Choice Questions

1. The flow in Kaplan turbine is
   (a) tangential
   (b) radial
   (c) axial
   (d) mixed flow
2. The most efficient turbine at part load operation is
   (a) Kaplan (b) propeller
   (c) Francis (d) Pelton wheel
3. A Kaplan turbine is
   (a) radial flow reaction turbine
   (b) low head axial flow turbine
   (c) high head axial flow turbine
   (d) high head radial flow turbine.

4. Which of the following advantage does a Kaplan turbine has over the Francis turbine?
   (a) compactness (b) high part load efficiency
   (c) low frictional losses (d) All of the above.
5. The number of runner blades in Kaplan turbine are:
   (a) 2 or 3 (b) 3 to 6
   (c) 8 to 16 (d) 18 to 24
6. Specific speed of Kaplan turbine ranges from
   (a) 50 – 150 (b) 250 – 1050
   (c) 1200 – 1500 (d) 150 – 250

## Answers

1. (c) 2. (a) 3. (b) 4. (d) 5. (b) 6. (b)

## Review Questions

1. What is a reaction turbine?
2. What is reaction pressure?
3. What is the difference between a propeller turbine and a Kaplan turbine?
4. What is the maximum head under which a Kaplan turbine works?
5. What is the range of ratio of hub diameter to runner external diameter?
6. What is the type of flow in Kaplan turbine?
7. What is the range of runner of blades on the Kaplan turbine and propeller turbine?
8. What is the range of specific speed of a Kaplan turbine?

## Numerical Problems

1. The following data pertain to a Kaplan turbine: Power available at shaft = 8850 kW; net available head = 5.5 m;

   Speed ratio = 2.1; flow ratio = 0.67, overall efficiency = 85%.

   Assuming that hub diameter of the wheel is a 0.35 times the outside diameter, determine:

   (a) Runner diameter;

   (b) Runner speed.

   [**Ans.** (i) 6.34 m, (ii) 65.7 r.p.m.]
2. A Kaplan turbine produces 44000 kW under a head of 34.7 m, with an overall efficiency of 90 per cent. Taking the value of speed ratio

as 10.6, flow ratio as 0.5 and the hub diameter as 3.5 times the outside diameter, find the runner diameter and speed of the turbine.

[**Ans.** 5.16 m, 130.7 r.p.m]

3. The following data pertain to a Kaplan turbine: Shaft power = 13230 kW; speed = 75 r.p.m, head = 8 m; diameter of boss of runner = 0.35 times the external diameter; speed ratio = 2, flow ratio = 0.6, Find the efficiency of the turbine.

[**Ans.** 80.7 %]

4. A Kaplan turbine has the following specifications:

Net working head = 10 m

Discharge = 84 $m^3/s$

Output = 7.5 MW

Speed = 125 rpm

Diameter of runner = 4 m

Calculate:

(a) Overall efficiency,

(b) Specific speed, and

(c) Speed ratio.

5. The following data were obtained from a test on a Kaplan turbine:

Speed ratio = 2

Flow ratio = 0.6

Speed = 75 rpm

Head = 8 m

Brake power = 14 MW

Find the efficiency of the turbine.

6. A Kaplan turbine develops 1300 kW at an average head of 29 m. The speed and flow ratios are 2.1 and 0.62 respectively. The diameter of boss is 0.34 times the external diameter of the runner. Overall efficiency is 0.89. Calculate the diameter and speed of runner.

7. A Kaplan turbine is rated at 11,550 kW when working under an average head of 4.3 m. The diameter of the hub is 0.3 times the external diameter of runner. Overall efficiency of turbine is 0.91. find the speed and diameter of turbine runner. Take the values of speed ratio and flow ratio as 2 and 0.65 respectively.

# Draft Tube and Cavitation

## 8.1 Draft Tube

The draft tube is a conduit which connects the runner exit of the turbine to the tail race. It is welded steel plate pipe or a concrete tunnel, whose cross – section gradually increases towards the outlet. The water after doing work on the runner passes on to the tail race through the draft tube. The tube should be drowned below the tail race level by approximately one metre. The functions of the draft tube are as follows:

1. It allows the turbine to be set above tail race level, without loss of head, to facilitate inspection and maintenance.
2. It helps to convert the velocity head available at the runner outlet into pressure head. The Pelton turbine does not require a draft tube because all the available energy is converted into kinetic (or velocity) energy before it strike the buckets . In the case of reaction turbines, the use of a draft tube is essential, because a large amount of pine tic energy is available in water at runner exit. The draft tube converts this kinetic energy into pressure head.

### 8.1.1 Draft tube theory

Consider a reaction turbine fitted with a draft tube as shown in Fig. 8.1.

Let $z$ = distance of the bottom of draft tube from tail race

$P_a$ = atmospheric pressure at the surface of tail race.

$H_s$ = suction head of draft tube

Applying Burnoulli's equation at section 2-2 and 3-3, we have

$$\frac{p_2}{\gamma}+\frac{{v_2}^2}{2g}+z_2=\frac{p_3}{\gamma}+\frac{{v_3}^2}{2g}+0+h_f \quad ...(1)$$

Where $h_f$ = loss of head between section 2-2 and 3-3.

$$\frac{p_2}{\gamma}=\frac{p_3}{\gamma}-z_2-\left[\frac{v_2^2-{v_3}^2}{2g}-h_f\right] \quad ...(2)$$

Now $\frac{p_3}{\gamma}=\frac{p_a}{\gamma}+z$

$$\therefore \qquad \frac{p_2}{\gamma}=\frac{p_a}{\gamma}+(z-z_2)-\left(\frac{{v_2}^2-{v_3}^2}{2g}-h_f\right)$$

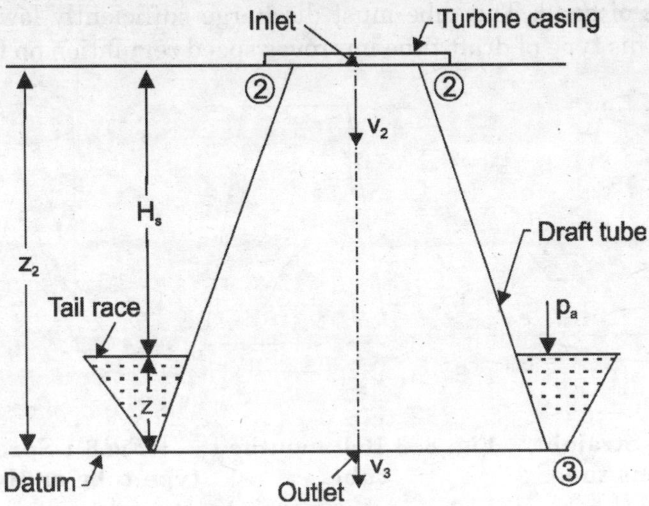


**Fig. 8.1 Draft tube theory.**

Now $z_2 - z = H_s$ = suction head

$$\frac{p_2}{\gamma} = \frac{p_a}{\gamma} - H_s - \left[\frac{v_2^2 - v_3^2}{2g} - h_f\right] \qquad \text{...(8.1)}$$

In Eq. (8.1), $\frac{p_2}{\gamma}$ is less than atmospheric pressure.

**Efficiency of a draft tube ($\eta_d$): It is defined as the ratio of net gain in pressure head to the velocity head at entrance of draft tube. Thus**

$$\eta_d = \frac{\left(\frac{v_2^2 - v_3^2}{2g} - h_f\right)}{\left(\frac{v_2^2}{2g}\right)} \qquad \text{...(8.2)}$$

Where $v_2, v_3$= velocity of water at inlet and outlet of draft tube respectively

$$h_f = \frac{v_2^2 - v_3^2}{2g} - \eta_d \times \frac{v_2^2}{2g} \qquad \text{...(8.3)}$$

### 8.1.2 Types of Draft Tubes

The various types of draft tubes employed are:

1. **Straight divergent tube**

The shape of this type of tube is that of a frustrum of a cone as shown in Fig. 8.2. It is employed for low specific speed, vertical shaft Francis turbine. The maximum cone angle is 8° (or $\alpha = 4°$). If this angle is more than 8°, the water detaches away from the inner wall of the tube forming vortices and causing loss of head. The tube must discharge sufficiently low under tail race level. This type of draft tube improves speed regulation on falling load.

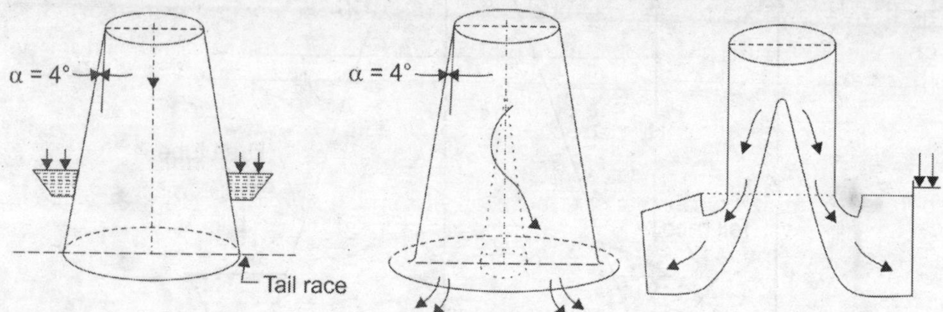


**Fig. 8.2 Straight divergent tube.** **Fig. 8. 3 Bell mouthed tube.** **Fig.8.4 Spreading type tube or Hydra cone.**

**2. Moody's Bell – mouthed tube**

This type of draft tube, as shown in Fig. 8.4, has a solid conical core in the entire central portion of the tube. This allows a large exit area without excessive length. It also reduces the whirl action of discharging water.

**3. Moody's spreading tube or Hydra cone**

This type of tube, as shown in Fig. 8.4, serves the same purpose as performed by bell-mouthed tube. It is suited particularly for helical flow which occurs when water leaves the runner with a whirl component.

**4. Simple Elbow tube**

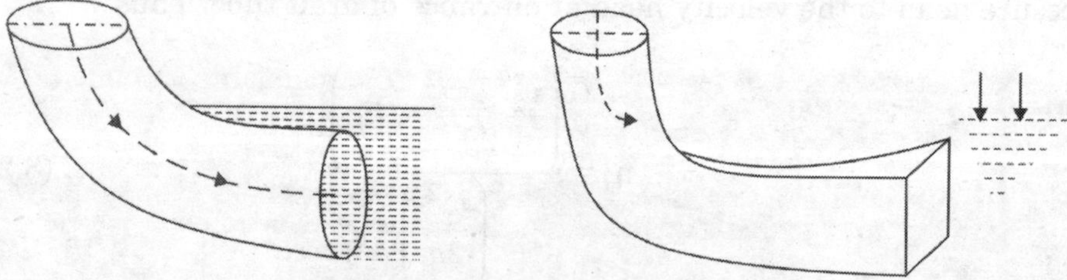

(a) Circular cross – section (b) circular cross-section at inlet and rectangular

**Fig. 8.5 Elbow type draft tubes.**

In order to reduce the cost of excavation, particularly in rocks , the vertical length of the tube should be minimum. To achieve this the tube is bent to keep its definite length the simple elbow type tube as shown in Fig 8.5 (a) serves such a purpose.

**5. Elbow tube with circular inlet and rectangular outlet cross - section**

This type of tube, as shown in Fig. 8.5 (b), has been designed to turn the water from the vertical to the horizontal direction with a minimum depth of excavation. At the same time it gives high efficiency. The transition from a circular section in the vertical leg to the rectangular section in the horizontal leg takes place in the bend. The horizontal portion of the tube is generally inclined upwards to lead the water gradually to the level of ant to prevent entry of air from the exit end. The exit end of the tail race tube must be totally immersed in water.

In order to avoid any whirl component of velocity of water at runner exit, one or two piers are constructed in the bend of the draft tube. Such piers behave similar to the central core of the spreading tube. Fig. 8.6 shows single and double piers elbow type draft tubes.

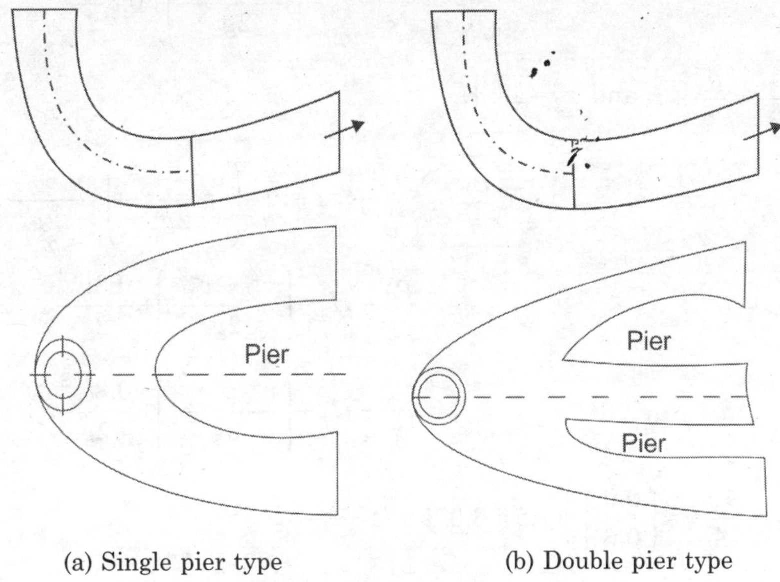


(a) Single pier type (b) Double pier type

**Fig. 8.6 Elbow type draft tube with piers.**

**Example 8.1** A Francis turbine is fitted with a vertical conical shaped draft tube. the top and bottom diameters are equal to 600 mm and 900 mm respectively. The tube is running full with water. It has a vertical height of 6m out of which 1.5 m is immersed in the tail race water. The friction head loss between the top and bottom points of tubc is 0.3 times the kinetic head at draft tube exit. The velocity at exit is 1.5 m/s. Determine:

(a) The pressure head at the top point of draft tube,

(b) The total head at the same point with reference to the tail race as a datum,

(c) The total head at the bottom point with reference to the tail race as a datum,

(d) The power in water at the top and bottom of tube , and

(e) The efficiency of draft tube.

**Solution.**

Given: $d_2 = 0.6m,\ d_3 = 0.9m,\ z_2 = 6m,\ h_f = 0.3\dfrac{v_3^{\ 2}}{2g}\ z = 1.5m, v_3 = 1.5m/s$

(a) Refer to Fig. 8.1. Applying Bernoulli's equation between points 2 and 3, we have

$$\frac{p_2}{\gamma} + \frac{v_2^{\ 2}}{2g} + z_2 = \frac{p_3}{\gamma} + \frac{v_3^{\ 2}}{2g} + 0 + h_f'$$

$$\frac{p_2}{\gamma} = \frac{p_3}{\gamma} - z_2 - \left(\frac{v_2^2 - v_3^{\ 2}}{2g}\right) + 0.3\frac{v_3^{\ 2}}{2g}$$

Now $\dfrac{p_3}{\gamma} = \dfrac{p_a}{\gamma} + z$ and $z_2 - z = \mathrm{H_s}$

$$\therefore \quad \frac{p_2}{\gamma} = \frac{p_a}{\gamma} - (z_2 - z) - \left(\frac{v_2^{\ 2} - v_3^{\ 2}}{2g}\right) + \frac{0.3v_3^2}{2g}$$

$$= \frac{p_a}{\gamma} - H_s - \left(\frac{v_2^2 - v_3^{\ 2}}{2g}\right) + \frac{0.3v_3^{\ 2}}{2g}$$

$$= \frac{p_a}{\gamma} - H_s - \left(\frac{v_2^2 - v_3^{\ 2}}{2g}\right) + \frac{0.3v_3^{\ 2}}{2g}$$

Now $v_2 = \dfrac{A_3 v_3}{A_2} = \left(\dfrac{0.9}{0.6}\right)^2 \times 1.5 = 3.375 m/s$

$$\therefore \quad \frac{p_2}{\gamma} = \frac{1.01325\times10^5}{9.81\times10^3} - (6-1.5)$$

$$- \left[\frac{(3.375)^2 - (1.5)^2}{2\times9.81}\right] + \frac{0.3\times(1.5)^2}{2\times9.81}$$

$$= 10.329 - 4.5 - 0.466 + 0.023$$

$$= 5.386m \text{ of water absolute}$$

or $\quad = -10.329 + 5.386 = -4.943m$ of water

(b) $$H_2 = \frac{p_2}{\gamma} + \frac{v_2^2}{2g} + (z_2 - z)$$

$$= -4.943 + \frac{(3.375)^2}{2\times9.81} + (6-1.5)$$

$$= 0.1376 \text{ m of water}$$

(c)
$$H_3 = \frac{p_3}{\gamma} + \frac{v^2{}_3}{2g} + h_f = H_2 - h_f$$

$$= 0.1376 - \frac{0.3(1.5)^2}{2 \times 9.81}$$

$$= 0.1032 \text{ m of water}$$

(d)
$$P_2 = \frac{\gamma Q H_2}{10^3}$$

$$= \frac{9.81 \times 10^3 \times \frac{\pi}{4}(0.6)^2 \times 3.375 \times 0.1376}{10^3}$$

$$= 1.288 \text{ kW}$$

$$P_3 = \frac{\gamma Q H_3}{10^3}$$

$$= \frac{9.81 \times 10^3 \times \frac{\pi}{4}(0.9)^2 \times 1.5 \times 0.1032}{10^3}$$

$$= 0.966 \text{ kW}$$

(e)
$$\eta_d = \frac{\left[\frac{v_2{}^2 - v_3^2}{2g} - \frac{0.3 v_3^2}{2g}\right]}{\left(\frac{v_2{}^2 - v_3{}^2}{2g}\right)}$$

$$= \frac{\left[\frac{(3.375)^2 - (1.5)^2}{2 \times 9.81} - \frac{0.3 \times (1.5)^2}{2 \times 9.81}\right]}{\left[\frac{(3.375)^2 - (1.5)^2}{2 \times 9.81}\right]}$$

$$= \frac{0.466 - 0.023}{0.466} = 0.9506$$

$$= 95.06 \ \%$$

**Example 8.2** A Kaplan turbine developing 3000 kW under a head of 6 m is provided with a draft tube. The inlet diameter of tube is 3m and it is set 2m above the tail race level. A vacuum gauge connected to the draft tube indicates a reading of 6m of water. The efficiency of draft tube is 80 percent. Determine the overall efficiency of Kaplan turbine.

**Solution.**

Given: $P = 3000$ kW, $H = 6$m, $h_s = 2m$, $\frac{p_2}{\gamma} = -\,6m$, $\eta_d = 0.8$

Applying Bernoulli's theorem between points 2 and 3, we have

$$\frac{p_2}{\gamma} = \frac{p_a}{\gamma} - H_s - \left(\frac{v^2{}_2 - v^2{}_3}{2g}\right), \text{neglecting } h_f$$

$$-6 = 0 - 2 - \frac{v^2{}_2 - v_3{}^2}{2g}$$

$$\frac{v^2{}_2 - v^2{}_3}{2g} = -2 + 6 = 4$$

$$\eta_d = \frac{\dfrac{v^2{}_2 - v_3{}^2}{2g}}{\dfrac{v^2{}_2}{2g}}$$

$$0.8 = \frac{4}{\dfrac{\left(v_2{}^2\right)}{2\times 9.81}}$$

$$v_2{}^2 = \frac{4\times 2\times 9.81}{0.8} = 98.1$$

$$v_2 = 9.9 \text{ m/s } v_2 = 9.9\ m/s$$

$$Q = \frac{\pi}{4}\times 3^2 \times 9.9 = 69.979\, m^3/s$$

$$\eta_{\text{Overall}} = \frac{10^3 P}{\gamma Q H} = \frac{10^3 \times 3000}{9.81\times 10^3 \times 69.979\times 6}$$

$$= 0.7283 \text{ or } 72.83\%$$

**Example 8.3** A conical draft tube having inlet and outlet diameters 2m and 3m discharges water at outlet with a velocity of 3m/s. The total length of draft tube is 7.5m and 1.5m of its length is immersed in water. If the atmospheric pressur we head is 10.3m of water and frictional head loss is equal to 0.25 × velocity head at outlet of tube, calculate: (a) pressur we head at inlet, and (b) efficiency of draft tube.

**Solution.**

Refer to Fig. 8.1

Given: $d_2 = 2\ m$, $d_3 = 3\ m$, $z_2 = 7.5\ m$, $z = 1.5\ m$, $h_f = 0.25\times\frac{v_3^2}{2g}$, $\frac{p_a}{\gamma} = 10.3m$ of $H_2O$, $v_3 = 3\ m/s$

(a)

$$Q = A_3 v_3 = \frac{\pi}{4} d^2{}_3 v_3 = \frac{\pi}{4}\times 3^2 \times 3$$

$$= 21.2 \text{ m}^3/\text{s}$$

$$v_2 = \frac{4Q}{\pi {d_2}^2} = \frac{4\times 21.2}{\pi \times 2^2} = 6.75\ m/s$$

$$h_f = 0.25\times \frac{3^2}{2\times 9.81} = 0.1147m \text{ of } H_2O$$

$$H_S = z_2 - z = 7.5 - 1.5 = 6m$$

Applying Bernoulli's equation, we have

$$\frac{p_2}{\gamma} = \frac{p_a}{\gamma} - H_s - \left(\frac{{v_2}^2 - {v^2}_3}{2g} - h_f\right)$$

$$= 10.3 - 6 - \left[\frac{(6.75)^2 - 3^2}{2\times 9.81} - 0.1147\right]$$

$$= 2.55m(abs)$$

(b)
$$\eta_d = \frac{\left[\frac{{v^2}_2 - {v_3}^2}{2g} - h_f\right]}{\frac{(v_2^2)}{2g}} = \frac{\frac{(6.75)^2 - 3^2}{2\times 9.81} - 0.1147}{\frac{(6.75)^2}{2\times 9.81}}$$

$$= \frac{1.7488}{2.322} = 0.7531 \text{ or } 75.31\%$$

**Example 8.4** A reaction turbine and its draft tube harve a vertical axis. The pressure head in the spiral casing at inlet is 50m above atmospheric pressure and velocity of water is 6m/s. The water flow through the tubes is 2m$^3$, and the hydraulic and overall efficiencies are 85 percent and 80 percent respectively. The top of the draft tube is 1.25m below the centre line of the spiral casing while the tail race is 4m below the top of draft tube. The diameter of draft tube at inlet is 1m and that at tail race level is 1.25m. Determine:

(a) Total head across the turbine

(b) Shaft power

(c) Head lost due to friction in turbine and tube

(d) Power lost in mechanical friction.

**Solution.**

Refer to Fig. 8.7.

Given: $d_2 = 1$m, $d_2 = 1.25$ m, $z_1 = 1.25$ m, $z_2 = 4$m, $Q = 2$ m$^3$/s, $\eta_{hyd} = 0.85$, $\eta_{overall} = 0.80$, $\frac{p_c}{\gamma} = 50$ m, $v_2 = 6$ m/s

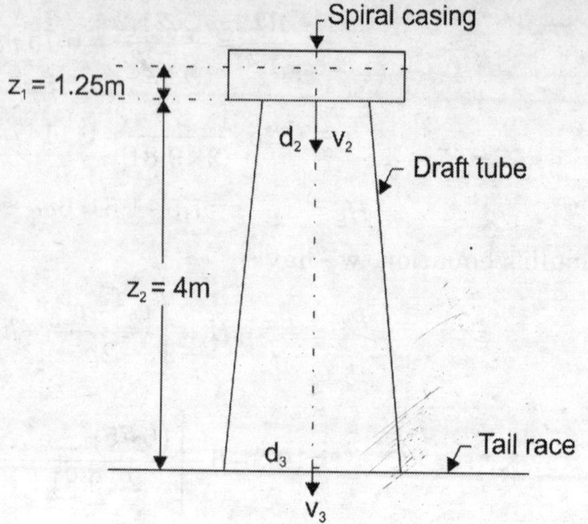


**Fig. 8.7**

(a) Total head across the turbine above tail race level,

$$H = \frac{p_c}{\gamma} + \frac{{v_2}^2}{2g} + z_1 + z_2$$

$$= 50 + \frac{6^2}{2\times 9.81} + 1.25 + 4$$

$$= 57.085\ m$$

(b) Shaft power, $P_s = \eta_{overall} \times \gamma Q H$

$$= \frac{0.80\times 9.81\times 10^3 \times 2\times 57.085}{10^3}$$

$$= 896\text{kW}$$

(c) $v_3 = \frac{4Q}{\pi {d_3}^2} = \frac{4\times 2}{\pi(1.25)^2} = 1.63\text{ m/s}$

Head utilised by turbine, $H_t = H - h_{ft} - h_{fd} - \frac{{v_3}^2}{2g}$

$$= 57.085 - \frac{(1.63)^2}{2\times 9.81} - (h_{ft} + h_{fd})$$

$$= 56.95 - (h_{ft} + h_{fd})$$

Where $h_{ft}, h_{fd}$ = frictional head lost in turbine and draft tube respectively.

Now $\eta_{hyd} = \frac{H_t}{H} = \frac{56.95 - (H_{ft} + H_{fd})}{57.085} = 0.85$

$$h_{ft} + h_{fd} = 8.428\ m$$

(d) Power developed by turbine runner, $P = P_s + P_f$

$$\eta_{mech} = \frac{P_s}{P}$$

But $\eta_{overall} = \eta_{mech} \times \eta_{hyd}$

$$\therefore \qquad 0.9412 = \frac{0.80}{0.85} = 0.9412$$

$$\therefore \qquad 0.9412 = \frac{896}{896 + P_f}$$

$$84303 + 0.9412\ P_f = 896$$

$$P_f = 55.97\ \text{kW}$$

## 8.2 Cavitation

According to Bernoulli's theorem, we have

$$\frac{p}{\gamma} + \frac{v^2}{2g} + z = \text{Constant}$$

or

Pressure head + velocity head + datum head = const.

If velocity head increases the pressure head will decrease. In case of liquids, the pressure cannot fall below vapour pressure. Whenever the pressure in any turbine part drops below the evaporation pressure, the liquid boils and a large number of small bubbles of vapour are formed. It may happen that a stream of water cuts short of its path giving rise to eddy and vortices which may contain voids or bubbles. These bubbles mainly formed on account of low pressure, are carried by the stream to higher pressure regions where the vapours condense and the bubbles suddenly collapse, as the vapours are condensed to liquid again. This results in the formation of a cavity and the surrounding liquid rushes to fill this cavity. The streams of liquid coming from all directions collide at centre of this cavity giving rise to a very high local pressure. Formation of cavity and high pressure are repeated many thousand times a second. This causes pitting on the metallic surface of runner blades or draft tube. The material then fails by fatigue and perhaps by corrosion. Some parts of runner blades of turbine may be torn away completely by this process.

The phenomenon which is responsible for pitting of the metallic surfaces of turbine parts is known as cavitation because of the formation of cavities. The cavities may be formed on the solid surface of blades of runner.

Now $\frac{p_2}{\gamma} = \frac{p_a}{\gamma} - \left[ H_s + \frac{v_2^2 - v_3^2}{2g} - \frac{kv_3^2}{2g} \right]$

Where $k$ = a fraction.

From the above equation, it is seen that the cavitation depends upon:

1. Vapous pressure which is a function of temperature of flowing water.
2. Absolute or barometric pressure due to the location of turbine above mean sea level.
3. Suction pressure $H_s$ which is the height of runner outlet above tail race level.
4. Effective dynamic suction head and absolute velocity of water at runner exit, i.e., $\eta_d \left( \frac{v_2^2 - v_3^2}{2g} \right)$ and $v_2$ respectively.

### 8.2.1 Tahoma's Cavitations Factor σ (Sigma)

**It is a factor suggested by Prof. Dietrich Thoma to determine the zone where turbine can work without being affected from cavitation. Its critical value is given by,**

$$\sigma_{crit} = \frac{H_b - H_s}{H} = \frac{(H_a - H_v) - H_s}{H}$$

Where $H_b$ = barometric pressure in m of water.

$= H_a - H_v$

$H_a$ = average atmospheric pressure in m of water,

$H_v$ = vapour pressure in m pf water.

$H_s$ = section pressure head in m of water

= height of runner outlet above tail race level.

$H$ = working head of turbine in m of water.

The cavitation factor $\sigma_{crit}$ depends upon the specific speed of the turbine. According to Prof. Thoma, cavitation can be avoided if the values of σ are not less than the value of $\sigma_{crit}$.

Thus the deciding factors in the selection of reaction turbine are specific speed $N_S$ and the cavitation factor σ. Prof F.H Roger has suggested the following empirical relation for Francis turbine:

$$\sigma_{crit} = 0.625 \left( \frac{N_s}{380.78} \right)^2 = \frac{N_s^2}{231989} \qquad \text{...(8.5)}$$

The maximum permissible specific speed then becomes,

$$N_s = 380.78\sqrt{\frac{\sigma_{crit}}{0.625}} = 481.65\sqrt{\sigma_{crit}} \quad ...(8.6)$$

$$= 481.65\sqrt{\frac{H_b - H_s}{H}} \quad ...(8.7)$$

The effect of cavitation σ on the turbine efficiency η is shown in Fig.8.8

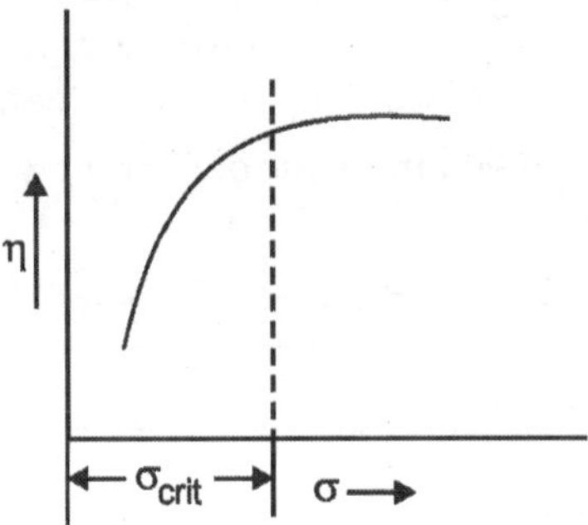


**Fig. 8.8 Turbine efficiency η v's cavitation factor σ.**

For propeller turbine: $\sigma_{crit} = 0.28 + \left[\frac{1}{7.5}\left(\frac{N_s}{380.78}\right)^3\right]$ ...(8.8)

For Kaplan turbine: $\sigma_{crit} = 1.1 \times \sigma_{crit}$ for propeller turbine.

Specific speed, $N_s = N\sqrt{\frac{P_t}{H^{5/4}}}$

Where $N$ = rpm, $P_t$ in kW and H in m.

### 8.2.2 Methods to Avoid Cavitation

The following methods may be adopted to avoid cavitation:

1. Installation of turbine below tail race level: if the value of $H_s$ is less, the pressure $p_2$ and σ will increase, so that turbine may work in safe zone. In order to keep $p_2$ within cavitation limits, the value of $H_s$ is made negative. This shall require the turbine to be installed below the tail race level such that the turbine remains always under water. This shall require pumping out of the whole water during maintenance and inspection. Therefore, this method is not advisable.
2. Use runner of low specific speed.
3. Use cavitation free runner of reaction turbines.
4. Cavitation can be reduced by selecting materials which offer better resistance to cavitation effect. Cast steel is batter than cast iron and stainless or alloy steel is still better than cast steel. There is a preference for chrome nickel stainless steels.

5. The highly polished turbine runner blades are less prone to cavitation.
6. The cavitation may be avoided by selecting a runner of proper specific speed for a given head.

## 8.3 Runaway Speeds

The runaway speed of a turbine is the maximum speed attained by the runner under maximum head at full gate opening when the generator is disconnected from the system and the governor ceases to function.

The range of runaway speed in terms of normal working speed for turbines is given in Table 8.1.

**Table 8.1 Range of runaway speed for turbine**

| Type of turbine | Runaway speed, $N_r$/working speed, $N$ |
|---|---|
| 1. Pelton turbine | 1.8 – 1.9 |
| 2. Francis turbine | 2 – 2.2 |
| 3. Kaplan turbine | 2.5 – 3 |

**Example 8.5** A Francis turbine working under a head of 30m produces 12 MW while running at 120 rpm. The atmospheric pressure is 10m of water and vapour pressure is 0.2m of water.

Determine the maximum height of straight draft tube for the turbine.

**Solution.**

Given: $H = 30$m, $P_t = 12$ MW, $N = 120$ rpm, $H_a = 10$ m of $H_2O$, $H_v = 0.2$ m of $H_2O$.

Specific speed,
$$N_s = \frac{N\sqrt{P_t}}{H^{S/4}} = \frac{120\sqrt{12000}}{30^{S/4}} = 187.2 \text{ rpm}$$

Critical value of thoma's cavitation factor for Francis turbine,

$$\sigma_{crit} = 0.625\left(\frac{N_s}{380.78}\right)^2 = 0.625\left[\frac{187.2}{380.78}\right]^2$$

$$= 0.151$$

$$\sigma_{crit} = \frac{H_a - H_v - H_s}{H}$$

$$0.151 \times 30 = 10 - 0.2\ H_s$$

$$H_s = 5.27\text{m}$$

Height of draft tube = 5.27m

## Summary

1. A draft tube is a conduit which connects the runner exit of the turbine to the tail race.
2. Governing equation for draft tube:

$$\frac{p_2}{\gamma} = \frac{p_a}{\gamma} - H_s - \left(\frac{v^2_2 - v_3^2}{2g} - h_f\right)$$

3. Efficiency of draft tube,

$\eta_d$ = Net gain in pressure head

$$\text{Velocity head at entrance} = \frac{\left(v^2_2 - v^2_3 - h_f\right)}{\left(\frac{v_2^2}{2g}\right)}$$

4. The various types of draft tubes are:

Straight divergent tube, bell mouthed tube, hydracone or spreading tube, simple elbow tube, and elbow tube with circular inlet and rectangular outlet.

5. The formation, growth and collapse of vapour filled cavities or bubbles in a flowing liquid due to local fall in pressure is called cavitation. Cavitation is responsible for pitting of the metallic surface of turbine runner.
6. Thoma's critical cavitation factor,

$$\sigma_{crit} = \frac{H_a - H_v - H_s}{H}$$

7. $\sigma_{crit} = 0.625\left(\frac{N_s}{380.78}\right)^2$ for Francis turbine

$= 0.28 + \left[\frac{1}{7.5}\left(\frac{N_s}{380.78}\right)^3\right]$ for propeller turbine

$= 1.1(\sigma_{crit})_{prop}$ for Kaplan turbine

## Multi-Choice Questions

1. The function of a draft tube is to
   (a) provide safety to turbine
   (b) prevent air from entering
   (c) convert kinetic energy into pressure energy
   (d) increase the rate of flow

2. Cavitation damage in turbine runner occurs near the
   (a) inlet on the convex side of blades
   (b) outlet on the convex side of blades
   (c) inlet on concave side of blades
   (d) outlet on concave side of blades.

3. Which of the following problems arise from cavitation ?
   (a) noise and vibration
   (b) damage to blade surface
   (b) fall in efficiency
   (d) all of the above

4. Which of the following draft tubes is suitable for helical flow?
   (a) comical type
   (b) elbow type
   (c) moody's spreading type
   (d) none of the above

5. Critical value of cavitation factor ($\sigma_{crit}$) is given by

   (a) $\dfrac{H_a + H_v - H_s}{H}$

   (b) $\dfrac{H_a - H_v + H_s}{H}$

   (c) $\dfrac{H_a - H_v - H_s}{H}$

   (d) $\dfrac{H_a + H_v + H_s}{H}$

**Answers**

1. (c) 2. (b) 3. (d) 4. (c) 5. (c)

## Review Questions

1. What are the functions of a draft tube?
2. What is Hydra cone? What are its advantages?
3. How do the losses in the draft tube effect the pressure at runner exit?
4. What is cavitation? How and where does it occur in reaction turbines?
5. On what factors does the cavitation in water turbines depend?
6. What is Thoma's factor of cavitation?
7. Describe some methods to avoid cavitation in water turbines.
8. Define runaway speed of a turbine.

## Numerical Problems

1. A turbine develops 400 kW at an overall efficiency of 82% when working under a static head of 5m. A tapered draft tube having inlet and outlet diameters of 2m and 3m respectively and efficiency of 90% is fitted at the exit of turbine. What increase in power and efficiency of turbine would you expect?

2. A reaction water turbine is fitted with a straight flaring draft tube having top and bottom diameters of 0.5m and 0.75m respectively. The water velocity at the top is 3m/s, where the elevation is 5m above the level of tail water. Assuming a loss in the draft tube equal to half the velocity head at exit, determine:
   (a) The pressure head at the top
   (b) The total head at the top with reference to the tail race as a datum
   (c) the total head at the exit, and
   (d) the power in water at top and at exit.

3. A turbine runner has an exit velocity of 10m/s. the loss of head due to friction and other causes in the draft tube should not exceed 1.5m. What maximum height of setting for turbine would you recommend for the turbine to avoid cavitation?

   Assume (a) the velocity of water at the outlet of draft tube as 2.5m/s, and (b) The cavitation commences when the pressure is 2.5 m water absolute.

4. A Kaplan turbine develops 1471 kW under a head of 6m. The turbine is set 2.5 m above the tail race level. A vacuum gaunge inserted at

the turbine outlet records a suction head of 3.1m. If the hydraulic efficiency is 85 per cent, what would be the efficiency of draft tube having inlet diameter if 3m? What will be the reading of suction gauge if power developed is reduced to half (735.5 kW ), the head and speed remaining constant.

[**Ans.** 68%; 2.6496 m]

5. Calculate the efficiency of a Kaplan turbine developing 2900kW under a net head of 5m. It is provided with a draft tube with its inlet (diameter 3 m) set 1.6 m above the tail race level. A vacuum gauge connected to the draft tube indicates a reading of 5 m of water. Assume draft tube efficiency as 78 per cent.

[**Ans.** 90.4%]

CHAPTER 9

# Governing, Selection and Performance of Hydraulic Turbines

## 9.1 Introduction

Governing means regulation of speed. The main function of governing is to maintain a constant speed when load on the turbine fluctuates. It is essential to keep the speed of turbine constant, because the turbine is directly coupled to a generator, which must run at a constant speed to generate power at a constant frequency. This is accomplished by controlling the rate of flow of water leading to the turbine runner, in proportion to the load. The device used for governing is called the governor.

The governing of Pelton turbines is accomplished by closing or opening the needle valve. In case of prance turbines, the governing function is achieved by closing or opening the wicket gates. In case of Francis turbine, the runner blades are moved in addition to closing and opening of wicket gates.

The turbine governor must satisfy the following conditions:

1. It should be quick acting, *i.e*., it should restore the speed to normal as soon as possible after it has been disturbed due to variation of load.
2. It should not close the pipe line (penstock) so quickly as to cause a serious hammer flow.

## 9.2 Functions of a Water Turbine Governor

The functions of a water turbine governor are:

1. To control the speed of turbine – alternator set and match it to the hydro - electric system for synchronizing.
2. To control the speed and frequency of turbine unit to maintain a desired frequency and voltage.
3. To set the amount of load a turbine unit has to carry.
4. To help in starting and shutting down the turbo – unit by opening and closing the nozzle of Pelton wheel and wicket gates of Francis and Kaplan turbines.

## 9.3 Types of Water Turbine Governors

The types of governors used for water turbines are:

1. Mechanical governor – such as ball and spring type.
2. Oil pressure governors.

### 9.3.1 Governing of Impulse Turbines

The quantity of water rejected from the turbine nozzle and from striking the buckets may be regulated by any one of the following ways:

1. Spear regulation,
2. Deflector regulation, and
3. Combined spear and deflector control.

**1. Spear Regulation**– The rate of flow of water according to the load is regulated by the to and fro movement of a spear inside the nozzle, which alters the cross – sectional area of stream. Spear regulation is satisfactory when a relatively large penstock feeds a small turbine, where fluctuations of load are small. With the sudden fall of load, the turbine nozzle has to be closed suddenly, which may create water hammer in the penstocks.

**2. Deflector regulation**– The deflector is generally a curved plate connected to the oil pressure governor by means of levers. When it is required to deflect the jet, the plate can be brought in between the nozzle and buckets, thereby diverting the water from the runner and directing into the tail race.

The deflector control is adopted when supply of water is constant but load fluctuates. The spear position can be adjusted by hand. As the nozzle has always a constant opening, it involves considerable wastage of water. It is used only when supply of water is abundant.

**3. Combined Spear Deflector Regulation**– The combined spear – deflector regulation controls the speed by spear and the pressure by deflector arrangement.

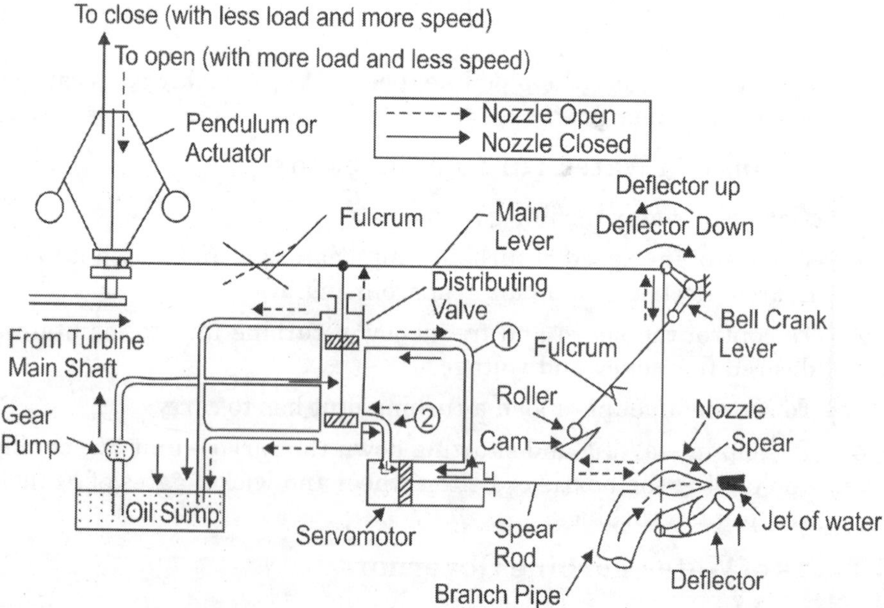


**Fig. 9.1 Combined spear – deflector regulation (in neutral position).**

The jet deflector is controlled directly by the governor, which deflects the water jet from runner within a very short period until the spear has been adjusted to a new position of equilibrium. The close of the spear can thus be retarded to avoid undue pressure rise.

A simple line diagram of combined spear – deflector regulation is shown in Fig. 9.1 in the neutral position. As the load on the turbine is decreased, the pendulum balls occupy the top position, thus moving the bell – crank lever towards bottom with which the deflector is brought in front of the nozzle, diverting whole or part of the water jet direct to the tail race. With movement of bell – crank lever, the roller on the cam is also raised. The distributing valve also functions simultaneously thus moving the servomotor piston rod.

The working of the system is as follows:

1. When the load on the turbine increases the speed of the runner decreases and consequently the balls of the centrifugal governor moves inwards and sleeve moves downwards.
2. The downward movement of sleeve is transmitted to distributing valve through linkages which admits oil under pressure to a servomotor. The oil exerts a force on the piston of servomotor, and that pushes the spears to a position which increases the annular area of nozzle flow passage. This results in quantity of water striking the buckets increased for the turbine to regain the normal speed.
3. When the load on the turbine decreases the direction of movement of servomotor is such that the nozzle area decreases and that allows a smaller quantity of water to strike the buckets of the turbine.
4. The deflector arrangement safe guards against excessive water hammer pressure.

### 9.3.2 Governing of Reaction Turbines

A simple line diagram for the governing of a reaction turbine is shown in Fig.3.2. The guide blades of a reaction turbine are pivoted and connected by leversand links to the regulation ring. To the regulating ring are connected two long regulating rods attached to a regulating lever.

The regulating lever is keyed to a regulating shaft which is turned by the servomotor piston of oil pressure governor.

The penstock, feeding the turbine inlet, has a relief valve (or pressure regulator). When the guide vanes have to be suddenly closed, the relief valve opens and diverts water direct to the tail race. Thus the double regulation is accomplished by moving the guide vanes and relief valve in Francis turbine, by the governor.

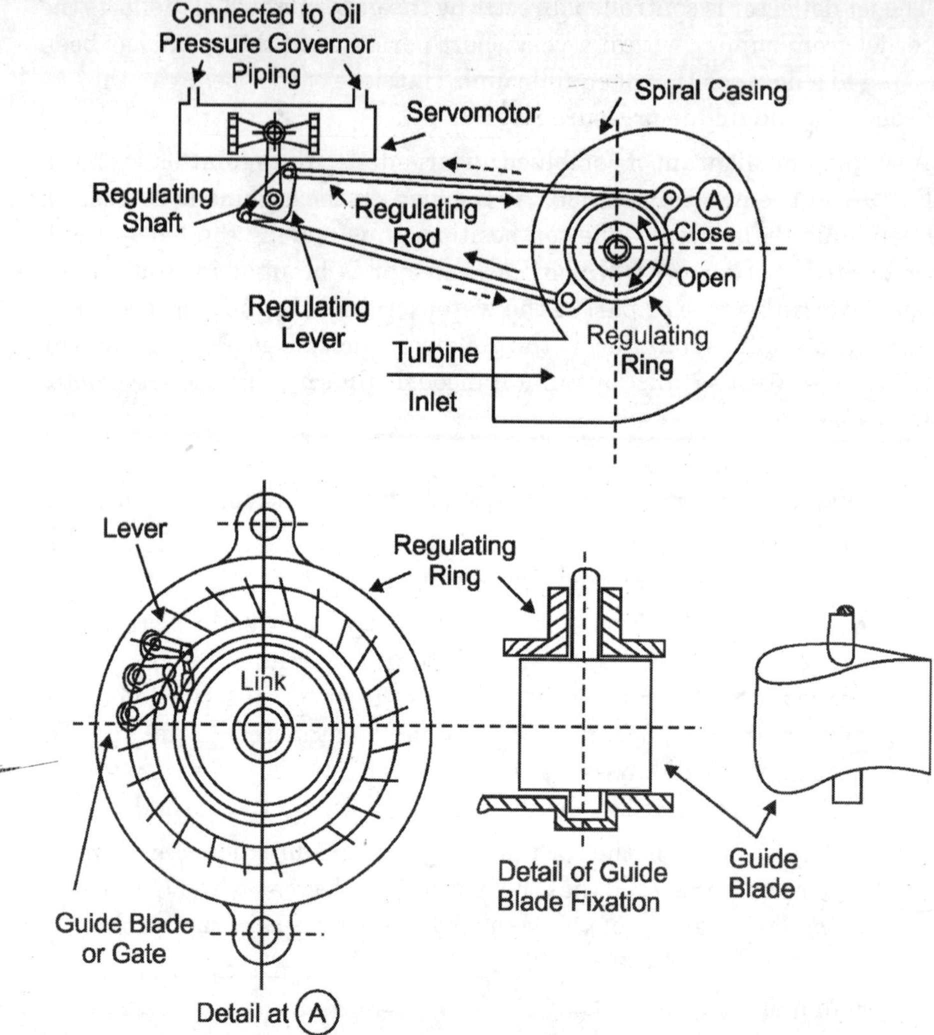


**Fig. 9.2 Simple line diagram for governing of reaction turbines.**

## 9.4 Safety Devices for Penstock

As the load on the generating units varies, openings of turbine gates are to be changed, depending on the load, adjusting the flow of water entering the runner. Whenever the turbine gates are suddenly closed the water flowing in the long penstock towards the turbine has to be retarded. This leads to water hammer which causes excessive pressure in the pipe line with which the pipe may burst. It is therefore necessary to provide some safety device for the penstock. To relieve the penstock of its excessive pressure the following devices are employed:

### 9.4.1 Surge tank

It is used in case of high and medium head power plants where the penstock is very long. A surge tank is a storage reservoir fitted at some opening made on a long penstock to relive the rejected flow when the penstock is suddenly closed by a valve fitted at its steep end. Surge tank, therefore, relieves the pipe line of excessive pressure produced due to closing of penstock, thus eliminating positive water hammer effect. Surge tank also serves the purpose of supplying initial water for an increasing load on the turbine while the water in the pipe line is being accelerated.

#### *Functions of Surge Tank*

(i) Control of pressure variations resulting from rapid changes of flow in penstock, relieving the pipe line of excessive pressure, thus eliminating water hammer effects.

(ii) Regulation of flow in power and pumping plants by providing necessary accelerating or retarding head.

(iii) Regulation of turbine speed with the help of(ii) above.

#### *Location of surge tank*

Theoretically a surge tank should be located as close to a power house as possible. The ideal place is at the turbine inlet. But it is seldom possible in case of medium and high head plants because it will have to be made very high. In order to reduce its height, it is generally located at the junction of pressure tunnel and penstock or on the side of the mountain.

#### *Types of Surge Tanks*

There are mainly three types of surge tanks: Simple, restricted orifice, and differential

(i) *Simple surge tank*– It is a reservoir directly connected to the penstock, as shown in fig.9.3. In this type of surge tank, the accelerating and retarding heads accumulate slowly, therefore its action is sluggish. It is liable to set up considerable oscillations unless it is of very large size. It is not used in practice.

(ii) *Restricted orifice type surge tank*– It consists of a restricted orifice (Fig 9.4)installed between the penstock and the simple tank. With the help of orifice the water storage function of surge tank can be separated from accelerating and retarding function. The retarding and accelerating heads are developed by the orifice more quickly than with a simple surge tank. However, this results in fluctuation on the turbine head, making the governor mechanism more cumbersome and costly

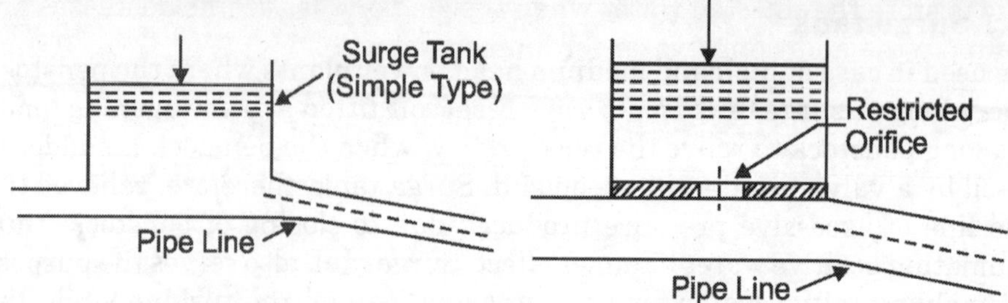


**Fig. 9.3 Simple surge tank.** **Fig. 9.4 Restricted orifice type surge tank.**

(iii) *Differential surge tank*– It is a combination of simple and restricted orifice type, with an addition of an internal riser , as shown in Fig.9.5. The diameter of the riser is smaller than the connection of the penstock. At the base of the riser there is an annular port, which communicates with the tank. The water enters the tank through the annular port. When there is a sudden drop in load on the turbine, the water from the penstock rises immediately in the riser and spills over from its top into the tank. This establishes an immediate retarding head on the penstock, as well as a differential head on the port. The differential head is responsible for forcing the water by the turbine, through the port into the tank. When there is a sudden demand of water by the turbine, the water level in the riser falls rapidly, there by developing an immediate considerable acceleration head in the penstock. The additional water required is supplied by the tank slowly through the ports.

$oaa_1$ : steady turbine load.

$obb_1$: falling turbine load

$occ_1$ :increasing turbine load.

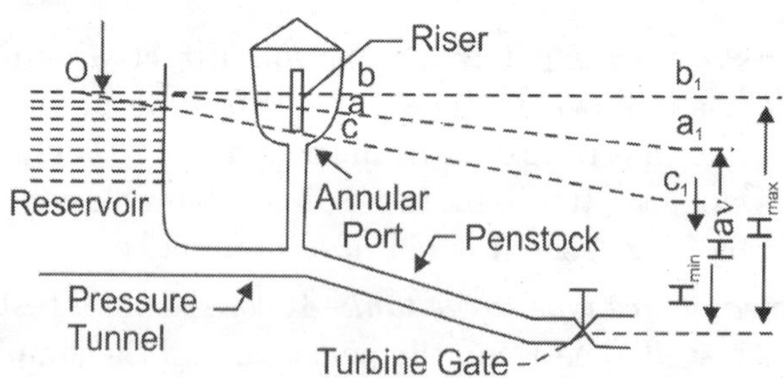


**Fig. 9.5 Differential surge tank.**

### 9.4.2 Forebay

A fore bay is a storage reservoir at the beginning of a penstock. When a hydro – power plant is located at the end of a canal, a forebay is made by enlarging the canal at the intake to the penstock. The function of the fore by is to store the rejected water when the load on the turbine decreases and to supply the additional water for an increasing load.

## 9.5 Selection of Hydraulic Turbines

The following points should be considered for the selection of right type of turbine:

1. Specific speed
2. Rotational speed
3. Efficiency
4. Part load operation
5. Overall cost
6. Cavitation
7. Disposition of turbine shaft
8. Number of units, and
9. Head.

**1. Specific Speed**– The specific speed of a turbine is given by:

$$N_s = \frac{N\sqrt{P_t}}{H^{5/4}}$$

It is a common practice to select a high specific speed runner which is always economical because the size of the turbo – generator as well as that of power house will become smaller. High specific speed is essential where head is low and output is large. Otherwise the rotational speed will be very low which will increase the cost of turbo – generator and powerhouse. However for high head installations low values of specific speed can be chosen to attain high rotational speed.

**2. Rotational Speed**– The specific speed is directly proportional to the rotational speed of the turbine. Also the rotational speed of an electric generator with which the turbine is to be directly coupled, depends on the frequency and number of pair of poles. Therefore, the value of specific speed adopted should be such that it will give the synchronous speed of the generator.

**3. Efficiency**– The selected turbine should give the highest overall efficiency for various operating conditions. The turbine is normally designed to give maximum efficiency at the annual mean net head available.

**4. Part – load operation**– The efficiency at part – loads and over – loads is less than normal. The turbine should always run with maximum possible efficiency to get more revenue. Deriaz turbine should be preferred over Francis turbine when it is required to run the turbine at part or overload conditions.

**5. Overall Cost**– the overall cost of turbine comprises of initial cost and running cost. The initial cost includes manufacture and installation, where as the running cost includes the maintenance and overhead charges. consisting of operation, repair and replacement. These costs must be considered properly before selecting a turbine.

**6. Cavitation**– While selecting reaction turbine, the critical value of cavitation factor must be calculated to see that turbine works in the safe zone.

**7. Disposition of Turbine Shaft**– A vertical shaft turbine will require deeper foundations and high building. On the other hand, horizontal shaft turbine will need greater floor area. The vertical shaft arrangement is better for large – sized reaction turbines and horizontal shaft arrangement for impulse turbines.

The summary of criterias for selection of turbines is given in Table 9.1

**Table 9.1 Selection criterias for turbines.**

| Sl. No. | Type of turbine | Head, H m | Specific speed $N_s$ | Speed ratio $K_u$ | Hydraulic Efficiency % (max) | Remarks |
|---|---|---|---|---|---|---|
| 1. | Pelton:<br>1 jet<br>2 jets<br>3 jets | Upto 2000m<br>Upto 1500m<br>Upto 350m | 8.5 – 30<br>15 – 50<br>25 – 70 | 0.43-0.48 | 89 | Employed for very high heads. |
| 2. | Francis:<br>High head<br>Medium head<br>Low head | 150-350m<br>60-150m<br>Upto 60m | 80 – 150<br>150 – 250<br>250 – 340 | 0.6-0.9 | 93 | High full load efficiency. Part-load efficiency lower than Pelton turbine. |
| 3. | Propeller<br>Kaplan | 15m<br>4 – 30m | 250 – 1025 | 1.4 – 2 | 93 | High part load efficiency. High discharge with low head |
| 4. | Bulb or Tubular turbines | 1 – 15m | 1000 - 1200 | 6 -8 | 91 | Employed for tidal power plants. |

**8. Number of Units:** The number of units depend on the connected load and the generation capacity of a turbine set. When more than one unit is installed they should be of the same capacity so that their parallel running is more stable and to achieve economy in the use of spare parts. The multi – unit plant is useful for peak - load conditions.

**9. Head**

Very high head – above 350 m:Pelton turbine

High head– 150m to 350 m:Pelton or Francis turbine

Medium head – 60 m to 150 m:Francis turbine

Low heads – under 60 m:Francisand Kaplan turbines

Very low heads – 1 m to 15 m: Bulb turbine and Kaplan turbine

## 9.6 Similarity Relations

Water turbines are manufactured for specific conditions and are usually large – sized units. It is not possible to manufacture them on mass scale. The limitations of testing the turbine after manufacturing the prototype are the following:

1. The extensive tests on prototype machines are very costly and time consuming.
2. The test results on prototype machines may not tally with the design data. To change the dimensions and size of the turbine at this stage will be very costly.
3. The limitations of tests performed on the prototype are:

(i) It is not possible to vary the head.

(ii) The speed of the unit cannot be varied.

(iii) The load available on the unit may not be as steady as is desired for accurate observations.

Due to these limitations, it is essential to make a turbine model, which is geometrically similar for carrying out the tests in order to predict the behavior and working conditions of the full sized turbine.

To ensure similarity in the model and prototype flows, the following similarity laws are used.

**(a) Geometric Similarity**– It implies that the model must be similar to the prototype turbine, *i.e.*, all linear dimensions of the two machines compared are proportional. Thus,

$$\left(\frac{D_2}{D_1}\right)_m = \left(\frac{D_2}{D_1}\right)_p$$

**(b) Kinematic Similarity**– It requires the simlitudes of flow fields in the passages of model and prototype turbine. It mean that velocity at the corresponding points in flows compared have to be proportionate and have the same directions.

Thus $$(K_u)_m = (K_u)_p \quad ...(9.2a)$$

$\therefore$ $$(K_v)_m = (K_v)_p \quad ...(9.2b)$$

$$(K_{vf})_m = (K_{vf})_p \quad ...(9.2c)$$

$$(K_w)_m = (K_w)_p \quad ...(9.2d)$$

**(c) Dynamic similarity–** It demands the prototype of force acting on the blades of the two similar turbines. Thus

$$F_m = F_p$$

The force which act simultaneously are the inertia, gravity, pressure and friction forces. They include body and surface forces. For complete dynamic similarity the relative losses in the flows and efficiencies should be same Dynamic similarity requires that the dimensionless numbers like Reynolds number (*Re*), Froude number (*Fr*), Euler number (*Eu*) and Weber number (*W*) should be same for the model and prototype. In addition to these the Strouhal number (*Sh*) should also be equal.

### 9.6.1 Conditions of Similarity for Turbine Models

1. $$\left(\frac{D_2}{D_1}\right)_m = \left(\frac{D_2}{D_1}\right)_p \quad ...(9.4a)$$

Since $\frac{D_2}{D_1} = \frac{K_{u2}}{K_{u1}}$, i.e., speed ratios are proportionate.

$\therefore$ $$\left(\frac{K_{u2}}{K_{u1}}\right)_m = \left(\frac{K_{u2}}{K_{u1}}\right)_p \quad ...(9.4b)$$

2. $$(K_u)_m = (K_u)_p = \text{speed ratio} \quad ...(9.5)$$

3. $$(E_u)_m = (E_u)_p \quad ...(9.6a)$$

$\therefore$ $$(Q_{11})_m = (Q_{11})_p = \frac{Q}{D_1^2\sqrt{H}} \quad ...(9.6b)$$

4. $$(N_s)_m = (N_s)_p \left(\frac{\eta_m}{\eta_p}\right)^{5/4} \quad ...(9.7)$$

5. Friction factor, $f_m > f_p$ and $(R_e)_m < (R_e)_p$

$$(\eta_H)_m < (\eta_H)_p \quad ...(9.8)$$

$$(\eta_Q)_m < (\eta_Q)_p \qquad ...(9.9)$$

$$(\eta_{mech})_m < (\eta_{mech})_p \qquad ...(9.10)$$

$$\therefore \quad (\eta_t)_m < (\eta_t)_p \qquad ...(9.11)$$

### 9.6.2 Strouhal number

For unsteady flow, Strouhal number for the model and prototype must be same. The Strouhal number is defined as:

$$Sh = \left[\left(v^2 / L\right) / \left(v / T\right)\right]_m = \left[\left(v^2 / L\right) / \left(v / T\right)\right]_p$$

Or

$$(vT / D_1)_m = (vT / D_2)_p$$

$$= Sh = \text{constant} \qquad ...(9.12)$$

Where $T$ = characteristic interval of time for the motion considered.

$D_1$ = inlet runner diameter, which is the linear dimension $L$.

Now $T = \dfrac{1}{N}$

$N$ = speed of the runner, rpm.

$$Sh = \frac{vT}{D_1}$$

$$= \frac{K_{v1}\sqrt{2gH}}{D_1 N} \quad \alpha \quad \frac{\sqrt{H}}{D_1 N}$$

$$= \frac{1}{N_{11}} \qquad ...(9.13)$$

Or

$$N_{11} = \frac{ND_1}{\sqrt{H}} \, m^{1/2} / s \qquad ...(9.14)$$

Thus

$$(N_{11})_m = (N_{11})_p \qquad ...(9.15)$$

The Strouhal's number is used when there are vibrations duc to the unsteady flow.

### 9.6.3 Scale effect

It is not possible to maintain the Reynolds number of the model and prototype constant. The model tests are carried out with $(Re)_m < (Re)_p$, in which case the complete geometrical similarity will not be maintained. In actual practice, the relative roughness $\delta_m > \delta_p$ so that friction factor $f_m > f_p$. Therefore, the relative losses of energy of model are more. Also $\eta_m < \eta_p$. In order to decrease the influence of Re on $\eta$, the model experiments are carried with very high value of $Re > 10^6$.

The difference in efficiency of model and prototype is takes into account by making necessary corrections in the formulae. The size of the model is much smaller than the prototype. This is known as scale effect. Then

$$(N_{11})_p / (N_{11})_m = \sqrt{\eta_p / \eta_m} \qquad ...(9.16)$$

And $$(Q_{11})_p / (Q_{11})_m = \sqrt{\eta_p / \eta_m} \qquad ...(9.17)$$

Where $$N_{11} = \frac{ND_{11}}{\sqrt{\eta_{hyd}.H}} \qquad ...(9.18)$$

And $$Q_{11} = \frac{Q}{D_1^2\sqrt{\eta_{hyd}.H}} \qquad ...(9.19)$$

Now $$u_1 = K_{u1}\sqrt{2gH\eta_{hyd}}$$

$$= \frac{\pi D_1 N}{60}$$

If $D_1 = 1$ m and $H = 1$m, then

$$N_{11} = \frac{ND_1}{\sqrt{H}} = \frac{60K_{u1}\sqrt{2g\eta_{hyd}}}{\pi} \qquad ...(9.20)$$

Rate of flow through the runner,

$Q$ = acrea across flow × relative velocity of flow

$$= K_t . \frac{\pi}{4} D_1^2 v_{f1}$$

Where $K_t$ = blade thickness coefficient

$$K = \frac{\pi}{4} K_t$$

and $K_f$ = relative flow velocity coefficient

$$Q = \frac{KD_1^2 K_f\sqrt{2gH\eta_{hyd}}}{\eta_Q}$$

Where $\eta_Q$ = volumetric efficiency

$$\therefore \quad Q_{11} = \frac{Q}{D_1^2\sqrt{H}}$$

$$= \frac{K.K_f\sqrt{2g\eta_{hyd}}}{\eta_Q} \qquad ...(9.21)$$

And turbine specific power,

$$P_{11} = \frac{P}{H^{3/2} D_1^2} = \frac{\gamma Q H \eta_t}{10^3 H^{3/2} D_1^2}$$

$$= \frac{\gamma Q.\eta_t}{10^3 H^{1/2} D_1^2} = \frac{\gamma Q_{11} \eta_t}{10^3}$$

$$= \frac{\gamma K K_f \sqrt{2g\eta_{hyd}}}{10^3 \eta_Q}.\eta_t \qquad ...(9.22)$$

### 9.6.4 Relationships Between Model and Prototype

Let $n$ = scale of model

$$= \frac{(D_1)_p}{(D_1)_m} = \frac{H_p}{H_m}$$

**1. Speed of Turbine:**

$$(K_{u1})_p = (K_{u1})_m$$

or

$$\frac{\pi (D_1 N)_p}{60\sqrt{2g}\sqrt{H_p}} = \frac{\pi (D_1 N)_m}{60\sqrt{2g}\sqrt{H_m}}$$

or

$$N_p = N_m \frac{(D_1)_m}{(D_1)_p} . \sqrt{\frac{H_p}{H_m}} \qquad ...(9.23)$$

**2. Rate of Flow:**

$$(Q_{11})_p = (Q_{11})_m$$

$$\left(\frac{Q}{D_1^2 \sqrt{H}}\right)_p = \left(\frac{Q}{D_1^2 \sqrt{H}}\right)_m$$

$$Q_p = Q_m \frac{\left(D_1^2\right)_p}{\left(D_1^2\right)_m} . \sqrt{\frac{H_p}{H_m}} \qquad ...(9.24)$$

**3. Available Power:**

$$(P_a)_p = (\gamma Q H)_p / 10^3$$

$$(P_a)_m = (\gamma Q H)_p / 10^3$$

$$\frac{(P_a)_p}{(P_a)_m} = \frac{\gamma_p}{\gamma_m} \times \frac{Q_p}{Q_m} \times \frac{H_p}{H_m}$$

$$= \frac{\left(D_1^2\right)_p}{\left(D_1^2\right)_m} . \left(\frac{H_p}{H_m}\right)^{3/2} \text{ for } \gamma_p = \gamma_m \qquad ...(9.25)$$

**4. Brake power:**

$$P_t = P_a \eta_t = P_a \eta_H . \eta_Q . \eta_{mech}$$

$$\frac{(P_t)_p}{(P_t)_m} = \frac{(P_a)_p}{(P_a)_m} \cdot \frac{(\eta_H)_p}{(\eta_H)_m} \cdot \frac{(\eta_Q)_p}{(\eta_Q)_m} \cdot \frac{(\eta_{mech})_p}{(\eta_{mech})_m}$$

$$= \frac{(D_1^2)_p}{(D_1^2)_m} \cdot \left(\frac{H_p}{H_m}\right)^{3/2} \times . \frac{(\eta_H)_p}{(\eta_H)_m} \cdot \frac{(\eta_Q)_p}{(\eta_Q)_m} \cdot \frac{(\eta_{mech})_p}{(\eta_{mech})_m} \quad \text{...(9.26)}$$

**5. Efficiency**

(a) *Head efficiency:*

(i) *Dr. J. Ackeret formula:*

$$(\eta_H)_p = 1 - \left[1 - (\eta_H)_m\right] \times \left[0.5 - 0.5\left\{\frac{(R_e)_m}{(R_e)_p}\right\}^{1/4}\right] \quad \text{...(9.27)}$$

(ii) *Prof. Moody modified formula:*

$$(\eta_H)_p = 1 - \left[1 - (\eta_H)_m\right]\left[\frac{(D_1)_m}{(D_1)_p}\right]^{1/5} \quad \text{...(9.28)}$$

(iii) *Dr. R. Gregorig formula:*

$$(\eta_H)_p = 1 - \left[1 - (\eta_H)_m\right]\left[\frac{(R_e)_m}{(R_e)_p}\right]^{1/4} - K_{vf3}\left[1 - \left\{\frac{(R_e)_m}{(R_e)_p}\right\}^{1/4}\right] \quad \text{...(9.29)}$$

(iv) *Dr. S.P. Hutton formula:*

$$(\eta_H)_p = 1 - \left[1 - (\eta_H)_m\right]\left[0.3 - 0.7\left\{\frac{(R_e)_m}{(R_e)_p}\right\}^{1/5}\right] \quad \text{...(9.30)}$$

Where $Re = \frac{\sqrt{2gH}.D_1}{\upsilon}$,

$\upsilon$ = kinematic viscosity

$$K_{vf3} = \frac{v_{f3}}{\sqrt{2gH}},$$

$v_{f3}$ = velocity of flow of water at the outlet of draft tube.

Leakage losses, $\eta_Q = 1 - \frac{\Delta Q}{Q}$

Mosony formula for leakage loss,

$$\Delta Q = 0.9\sqrt{2gH}.D_1\pi S \quad ...(9.31)$$

Where $S$ = clearance width between runner blades and turbine casing

(b) Volumetric efficiency:

**Dubs-Moody formula:** $$(\eta_t)_p = \left[1-\{1-(\eta_t)_m\}\sqrt{\frac{(D_1)_m}{(D_1)_p}}\right]^2 \quad ...(9.33)$$

**Example 9.1** A turbine model having the following specifications was made to predict the behavior of the actual turbine:

$P_t$ = 12 kW, $H$ = 10 m, $N$ = 450 rpm, model scale, $n$ = 10

Determine: (a) the speed of the actual turbine runner, and (b) power developed.

Assume the overall efficiency of model and prototype to be equal. Find the type of turbine of which the model was made.

**Solution.**

Let $$n = \frac{(D_1)_p}{(D_1)_m}$$

$$= \frac{H_p}{H_m} = 10$$

Then $H_p = 10 \times H_m = 10 \times 10 = 100\text{m}$

(b) For $(\eta_o)_m = (\eta_o)_p$, the specific powers of model and prototype must be equal.

$$(P_{t11})_p = (P_{t11})_m$$

$$\frac{(P_t)_p}{(D_1^2)_p .H_p{}^{3/2}} = \frac{(P_t)_m}{(D_1^2)_m .H_m{}^{3/2}}$$

$$(P_t)_p = (P_t)_m \cdot \left[\frac{(D_1)_p}{(D_1)_m}\right]^2 \times \left[\frac{H_p}{H_m}\right]^{3/2}$$

$$12\times(10)^2\times(10)^{3/2} = 12\times(10)^{7/2} = 37947.3 \text{ kW}$$

(a) $$(K_{u1})_p = (K_{u1})_m$$

$$\frac{\pi(D_1)_p N_p}{60\sqrt{2g}\sqrt{H_p}} = \frac{\pi(D_1)_m N_m}{60\sqrt{2g}\sqrt{H_m}}$$

$$N_p = N_m \times \frac{(D_1)_m}{(D_1)_p} \times \sqrt{\frac{H_p}{H_m}}$$

$$= 450 \times \frac{1}{10} \times \sqrt{10} = 142.3 \text{ rpm}$$

(c) $$(N_s)_m = (N_s)_p = \frac{N\sqrt{P_t}}{H^{5/4}} = \frac{450\sqrt{12}}{10^{5/4}} = 87.66$$

∴ It is a Kaplan turbine.

**Example 9.2** The runner of a Pelton turbine was built according to the following specifications.

$P_t$ = 12,500 kW, $H$ = 500 m, $Q$ = 2950 lps, $N$ = 300 rpm

Find the specifications of a one-eighth scale model of this turbine, assuming the efficiencies of the two turbines to be same.

**Solution.**

Let $$\frac{(D_1)_p}{(D_1)_m} = \frac{H_p}{H_m} = n = 8$$

(a) $$\frac{H_p}{H_m} = 8$$

$$H_m = \frac{500}{8} = 62.5\text{m}$$

(b) $$(K_{u1})_p = (K_{u1})_m$$

$$N_m = N_p \cdot \frac{(D_1)_p}{(D_1)_m} \cdot \left(\frac{H_m}{H_p}\right)^{1/2}$$

$$= 300 \times 8 \times \left(\frac{1}{8}\right)^{1/2} = 848.5 \text{ rpm}$$

(c) $$(Q_{11})_p = (Q_{11})_m$$

$$\frac{Q_p}{(D_1^2)_p \sqrt{H_p}} = \frac{Q_m}{(D_1^2)_m \sqrt{H_m}}$$

$$Q_m = Q_p \cdot \left[\frac{(D_1)_m}{(D_1)_p}\right]^2 \cdot \left(\frac{H_m}{H_p}\right)^{1/2}$$

$$2950 \times \left(\frac{1}{8}\right)^2 \times \left(\frac{1}{8}\right)^{1/2} = 2950 \times \left(\frac{1}{8}\right)^{5/2} = 16.297 \text{ lps}$$

(d) $$(P_{t11})_p = (P_{t11})_m$$

$$\frac{(P_t)_p}{(D_1^2)_p H_p^{3/2}} = \frac{(P_t)_m}{(D_1^2)_m H_m^{3/2}}$$

$$(P_t)_m = (P_t)_p \times \left[\frac{(D_1)_m}{(D_1)_p}\right]^2 \times \left(\frac{H_m}{H_p}\right)^{3/2}$$

$$= 12{,}50\left(\frac{1}{8}\right)^2 \times \left(\frac{1}{8}\right)^{3/2}$$

$$= 12{,}50\left(\frac{1}{8}\right)^{7/2} = 8.63 \text{ kW}$$

**Example 9.3** A model turbine is provided with a runner of 2m diameter. It develops 45 kW under a head of 100 m when running at a speed of 4000 rpm. Calculate the specific speed and unit speed for this model. It is required to build a similar turbine to develop 150 k W under a head of 10 m. Calculate the required diameter.

**Solution.**

Given: $D_m = 2\text{m}$, $(P_t)_m = 45$ kW, $H_m = 100$ m, $N_m = 4000$ rpm, $(P_t)_p =$ 150 kW, $H_p = 120$ m

Specific speed of model, $$(N_s)_m = \frac{N_m\sqrt{(P_t)_m}}{H_m^{5/4}}$$

$$= \frac{4000\sqrt{45}}{100^{5/4}} = 84.85$$

Unit speed of model, $$(N_1)_m = \frac{N_m}{\sqrt{H_m}} = \frac{4000}{\sqrt{100}} = 400$$

Specific powers of model and prototype are equal.

$$\frac{(P_t)_p}{D_p^2.H_p^{3/2}} = \frac{(P_t)_m}{D_m^2.H_m^{3/2}}$$

$$D_p = D_m.\left[\frac{(P_t)_p}{(P_t)_m} \times \left(\frac{H_m}{H_p}\right)^{3/2}\right]^{1/2}$$

$$= 2\left[\frac{150}{45} \times \left(\frac{100}{120}\right)^{3/2}\right]^{1/2} = 3.158 \text{ m}$$

**Example 9.4** A quarter scale turbine model is tested under a head of 40 m. The full scale turbine is required to work under a head of 100 m and to run at 425 rpm. At what speed must the model be run if it develops 100 kW and uses 0.35 m³/s of water at this speed. What power will be admitted from the full scale turbine assuming that its efficiency is 3 % better than that of the model?

**Solution.**

Given: $n = \dfrac{D_p}{D_m} = 4$, $H_m = 40\text{m}$, $H_p = 100$ m, $N_p = 425$ rpm, $(P_t)_m = 100$ kW,

$Q_m = 0.35$ m³/s, $(\eta_t)_p = (\eta_t)_m + 3\%$

(a) $$(K_{u1})_m = (K_{u1})_p$$

or $$N_m = N_p \times \frac{D_p}{D_m} \times \sqrt{\frac{H_m}{H_p}} = 425 \times 4 \times \sqrt{\frac{40}{100}}$$

$$= 1075 \text{ rpm}$$

(b) $$(P_t)_m = \frac{\gamma Q_m H_m}{10^3} \times (\eta_t)_m$$

$$(\eta_t)_m = \frac{100 \times 10^3}{9.81 \times 10^3 \times 0.35 \times 40} = 0.728$$

$$(\eta_t)_p = 0.728 + 0.03 = 0.758 \text{ or } 75.8\ \%$$

Now $$(Q_{11})_p = (Q_{11})_m$$

or $$Q_p = Q_m \left(\frac{D_p}{D_m}\right)^2 \left(\frac{H_p}{H_m}\right) H^{1/2}$$

$$= 0.35(4)^2 \left(\frac{100}{40}\right)^{1/2} = 8.85 \text{ m}^3/\text{s}$$

$$(P_t)_p = \frac{\gamma Q_p H_p \times (\eta_t)_p}{10^3}$$

$$= \frac{9.81 \times 10^3 \times 100 \times 0.758}{10^3} = 743.6 \text{ kW}$$

(c) Specific speed, $$(N_s)_m = \frac{N_m \sqrt{(P_t)_m}}{H_m^{5/4}} = \frac{1075\sqrt{100}}{40^{5/4}} = 106.86$$

**Example 9.5** A Kaplan turbine hydroelectric plant works under a head of 1.8 m and consume 250s m³/s of water. Assuming the speed of the turbines to be 50 rpm and their efficiency 82 % determine the least number of turbines, of same size, required if their specific speed does not exceed 800. What would be the output of each turbine?

**Solution.**

Given: $Q_{total} = 250$ m³/s, $H = 1.8$ m, $N = 50$ rpm, $N_s = 800$, $\eta_t = 82$ %

Specific speed,
$$N_s = \frac{N\sqrt{P_t}}{H^{5/4}}$$

$$800 = \frac{50\sqrt{P_t}}{(1.8)^{5/4}}$$

$$P_t = 1112.8 \text{ kW}$$

Now
$$P_t = \frac{\gamma QH}{10^3} \times \eta_t$$

$$Q = \frac{1112.8 \times 10^3}{9.81 \times 10^3 \times 1.8} = 63 \text{ m}^3\text{/s}$$

$$\text{Number of turbines} = \frac{Q_{total}}{Q} = \frac{250}{63} = 3.97 \simeq 4$$

**Example 9.6** A model turbine constructed to a scale of 1:10 when tested under a head of 8m at 400 rpm gave an efficiency of 77 %. Determine the rpm of the prototype and the ratio of powers developed by the model and prototype if the prototype works under a head of 100 m. What will be the efficiency of prototype if scale effect is considered?

**Solution.**

Given: $n = 10$, $H_m = 8$m, $N_m = 400$ rpm, $(\eta_t)_m = 0.77$, $H_p = 100$ m, $\frac{D_p}{D_m} = n = 10$

$$(K_{u1})_m = (K_{u1})_p$$

$$N_p = N_m \times \frac{D_m}{D_p} \times \left(\frac{H_p}{H_m}\right)^{1/2}$$

$$= 400 \times \frac{1}{10} \times \left(\frac{100}{8}\right)^{1/2} = 141.42 \text{ rpm}$$

***Ratio of power developed,***

$$\frac{(P_t)_p}{(P_t)_m} = \left(\frac{N_p}{N_m}\right)^3 \times \left(\frac{D_p}{D_m}\right)^5$$

$$= \left(\frac{141.42}{400}\right)^3 \times 10^5 = 4419.4$$

Now $$\frac{N_p}{N_m} = \frac{D_p}{D_m}\sqrt{\frac{H_p}{H_m}}.\sqrt{\frac{(\eta_t)_p}{(\eta_t)_m}}$$

$$(\eta_t)_p = (\eta_t)_m\left(\frac{N_p}{N_m}\right)^2\left(\frac{D_p}{D_m}\right)^2\left(\frac{H_m}{H_p}\right)$$

$$= 0.77\left(\frac{141.42}{400}\right)^2(10)^2\left(\frac{8}{100}\right)$$

$$= 0.77 \text{ or } 77\ \%$$

## 9.7 Characteristics of Water Turbines

The graphical representation of the behavior of water turbines under varied conditions by means of curves are known as characteristics. The practical data for these curves are obtained from experiments on models or actual sized models. To draw the characteristic curves, the following data have to be obtained from the tests:

1. Rate of flow of water
2. Net head
3. Available power
4. Brake power, and
5. Overall efficiency

The behavior of a water turbine may be exhibited by the following curves:

(i) Main characteristics (constant head curves)

(ii) Operating characteristics (constant speed curves)

(iii) Muschel curves (constant efficiency curves )

***(i) Main Characteristics (or constant head curves)***

To plot these curves, the head is kept constant and speed is varied by allowing available quantity of water of flow through the inlet opening. Thus a series of values of N and Q are obtained. For each value, brake power is measured by applying brakes mechanically. All the measured and calculated quantities are tabulated for inlet opening percentage. All the quantities are then reduced to unit head. The following curves are then drawn:

$Q_1 = f(N_1)$, $P_{t1} = f(N_1)$ and $M_1 = f(N_1), \eta_t = f(N_1)$

Fig. 9.6 and 9.7 show the typical main characteristics of Peloton and Kaplan turbines respectively.

The main characteristics of Francis turbine are similar except $Q_1 = f(N_1)$ curves, shown, in Fig. 9.8.

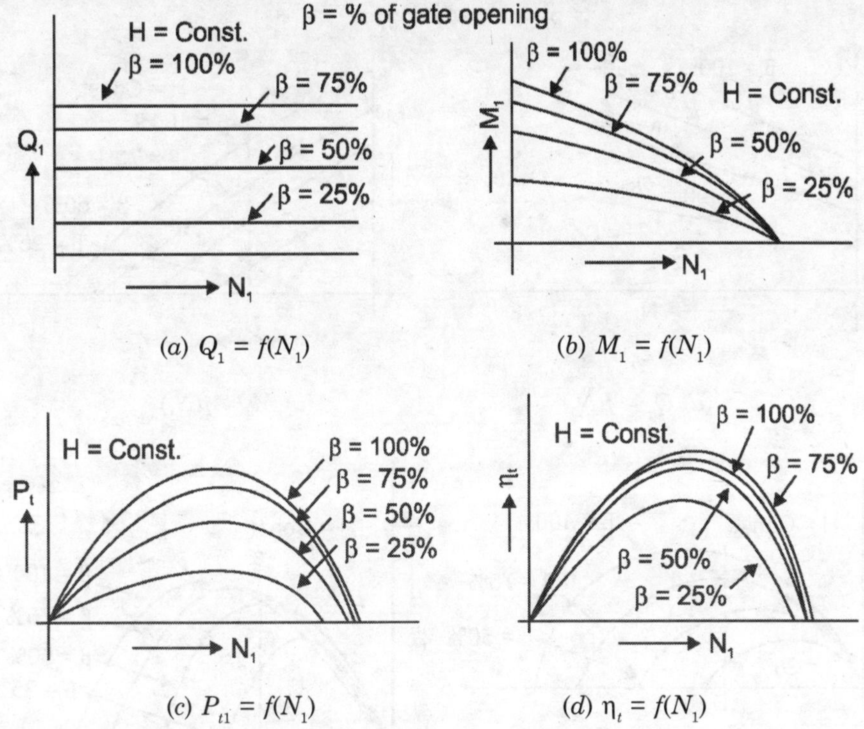


(a) $Q_1 = f(N_1)$ (b) $M_1 = f(N_1)$

(c) $P_{t1} = f(N_1)$ (d) $\eta_t = f(N_1)$

**Fig. 9.6 Main characteristics of Pelton turbine.**

The main characteristic curves yield the following results:

1. The discharge $Q_1$ for a Pelton turbine depends only upon gate opening and is independent of $N_1$. The curves for $Q_1$ are horizontal.
2. The curves between $Q_1$ and $N_1$ for a Francis turbine are falling curves. This is due to the fact that centrifugal head developed which acts outwards and opposes the external head causing flow. Consequently the discharge decreases as speed increases.
3. The curves between $Q_1$ and $N_1$ for a Kaplan turbine are rising curves, the discharge increases with increasing speed.
4. The curves between $P_{t1}$ and $N_1$, and those between $\eta_t$ and $N_1$ indicate that at a particular speed the efficiency is maximum. For the Pelton wheel $(\eta_t)_{max}$ occurs at the same speed for all gate openings.

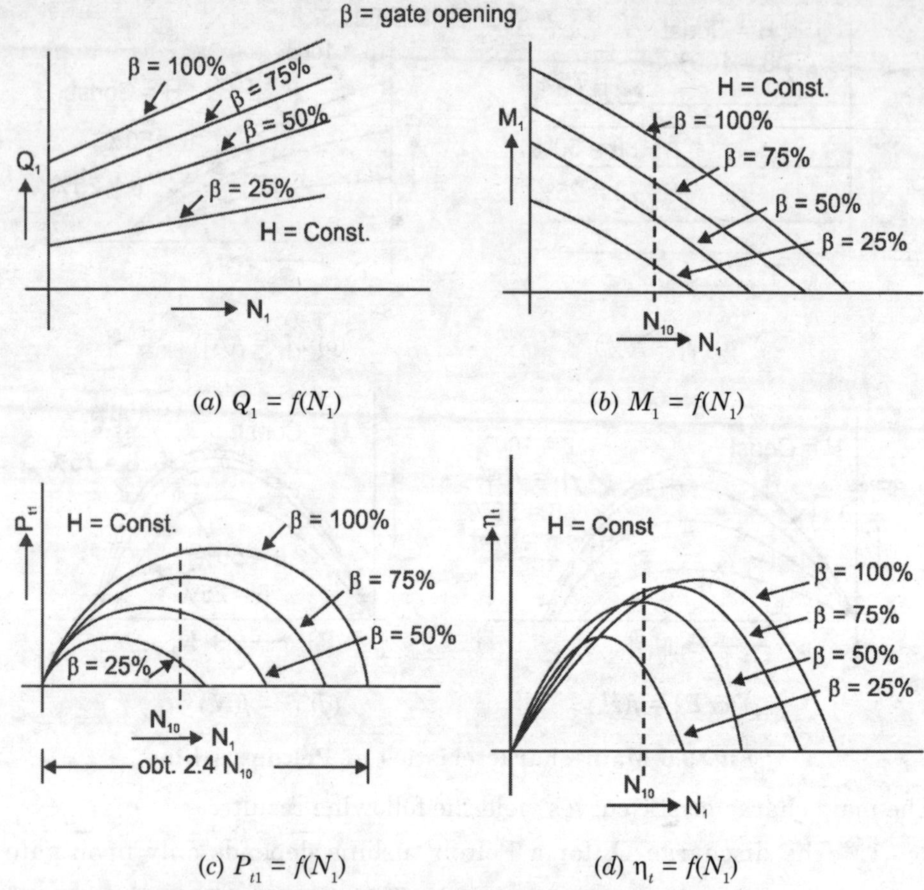


(a) $Q_1 = f(N_1)$ (b) $M_1 = f(N_1)$

(c) $P_{t1} = f(N_1)$ (d) $\eta_t = f(N_1)$

**Fig. 9.7 Main characteristics of Kaplan turbine.**

5. The maximum efficiency for Francis turbine accurse at different speeds for different gate opening

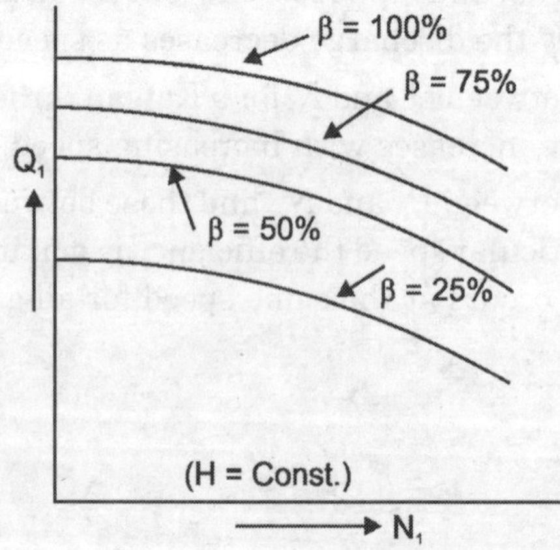


**Fig. 9.8 $Q_1 = f(N_1)$ for Francis turbines.**

### *(ii) Operating characteristics*

These characteristics are drawn by keeping the speed of the turbine constant and varying other conditions viz, $Q$ and $H$. The following curves are drawn, as shown in Fig. 9.9

$$P_t, \eta_t = f(Q)$$
$$\eta_t = f(P_t)$$
$$\% \text{ of } (\eta_t)_{max} = f(\% \text{ full load})$$

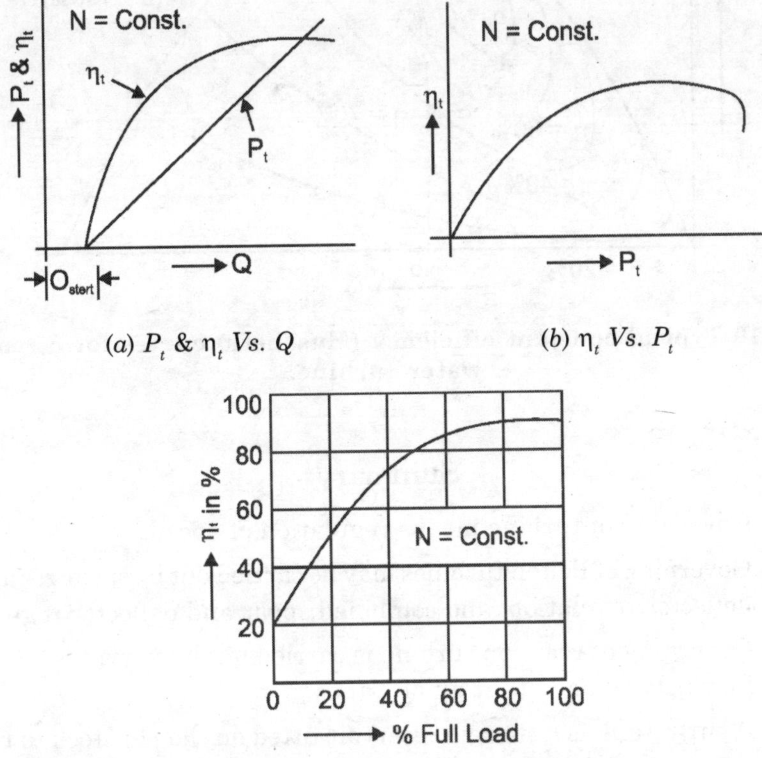


(a) $P_t$ & $\eta_t$ *Vs.* $Q$ (b) $\eta_t$ *Vs.* $P_t$

(c) $\eta_t$ (in %) *Vs.* % full load

**Fig. 9.9 Typical opereting chatacteristics (constant speed curves) of a water turbine.**

**The following points are worth noting:**

1. The output power *v's* discharge is a straight line, since $P_t \propto Q$, if $H$ is constant.
2. $\eta_t$ *v's* $Q$ curves are curvilinear and $\eta_t$ increases as $Q$ increases and becomes constant after a particular discharge.
3. $\eta_t$ increases as $P_t$ increases, becomes nearly constant after some value of $P_t$ and then starts decreasing.
4. Percentage efficiency increases as percentage full gate opining increases.

***(iii) Muschel curves***

The data for theses curves are obtained from the constant head and constant speed curves. These are constant efficiency curves shown in Fig. 9.10. The peak of these curves gives the maximum efficiency curves.

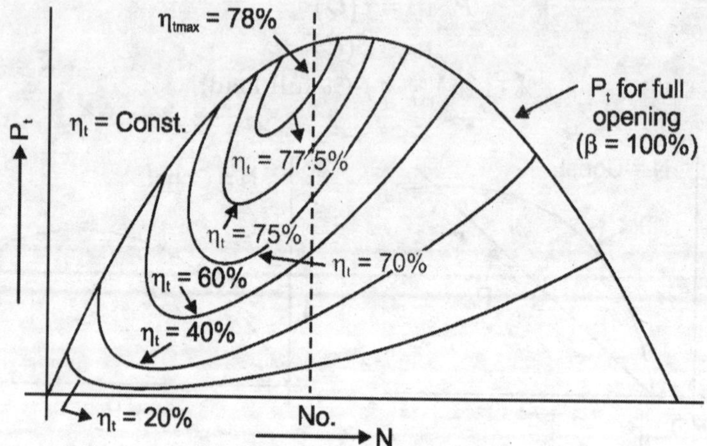


**Fig. 9.10 Typical constant efficiency (Muschel's) curves for a reaction water turbine.**

## Summary

1. Governing of turbine means regulation of speed.
2. Governing of Pelton turbines may be carried out by spear regulation, deflector regulation, and combined spear and deflector regulation.
3. Governing of reaction turbine is carried out by varying the area of flow between adjacent guide vanes
4. A surge tank is a storage reservoir fitted on the penstock to receive the rejected flow when the penstock is suddenly closed by a valve. It also supplies additional water when load increases.
5. A foreay is a storage reservoir at the beginning of a penstock made by enlarging the canal at the intake. Its function is to store the rejected water when the penstock is suddenly closed due to decrease of load. It also supplies the additional water when load increases.
6. Similarity relations:

$$\left(\frac{D_2}{D_1}\right)m = \left(\frac{D_2}{D_1}\right)p$$

$$(Q_{11})_m = (Q_{11})_p = \frac{Q}{D_1^2\sqrt{H}}$$

$$(N_s)_m = (N_s)_p \left(\frac{\eta_m}{\eta_p}\right)^{5/4}$$

7. $$N_{11} = \frac{ND_1}{\sqrt{N}}$$

$$(N_{11})_m = (N_{11})_p$$

8. $$\frac{N_p}{N_m} = \frac{(D_1)_m}{(D_1)_p} \cdot \sqrt{\frac{H_p}{H_m}}$$

$$\frac{Q_p}{Q_m} = \frac{(D_1^2)_p}{(D_1^2)_m} \sqrt{\frac{H_p}{H_m}}$$

$$\frac{(P_a)_p}{(P_a)_m} = \frac{(D_1^2)_p}{(D_1^2)_m} \times \left(\frac{H_p}{H_m}\right)^{3/2} \text{ for } \gamma_p = \gamma_m$$

$$\frac{(P_t)_p}{(P_t)_m} = \frac{(D_1^2)_p}{(D_1^2)_m} \times \left(\frac{H_p}{H_m}\right)^{3/2} \times \frac{(\eta_o)_p}{(\eta_o)_m}$$

## Review Questions

1. What kind of preliminary data is required for the selection of water turbines?
2. On what factors the selection of different types of water turbines depend?
3. How do the rotational speed and cavitation affect the selection of a water turbine?
4. Why is the turbine model manufactured?
5. What are the conditions of making the turbine models?
6. What are the various similarity considerations of model and prototypc?
7. What is scale effect?

## Numerical Problems

1. It is required to develop 75 MW by a number of water turbines. Each turbine runs at 166.7 rpm when working under a head of 30 m. Determine the number of turbines to be employed if the specific speed of each turbine is 350.

[**Ans.** 4]

2. The quantity of water available for a hydro – electric station is 250 $m^3/s$ under a head of 1.8 m. Assuming the speed of turbines to be 60 rpm and their efficiency 82%, determine the number of turbines required. Assume a specific speed of 900.

3. A hydro – electric station is installed to utilize 30 $m^3/s$ of water of a river at a head of 8 m. The turbines efficiency may be taken as 85 %. Determine:

   (a) Is it feasible to develop this power by two turbines with speed not less than 50 rpm and specific speed not greater than 540?

   (b) What type of runner will be used?

   (c) What is the diameter of runner if the speed ratio is 0.85?

4. In a water power site the available discharge is 340 $m^3/s$ under a net head of 30 m. Assuming a turbine efficiency of 88 % and rotational speed of 166.7 rpm, determine the least number of machines, all of the same size that may be installed if the selection is based on Francis turbine with specific speed not greater than 270.

5. A model of a water turbine develops 178 kW under a head of 5 m and running at 480 rpm. Determine the power of actual turbine if the working head is 40 m and model scale $\frac{1}{10}$ th of actual turbine. What is the speed of the actual turbine? Assume that the efficiency of model and actual turbine to be the same.

6. A model Francis turbine made to $\frac{1}{5}$ th scale, when tested gave 3 kW at a speed of 360 rpm under a head of 2 m. Calculate the equivalent speed and power of full size turbine when working under a head of 6 m. Find also the ratio of quantity of water flowing through the turbine per second.

7. A model having a scale ratio of $\frac{1}{10}$ is constructed to determine the best design for a Kaplan turbine to develop 7300 kW under a net head of 12 m when running at 100 rpm. If the head available in the laboratory is 8 m and the model efficiency if 80 %, find:

   (a) The correct running speed of model,

   (b) The flow required in laboratory in $m^3/s$,

   (c) The kW of model, and

   (d) The specific speed in each case.

# CHAPTER 10 Reciprocating Pumps

## 10.1 Introduction

A pump is a mechanical device to increase the pressure energy of a liquid. In most of the cases, a pump is used for raising fluids from a lower to a higher level. This is achieved by creating a low pressure at inlet or suction end and high pressure at the outlet or delivery end of the pump. Due to the low inlet pressure the fluid rises from a depth where it is available and the high outlet pressure forces it up to a height where it is required. Work has to be done by a prime-mover on the pump to enable it to impart energy to the fluid.

## 10.2 Classification of Pumps

The pumps may be classified a showing Fig. 10.1.

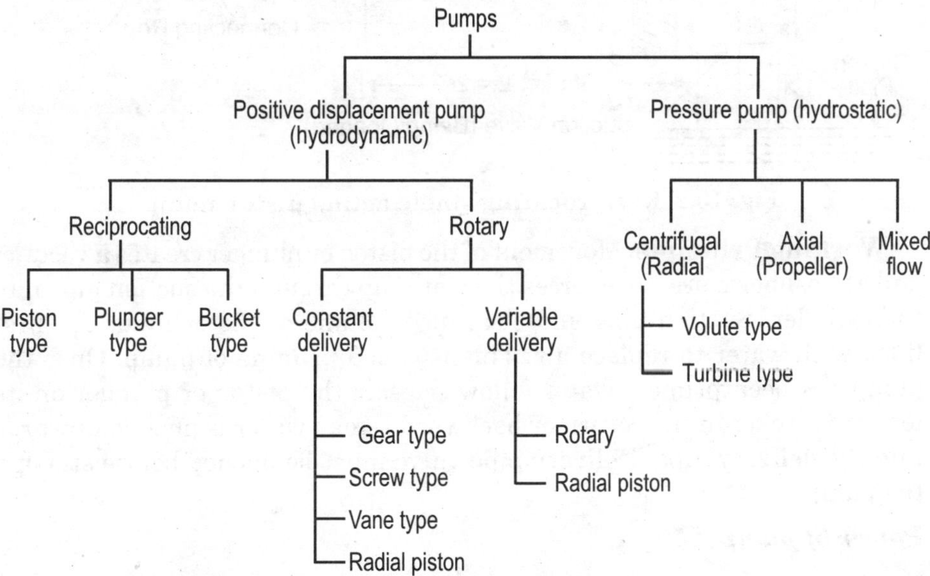


**Fig. 10.1 Classification of pumps.**

## 10.3 Reciprocating Pumps

### 10.3.1 Single Acting Piston Pump

A reciprocating pump shown in Fig. 10.2 consists primarily of a piston or a plunger reciprocating inside a close fitting cylinder, thus performing the suction and delivery strokes. It is a positive acting type which means that it is a displacement pump which creates lift and pressure by displacing liquid with a moving member or piston. The chamber or cylinder is alternately filled and emptied by forcing and drawing the liquid by mechanical motion. This type is called the positive displacement pump. Suction and delivery pipes are connected to the cylinder. Both the pipes are provided with non-return valves which ensure unidirectional flow of liquid. Volume of fluid delivered is constant regardless of pressure, and is varied only by speed changes. They operate at low speeds. They are used generally for pneumatic pressure systems, feeding small boilers condensate (steam condensed to water) return and light oil pumping.

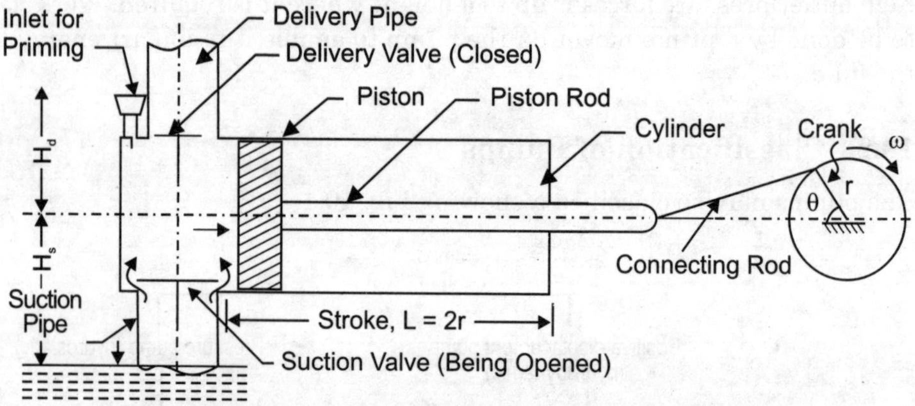


**Fig. 10.2 Reciprocating single acting piston pump.**

**Working Principle.** Movement of the piston or plunger creates a vacuum and atmospheric pressure forces the water up through the suction pipe into the cylinder. Suction pipe and clearance volume of the cylinder are first filled with water to replace air. This is called priming of pump. Once the pump has been primed, water follows closely the piston or plunger on its forward stroke. In the return or backward stroke water is pushed upwards into the delivery pipe. Delivery pipe valve must be opened before starting the pump.

***Power of pump***

Let $H_s$, $H_d$ = suction and delivery heads respectively of the pump, m

$A$, $a$ = cross-sectional area of piston and piston rod respectively, m$^2$

$L$ = length of stroke of piston, m

$N$ = speed of crank, rpm

Then average rate of flow, $Q = \frac{ALN}{60} \text{m}^3/\text{s}$

Force on the piston in forward stroke = $\gamma H_s A$ N

Force on the piston in backward stroke = $\gamma H_d A$ N

Static head, $H = H_s + H_d$

Neglecting head losses in transmission and at valves, the power required by the pump is,

$$P = \frac{\gamma QH}{10^3} = \frac{\gamma(ALN)(H_s + H_d)}{60 \times 10^3} \text{kW} \qquad ...(10.1)$$

### 10.3.2 Double Acting Piston Pump

In this pump, shown in Fig. 10.3, in every stroke, there is suction on one side and delivery on the other side. It has two suction and two delivery pipes.

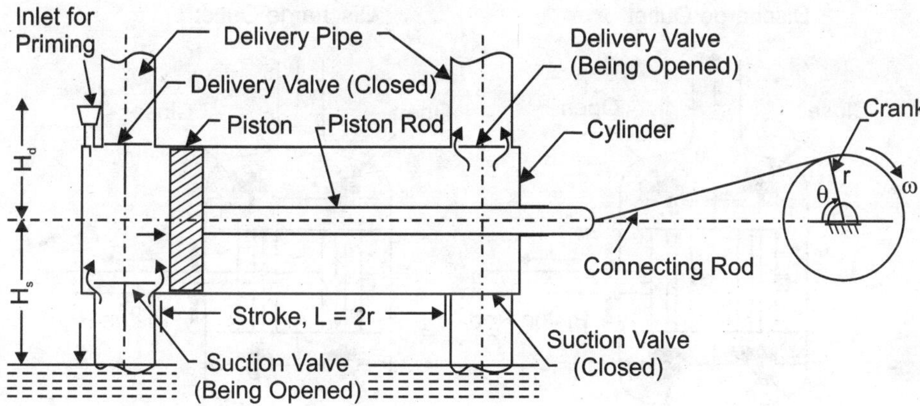


**Fig. 10.3 Double Acting Single Cylinder Piston Pump.**

$$Q = \frac{ALN}{60} + \frac{(A-a)LN}{60}$$

$$= \frac{(2A-a)LN}{60} \simeq \frac{2ALN}{60} \text{m}^3/\text{s} \qquad ...(10.2)$$

Force acting on piston during forward stroke

$$= \gamma H_s A + \gamma H_d (A-a) \text{N} \qquad ...(10.3)$$

Force acting on piston during backward stroke

$$= \gamma H_s (A-a) + \gamma H_d A \text{ N} \qquad ...(10.4)$$

### 10.3.3 Two–Throw Pump

It has two cylinders each equipped with one suction and one delivery pipe. The pistons reciprocating in the cylinders are moved with the help of connecting rods fitted with cranks at 180° , as shown in Fig. 10.4 (a). The operations of valves for two – throw pump is shown in Fig. 10.4 (b).

Rate of flow through two – throw pump, $Q = \frac{2ALN}{60} \text{ m}^3/\text{s}$ ...(10.5)

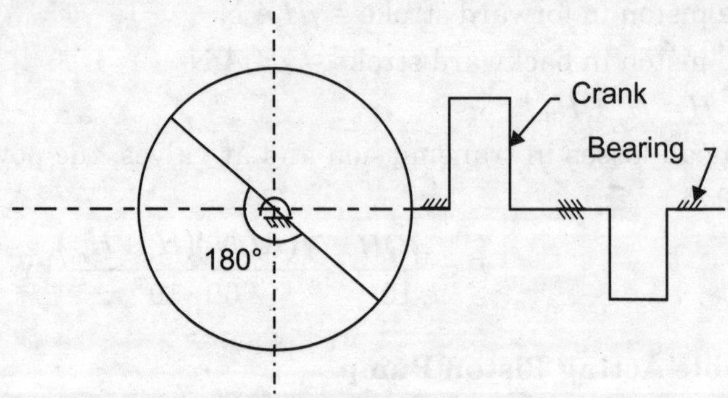


(a) Crank fitted at 180°

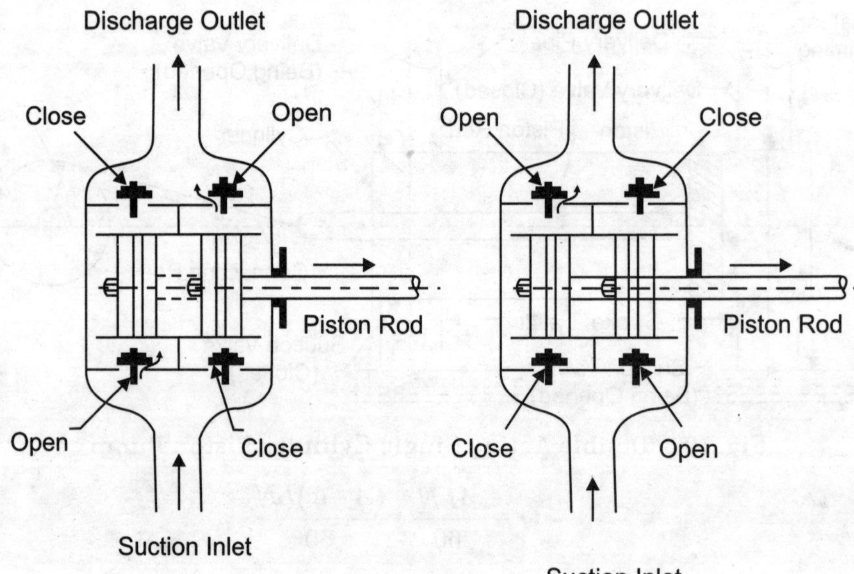


(b) Operation of Valves.

**Fig. 10.4 Two-Throw Reciprocating Pump.**

### 10.3.4 Three–Throw Pump

It has three cylinders and three pistons working with three connecting rods fitted to the crank at 120°. Such a pump is shown in Fig. 10.5.

Rate of flow through three – throw pump, $Q = 3 \times \frac{ALN}{60} \text{ m}^3/\text{s}$ ...(10.6)

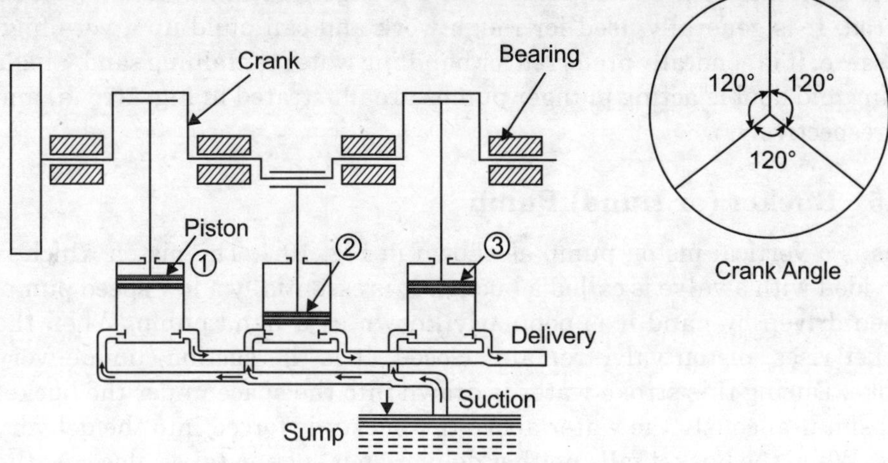


Fig. 10.5 Three – Throw Piston Pump.

## 10.4 Plunger Pump

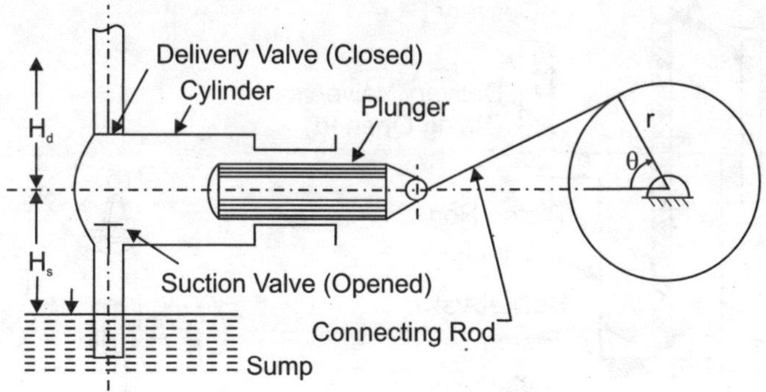


(a) Single Acting Plunger Pump.

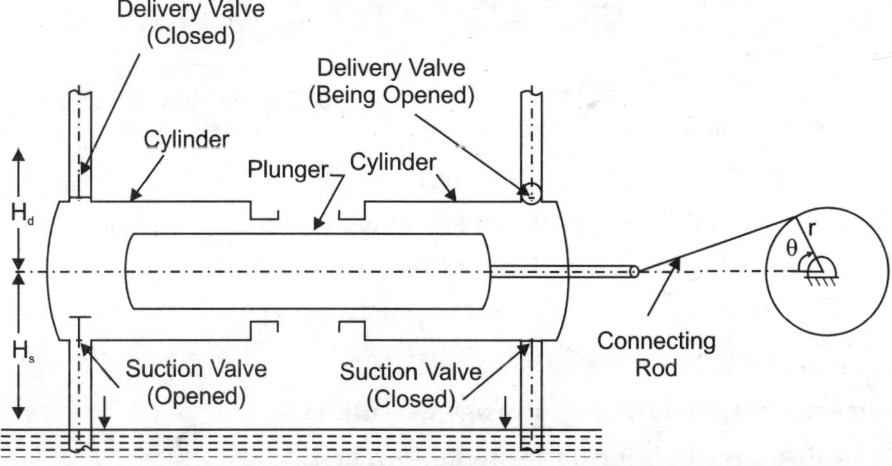


(b) Double Acting Single Cylinder Plunger Pump

Fig. 10.6 Piston or plunger pump.

It is similar to a piston pump with a plunger replacing the piston and its rod. It is generally used for rough work and can build up a very high pressure. It is generally preferred for handling water containing sand. Single acting and double acting plunger pumps are illustrated in Fig. 10.6 (a) and (b) respectively.

## 10.5 Bucket (or Hand) Pump

This is a vertical piston pump as shown in Fig. 10.7. The piston which is provided with a valve is called a bucket. It is essentially a low speed pump. When driven by hand it is popularly known as a hand pump. When the bucket rises, piston valve remains closed. It is the suction-cum-delivery stroke. During this stroke water is drawn into the space under the bucket and simultaneously the water above the bucket is forced into the delivery pipe. When the bucket falls neither delivery nor suction takes place. As the piston valve is opened, water flows across the piston from the suction to the delivery side of the cylinder.

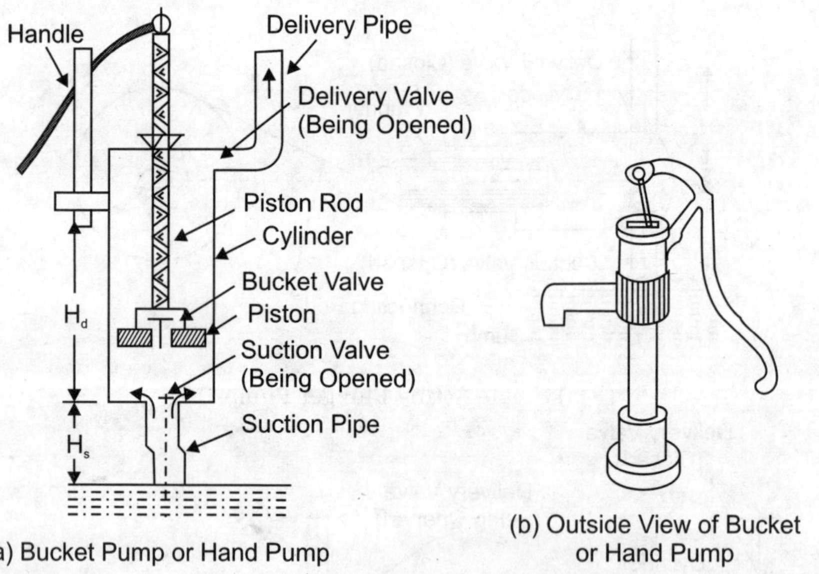


(a) Bucket Pump or Hand Pump

(b) Outside View of Bucket or Hand Pump

**Fig. 10.7**

Force required to raise the bucket = Force required to produce suction + Force required to force up the water

$$= \gamma A H_s + \gamma (A - a) H_d \text{ N}$$

Force required to lower the bucket = $\gamma a H_d$ N

Volume discharged during up stroke = $(A - a)$L m³/s

Volume discharged during down stroke = $aL$ m³/s

## 10.6 Rate of Delivery

The reciprocating pump is run by crank and connecting rod mechanism which gives the motion of piston as simple harmonic. The slider – crank mechanism is shown in Fig. 10.8. Let the crank be at an angular position θ from the inner dead centre at a time *t*. Then the displacement of crank pin from the *IDC* position, neglecting the obliquity of connecting rod for $\frac{\ell}{r}$ to be large, is given by

$$x = r(1 \cos \theta)$$
$$= r(1 - \cos wt)$$

where $\omega = \frac{2\pi N}{60} rad/s$

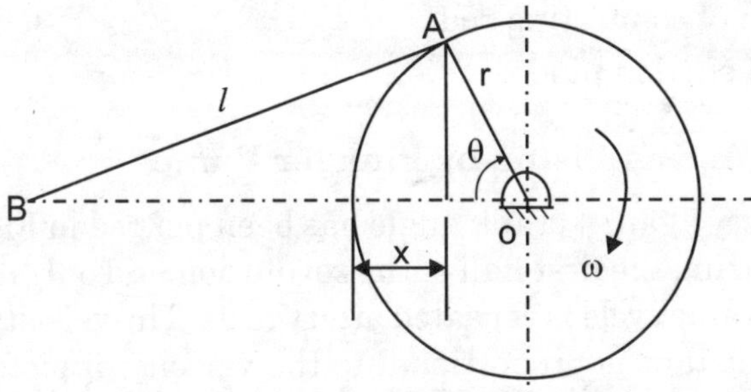


**Fig. 10.8 Slider – crank mechanism.**

Velocity of position, $$v_p = \frac{dx}{dt} = \omega r \sin\theta \quad ...(10.7)$$

Acceleration of piston, $$f_p = \frac{d^2x}{dt^2} = \omega^2 r \cos\theta \quad ...(10.8)$$

Acceleration of liquid in the suction pipe,

$$f_s = f_p \frac{A}{a_s} = \omega^2 r \cos\theta . \frac{A}{a_s} \quad ...(10.9)$$

where $a_s = \frac{\pi}{4} d_s^2$ = area of cross-section of suction pipe

$d_s$ = diameter of suction pipe

Acceleration head in suction pipe,

$$H_{a_s} = \frac{L_s}{g} \cdot f_s = \frac{L_s}{g} \cdot \omega^2 r \cos\theta \cdot \frac{A}{a_s} \quad ...(10.10)$$

At θ = 0° cos θ = 1, and

$$\left(H_{a_s}\right)_{max} = \frac{L_s}{g} \omega^2 r \cdot \frac{A}{a_s} \quad ...(10.11)$$

Considering the obliquity of connecting rod,

$$f_p = \omega^2 r\left[\cos\theta \pm \frac{\cos 2\theta}{n}\right]$$

where $n = \frac{\ell}{r}$

$$H_{a_s} = \frac{L_s}{g}\omega^2 r\left[\cos\theta \pm \frac{\cos 2\theta}{n}\right]\cdot\frac{A}{a_s}$$

$$\left(H_{a_s}\right)_{max} = \frac{L_s}{g}\omega^2 r\cdot\frac{A}{a_s}\left[1+\frac{1}{n}\right] \text{ at the beginning of suction stroke} \quad ...(10.\ 12a)$$

$$= \frac{L_s}{g}\omega^2 r\cdot\frac{A}{a_s}\left[1-\frac{1}{n}\right], \text{ at the end of suction stroke} \quad ...(10.12b)$$

where $\ell$ = length of connecting rod.

$L_s$ = length of suction pipe.

### 10.6.1 Single – Acting Piston or Plunger Pump

The rate of delivery against crank angle has been plotted in Fig. 10.9. There is only suction during the first half of the revolution and only delivery in the second half. The same cycle is repeated afterwards. The velocity of discharge of water at any instant is proportional to the velocity of piston or plunger at that instant. Therefore, the sine curve also represents the velocities of discharge to some scale.

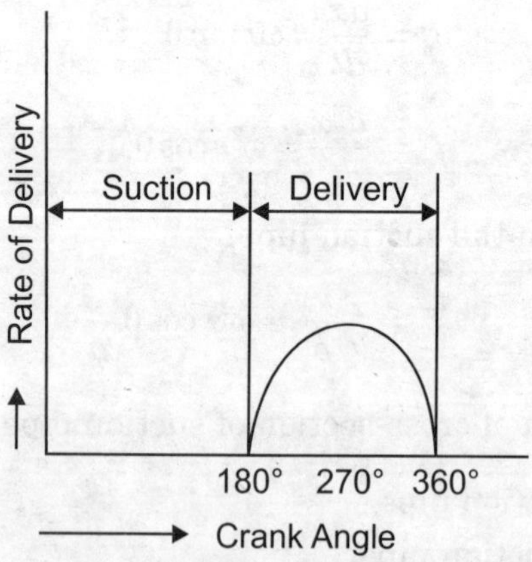


**Fig. 10.9 Rate of Delivery vs Crank Angle for single Acting Piston or Plunger Pump.**

### 10.6.2 Double–Acting, Single Cylinder Plunger Pump

The rate of delivery v's crank angle has been plotted in Fig. 10.10. Each stroke is suction cum delivery stroke. The rate of delivery against crank angle curve is a resultant of two sine curves drawn at a phase difference of 180°.

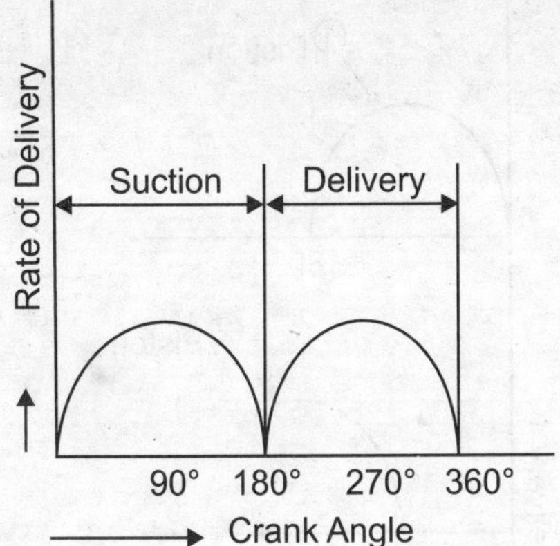


**Fig. 10.10 Rate of Delivery v Crank Angle for Double Acting Single Cylinder Piston or Plunger Pump.**

### 10.6.3 Two–Throw Pump

The rate of delivery v's crank angle has been plotted in Fig. 10.11. To make the delivery somewhat uniform, two equal cylinders with pistons connected to the perpendicular cranks of a common crank shaft are employed. The two – throw pump may be single acting or double acting. The rate of delivery curve for a double acting two – throw pump will be the resultant of two similar sine curves drawn at a phase difference of 90°.

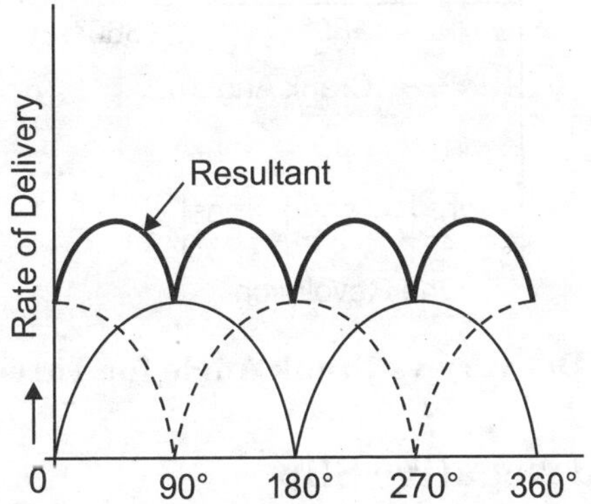


**Fig. 10.11 Rate of Delivery vs Crank Angle for Double Acting Driven by Two Cranks at Right Angles Piston or Plunger Pump.**

### 10.6.4 Three–Throw Pump

The rate of delivery curve is the resultant of three similar sine curves drawn at internals of 120°, as shown in Fig. 10.12. This gives more uniform rate of delivery.

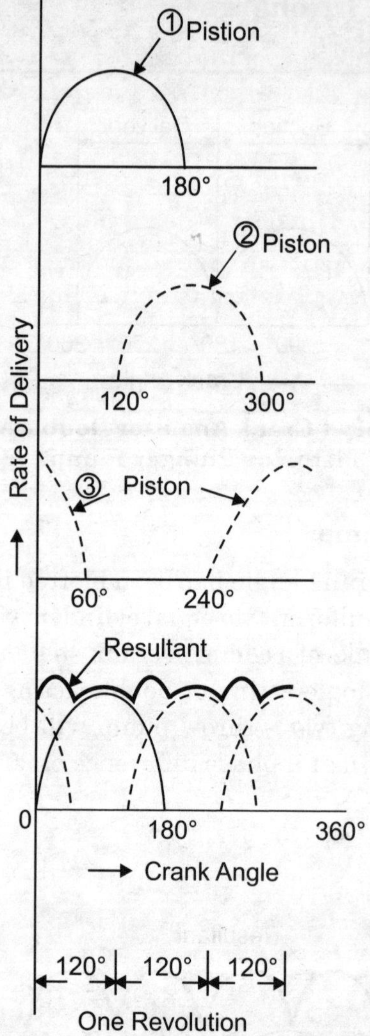


**Fig. 10.12 Rate of Delivery vs Crank Angle for Three-Throw Pump.**

## 10.7 Positive and Negative Slip

Slip is the difference of volume swept through the piston and the actual discharge per stroke.

Slip, $s$ = Volume swept per stroke – actual discharge per stroke ...(10.13)

The value of slip is generally positive. However in practice, sometimes the delivery valve opens before the suction stroke is completed, thus delivering a greater volume of water than actually swept by the piston. Hence, the slip wil be negative in such a case. Negative slip is possible in pumps having long suction pipe and short delivery pipe working at high speed.

## 10.8 Coefficient of Discharge

In a reciprocating pump, the actual discharge ($Q_{act}$) is always slightly different from theoretical discharge ($Q_{th}$) due to the following reasons:

1. Leakage through the valves, glands and piston packing.
2. Imperfect operation of the valves, glands and piston packing.
3. Partial filling of cylinder by the liquid.

The ratio between the actual discharge and theoretical discharge is called the coefficient of discharge ($C_d$) of the pump. Thus

$$\text{Coefficient of discharge,} \quad C_d = \frac{\text{Actual discharge}}{\text{Theoretical discharge}} = \frac{Q_{act}}{Q_{th}} \qquad ...(10.14)$$

The coefficient of discharge expressed as a percentage is called the volumetric efficiency of the pump.

Now
$$s = Q_{th} - Q_{act}$$
$$= Q_{th}\left(1 - \frac{Q_{act}}{Q_{th}}\right)$$
$$= Q_{th}(1 - C_d)$$

$$\therefore \quad Q_{th} = \frac{s}{1 - C_d} \qquad ...(10.15)$$

**Example 10.1** A single acting reciprocating pump, running at 50 rpm delivers 0.00736 m³/s of water. The diameter of the piston is 200 mm and stroke length 300 mm. If suction and delivery heads are 3.5m and 11.5 m respectively, calculate:

(a) Theoretical discharge,

(b) Coefficient of discharge,

(c) Percentage slip of the pump, and

(d) Power required to run the pump.

**Solution.**

Given: $N = 50$ rpm, $Q_{act} = 0.00736$ m³/s, $D = 200$ mm, $L = 300$ mm, $H_s = 3.5$ m, $H_d = 11.5$ m.

Are of piston,
$$A = \frac{\pi}{4}D^2 = \frac{\pi}{4}(0.2)^2 = 0.0314\text{m}^2$$

(a)
$$Q_{th} = \frac{ALN}{60} = 0.0314 \times 0.3 \times \frac{50}{60} = 0.00785 \text{ m}^3\text{/s}$$

(b)
$$C_d = \frac{Q_{act}}{Q_{th}} = \frac{0.00736}{0.00785} = 0.937$$

(c)
$$\% \text{ slip} = (1 - C_d) \times 100 = (1 - 0.937) \times 100 = 6.3\%$$

(d) Power required, $$P = \frac{\gamma ALN(H_s + H_d)}{60 \times 10^3}$$

$$= \frac{9.18 \times 10^3 \times 0.0314 \times 0.3 \times 50(3.5 + 11.5)}{60 \times 10^3}$$

$$= 1.155 \text{ kW}$$

**Example 10.2** A single acting reciprocating pump operating at 120 rpm has a piston diameter of 200 mm and stroke of 300 mm. The suction and delivery heads are 4m and 20m respectively. If the efficiency of both suction and delivery strokes is 75 percent, determine the power required by the pump.

**Solution.**

Given: $N = 120$ rpm, $D = 0.2$ m, $L = 0.3$ m, $\eta = 0.75$, $H_s = 4$m, $H_d = 20$ m

Power of pump, $$P = \frac{\gamma ALN(H_s + H_d)}{\eta \times 60 \times 10^3}$$

$$= \frac{9.81 \times 10^3 \times \frac{\pi}{4}(0.2)^2 \times 0.3 \times 120(4 + 20)}{0.75 \times 60 \times 10^3}$$

$$= 5.917 \text{ kW}$$

**Example 10.3** A single cylinder, single acting pump lifts 30 litres of water per second against a static head of 25 m. The cylinder bore is 350 mm and stroke is 450 mm. Determine the speed of pump if slip is 2 percent.

**Solution.**

Given: $H_s + H_d = 25$ m, $Q = 30$ litres/s $= 0.03$ m³/s, $D = 0.35$ m, $L = 0.45$ m, $s = 2\%$

Theoretical discharge,

$$Q_{th} = \frac{ALN}{60} = \frac{0.45 \times \frac{\pi}{4}(0.35)^2 \times N}{60}$$

$$= 7.216 \times 10^{-4} N \text{ m}^3/\text{s}$$

Slip, $$s = \frac{Q_{th} - Q_a}{Q_{th}}$$

$$0.02 = \frac{7.216 \times 10^{-4} \times N - 0.030}{7.216 \times 10^{-4} \times N}$$

$$0.1443 \times 10^{-4} \times N = 7.216 \times 10^{-4} \times N - 0.030$$

$$N = \frac{0.030}{7.0717 \times 10^{-4}} = 42.4 rpm$$

**Example 10.4** A single acting reciprocating pump has to deliver water against a total head of 50 m. The plunger diameter is 120 mm. What is the force exerted by the plunger during delivery stroke? Assume pump efficiency to be 80%.

**Solution.**

Given: $H = 50$ m, $D = 0.12$ m, $\eta_p = 0.8$

$$\text{Force on plunger during delivery stroke} = \frac{\gamma HA}{\eta_p}$$

$$= \frac{9.81\times10^3\times50\times\frac{\pi}{4}(0.12)^2}{0.8}$$

$$= 6934.3 \text{ N}$$

**Example 10.5** A double acting reciprocating pump delivering 80 litres per second has the following specifications:

Diameter of piston = 300 mm; Stroke = 600 mm; Speed = 80 rpm;

Suction head = 5m; Delivery head = 10m; and efficiency = 80%

Determine:

(a) the force required to operate the pump
(b) percentage slip
(c) power input.

**Solution.**

Given: $D = 0.3$m, $L = 0.6$ m, $N = 80$ rpm, $H_s = 5$m, $H_d = 10$m, $\eta_p = 0.8$, $Q_a = 80$ litres/s

Cross-sectional area of piston,

$$A = \frac{\pi}{4}D^2 = \frac{\pi}{4}(0.3)^2 = 0.0707 \text{ m}^2$$

(a) *Suction force,*

$$F_s = \gamma H_s A + \gamma H_d A = \gamma A(H_s + H_d)$$

$$= 9.81\times1000\times0.0707(5+10)$$

$$= 10403.5 \text{ N}$$

(b) *Theoretical discharge,*

$$Q_{th} = \frac{2\,LAN}{60} = \frac{2\times0.6\times0.0707\times80}{60}$$

$$= 0.11312 \text{ m}^2\text{/s or } 113.12 \text{ litres/s}$$

$$\text{Percentage slip} = \frac{Q_{th} - Q_a}{Q_{th}}\times100$$

$$= \left(\frac{113.12-80}{113.12}\right)\times100 = 29.28\%$$

(c) *Power input,*

$$P = \frac{\gamma Q_a(H_s + H_d)}{\eta_p\times10^3}$$

$$= \frac{9.81\times1000\times80\times10^{-3}\times10}{0.8\times10^3}$$

$$= 14.715 \text{ kW}$$

## 10.9 Separation and Water Hammer in Reciprocating Pump

After the pump has been primed, water follows the piston closely. For smooth working of the pump, there should be no discontinuity of flow. In other words, there should be no separation of flow of water in suction pipe, cylinder, and delivery pipe. If at any instant separation takes place, it will result in a sudden change of momentum of the moving water which has been separated from the rest. This causes impulsive force which is responsible for the phenomenon of "water hummer" in reciprocating pumps. The pump is liable to fracture under the heavy shocks sustained as a result of this.

To eliminate the cause of water hammer on the suction side of the pump, driving force must be sufficient to accelerate the mass of water following the piston at the same rate as the piston itself. Assuming that the pressure inside the cylinder is zero when the piston moves forward, total suction pressure is equal to atmospheric pressure and it has to work against the following forces:

1. Suction head, $H_s$.
2. Inertia forces equivalent to head, $H_{as}$
3. Friction forces equivalent to head, $H_f$
4. Force required to open the non-return valve, $H_{v_s}$
5. Friction in the valve, $H_{vfs}$.
6. Kinetic head due to velocity of water in suction pipe, $\frac{v_s^2}{2g}$.
7. Vapour pressure equivalent to head, $H_{vap}$

$$\therefore \quad H_{atm} = H_s + H_{as} + H_{fs} + H_{vs} + H_{vfs} + \frac{v_s^2}{2g} + H_{vap} \qquad \text{...(10.16)}$$

Separation will occur if the RHS in Eq. (10.16) becomes more than $H_{atm}$. The sum of terms on the RHS is maximum when the piston is at its inner dead centre, i.e. $\theta = 0°$, even though the velocity and frictional loss is zero at this point. The pump should be so designed that separation does not occur when the piston is at *IDC*. $H_s$ depends on the length of suction pipe ($L_s$) and speed of crankshaft ($N$). At $\theta = 0°$,

$$\left[H_{atm} - (H_s + H_{as})\right] < H_{vap}$$

or
$$\left[10.3 - (H_s + H_{as})\right] < 2.5$$

or
$$(H_s + H_{as}) < 7.8 \text{ m of water vacuum} \qquad \text{...(10.17)}$$

When the value of $(H_s + H_{as})$ is more than 7.8 m of water vacuum, the vapours are formed. Thus the flow will not remain continuous leading to separation or cavitation.

## 10.10 Maximum Speed of Reciprocating Pump

Now maximum head due to inertia forces,

$$(H_{as}) = \frac{L_s}{g} \cdot \frac{A}{a_s} \cdot \omega^2 r$$

$$= \frac{L_s}{g} \cdot \left(\frac{D}{d_s}\right)^2 \cdot \omega^2 r$$

$$= \frac{L_s}{g} \cdot \left(\frac{D}{d_s}\right)^2 \cdot \left(\frac{2\pi}{60}\right)^2 r.N^2$$

The variables $L_s$, $D$, $d_s$ and $r$ are fixed for a particular pump. Therefore,

$$(H_{as})_{m\omega} \propto N^2$$

Or

$$N \propto \sqrt{(H_{as})_{\max}} \qquad ...(10.18)$$

**Example 10.6** The piston in a reciprocating pump moves with SHM. The piston diameter is 240 mm and length of stroke is 450 mm. The suction pipe is 120 mm in diameter and 10 m long. The suction head is 3 m. Calculate the speed at which the pump can operate without separation occurring at the beginning of suction stroke. The barometer reads 9.25 m of water and $H_{vap}$ = 2.5 m of water.

**Solution.**

Given: $D$ = 0.24m, $L$ = 0.45m or $r$ = 0.225m, $d_s$ = 0.12m, $L_s$ = 10m, $H_{atm}$ = 9.25m, $H_s$ = 3m

$$H_{atm} = H_s + (H_{as})_{\max} + H_{vap}$$

$$= H_s + \frac{L_s}{g}\left(\frac{D}{d_s}\right)^2 \omega^2 r + H_{vap}$$

$$9.25 = 3 + \frac{10}{9.81}\left(\frac{0.24}{0.12}\right)^2 \left(\frac{2\pi N}{60}\right)^2 \times 0.225 + 2.5$$

$$3.75 = 0.01006\ N^2$$

$$N = 19.3 \text{ rpm}$$

**Example 10.6** A three throw pump has cylinders of 250 mm and stroke of 500 mm each. The pump is required to deliver 0.1 m$^3$/s at a head of 100 m. Friction losses are estimated to be 1 m in suction pipe and 20 m in delivery pipe. The velocity of water in delivery pipe is 1 m/s overall efficiency is 85% and the slip is 3%. Determine:

(a) Speed of the pump, and

(b) Power required to run the pump.

**Solution.**

Given: $D = 0.25$m, $L = 0.5$m, $Q_{act} = 0.1$ m³/s, $H_s + H_d = 100$m, $H_{fs} = 1$m, $H_{fd} = 20$m, $v_d = 1$ m/s, $\eta_o = 0.85$, $s = 0.03$

(a)
$$Q_{th} = \frac{3ALN}{60} = 3\times\frac{\pi}{4}(0.25)^2\times0.5\times\frac{N}{60}$$
$$= 1.2272\times10^{-3}\times N \text{ m}^3/\text{s}$$
$$Q_{act} = (1-s)Q_{th} = (1-0.03)\times1.2272\times10^{-3}\times N$$
$$= 1.19037\times10^{-3}\times N \text{ m}^3/\text{s}$$
$$= 0.1$$
$$N = 84 \text{ rpm}$$

(b) Total head,
$$H = (H_s + H_d) + H_{fs} + H_{fd} + \frac{v_d^2}{2g}$$
$$= 100 + 1 + 20 + \frac{1}{2\times9.81} = 121.05 \text{ m}$$

Power required,
$$P = \frac{\gamma QH}{\eta_o\times10^3} = \frac{9.81\times10^3\times\#\#\#\times121.05}{0.85\times10^3}$$
$$= 139.7 \text{ kW}$$

## 10.11 Indicator Diagrams

The indicator diagram is a representation of the stroke of pump piston along x-axis and pressure head along y-axis. The area of the indicator diagram represents the work done per revolution for a single acting pump. For a double acting pump, the work done per revolution is equal to twice the area of indicator diagram.

### 10.11.1 Theoretical Indicator Diagram

Fig. 10.13 shows the theoretical indicator diagram for a reciprocating pump under ideal conditions. Line $EF$ represents the atmospheric head. $AB$ line represents the suction stroke, and line $CD$ the delivery stroke.

Work done during suction stroke = area $ABFE$

Work done during delivery stroke = area $EFCD$

Work done by the pump per revolution = area $ABCD$

$$= AB\times BC$$
$$= AB\,(BF + FC)$$
$$= L\,(H_s + H_d)$$

Work done by the pump per second $= \dfrac{\gamma AN}{60}\times L(H_s + H_d)$ ...(10.19)

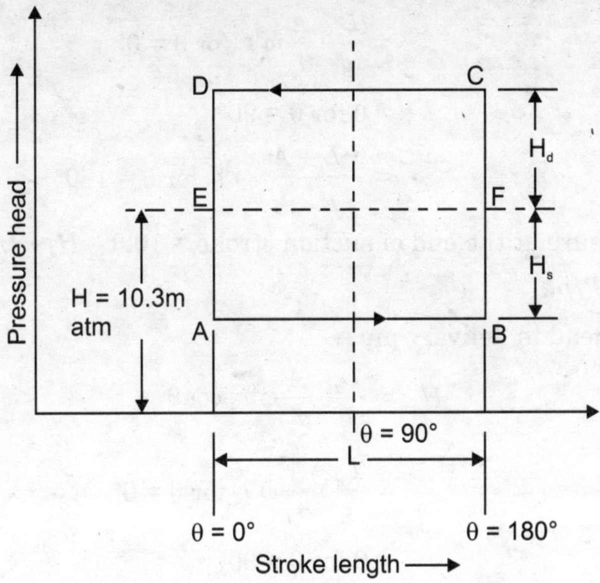


**Fig. 10.13 Theoretical indicator diagram.**

### 10.11.2 Effect of Acceleration in Suction and Delivery Pipes

#### *(a) Suction Pipe*

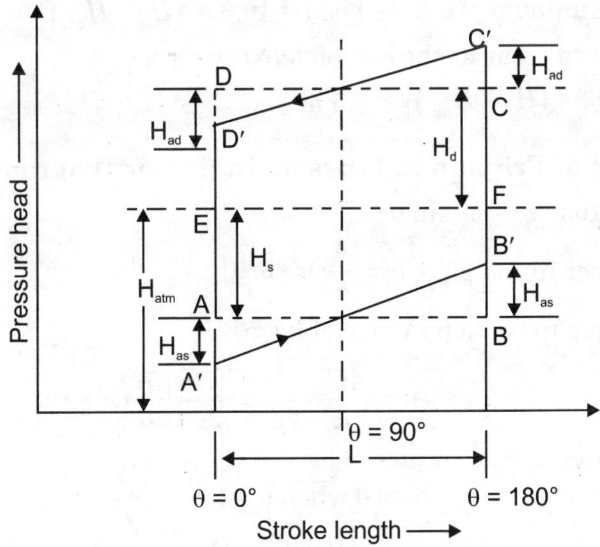


**Fig. 10.14 Effect of acceleration in pipes on indicator diagram.**

Acceleration head in suction pipe,

$$H_{as} = \frac{L_s}{g} \cdot \frac{A}{a_s} \omega^2 r \cos\theta \qquad \text{...(10.20a)}$$

$$= \frac{L_s}{g} \cdot \frac{A}{a_s} \omega^2 r \text{ for } \theta = 0° \quad ...(10.20b)$$

$$= 0 \text{ for } \theta = 90°$$

$$= -\frac{L_s}{g} \cdot \frac{A}{a_s} \cdot \omega^2 r \text{ for } \theta = 180° \quad ...(10.20c)$$

Absolute pressure at the end of suction stroke = $10.3 - H_s + H_{as}$

***(b) Delivery Pipe***

Acceleration head in delivery pipe,

$$H_{ad} = \frac{L_d}{g} \cdot \frac{A}{a_d} \omega^2 r \cos\theta$$

$$= \frac{L_d}{g} \cdot \frac{A}{a_d} \omega^2 r \text{ for } \theta = 0° \quad ...(10.21a)$$

$$= 0 \text{ for } \theta = 90° \quad ...(10.22b)$$

$$= \frac{A}{a_d} \omega^2 r \, for \, \theta = 180° \quad ...(10.21c)$$

The indicator diagram considering acceleration in suction and delivery pipes is shown in Fig. 10.14. The minimum pressure head occurs at the end of delivery stroke.

Absolute minimum pressure head = $10.3 + (H_d - H_{ad})$

For separation to occur at the end of delivery stroke,

$$[10.3 + (H_d - H_{ad})] < 2.5 \text{ m} \quad ...(10.23)$$

### 10.11.3 Effect of Friction in Pipes on Indicator Diagram

Velocity of piston, $v_p = \omega r \sin\theta$

Velocity of water in the pipe, $v = \frac{A}{a} \omega r \sin\theta$

Loss of head due to friction in pipe of length $L$,

$$H_f = \frac{4fL}{d} \cdot \frac{v^2}{2g} = \frac{4fL}{2gd}\left[\frac{A}{a}\omega r \sin\theta\right]^2 \quad ...(10.24)$$

Where $f$ = friction factor for pipe.

$$H_f = 0 \text{ when } \theta = 0$$

$$(H_f)_{max} = \frac{4fL}{2gd}\left(\frac{A}{a}\omega r\right)^2 \text{ when } \theta = 90° \quad ...(10.25)$$

$$H_f = 0 \text{ when } \theta = 180°$$

***(a) Suction Pipe***

Pressure head at the beginning of suction stroke

$$= H_{atm} - (H_s + H_{as}) \quad ...(10.27a)$$

Pressure head at the middle of suction stroke

$$= H_{atm} - (H_s + H_{fs}) \quad ...(10.26b)$$

Pressure head at the end of suction stroke

$$= H_{atm} - (H_s - H_{as}) \quad ...(10.26c)$$

***(b) Delivery Pipe***

Pressure head at the beginning of delivery stroke

$$= H_{atm} + (H_d + H_{ad}) \quad ...(10.27a)$$

Pressure head at the middle of delivery stroke

$$= H_{atm} + (H_d + H_{fd}) \quad ...(10.27b)$$

Pressure head at the end of delivery stroke

$$= H_{atm} + (H_d - H_{ad}) \quad ...(10.27c)$$

The modified indicator diagram considering accelerator and friction in pipe is shown in Fig. 10.15. The summary of acceleration and friction heads is given in Table 10.1.

**Table 10.1 Summary of accelerator and friction heads on pressure head.**

| Stroke | θ | Acceleration head | Friction head | Pressure head |
|---|---|---|---|---|
| 1. Suction | 0° | $H_{as} = \frac{L_s}{g}\frac{A}{a_s}\omega^2 r$ | 0 | $H_{atm} - (H_s + H_{as})$ |
| | 90° | 0 | $H_{fs} = \frac{4fL_s}{2gds}\left(\frac{A}{a_s}\omega r\right)^2$ | $H_{atm} - (H_s + H_{fs})$ |
| | 180° | $H_{as} = \frac{L_s}{g}\frac{A}{a_s}\omega^2 r$ | 0 | $H_{atm} - (H_s - H_{as})$ |
| 2. Delivery | 0° | $H_{ad} = \frac{L_s}{g}\frac{A}{a_d}\omega^2 r$ | 0 | $H_{atm} + (H_d + H_{ad})$ |
| | 90° | 0 | $H_{fd} = \frac{4fL_d}{2gds}\left(\frac{A}{a_d}\omega\ r\right)^2$ | $H_{atm} + (H_d + H_{fd})$ |
| | 180° | $H_{ad} = \frac{L_s}{g}\frac{A}{a_d}\omega^2 r$ | 0 | $H_{atm} + (H_d - H_{ad})$ |

Work done during suction stroke = area A′GB′FE

Work done during delivery stroke = area EFC'HD'
Total work done during on cycle = area A'GB'C'HD'

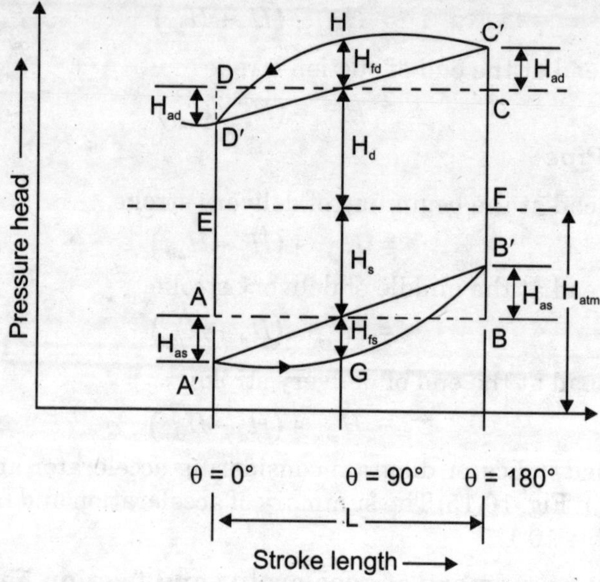


**Fig. 10.15 Effect of acceleration and friction in pipes on indicator diagram.**

## 10.12 Effect of Bent Delivery Pipe on Separation

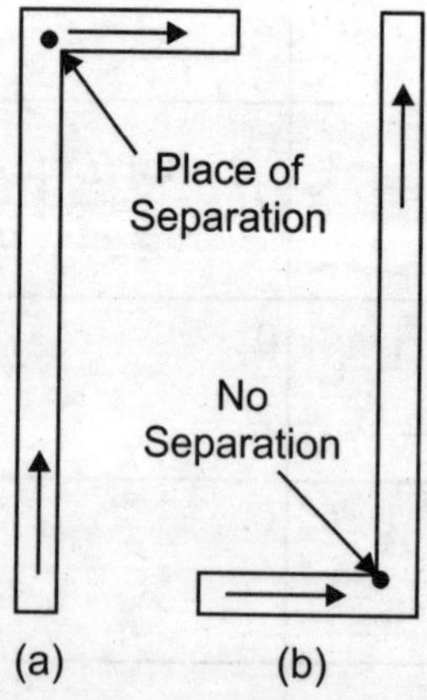


**Fig. 10.16 Separation in bend delivery pipe.**

The bent delivery pipe may be first vertical and then horizontal (Fig. 10.16a) or first horizontal and then vertical (Fig. 10.16b). In the first case, the delivery head will become zero at the bend. After the bend there is still a long horizontal pipe which will have a considerable value of acceleration head.

Therefore, separation will take place at the bend. In the second case, there is no possibility of separation at the bend because there is still considerable delivery head that is always greater than the acceleration head.

**Example 10.8** A single acting reciprocating pump has a plunger diameter of 200 mm and stroke of 300mm. The suction pipe is 100 mm in diameter and 10 m long. The water surface in the sump from which the pump draws water is 3 m below the pump cylinder axis. If the pump is working at 30 rpm, calculate the pressure head on the piston at the beginning, middle and end of suction stroke. Take $f = 0.01$, $H_{atm} = 10.3$ m.

**Solution.**

**Given:** $D = 0.2$ m, $L = 0.3$ m, or $r = 0.15$ m, $d_s = 0.1$ m, $L_s = 10$ m, $H_s = 3$ m, $N = 30$ rpm, $f = 0.01$

***(a) At the beginning of suction stroke,*** $\theta = 0°$

$$H_{as} = \frac{L_s}{g} \cdot \left(\frac{D}{d_s}\right)^2 \omega^2 r$$

$$= \frac{10}{9.81}\left(\frac{0.2}{0.1}\right)^2 \times \left(\frac{2\pi \times 30}{60}\right)^2 \times 0.15 = 6.036 \text{ m}$$

$$H_{as} = 0$$

$$H_{atm} = H_s + H_{as} + H_{fs} + H_{piston}$$

$$10.3 = 3 + 6.036 + 0 + H_{piston}$$

$H_{piston} = 1.264$ m of water absolute or 9.036 m of water vacuum

***(b) At the middle of suction stroke,*** $\theta = 90°$

$$H_{as} = 0$$

$$H_{fs} = \frac{4fL_s}{2gd_s}\left(\frac{A}{a_s}\omega r\right)^2$$

$$= \frac{4 \times 0.01 \times 10}{2 \times 9.81 \times .01}\left[\left(\frac{0.2}{0.1}\right)^2 \left(\frac{2\pi \times 30}{60} \times 0.15\right)\right]^2$$

$$= 0.724 \text{ m}$$

$$10.3 = 3 + 0 + 0.724 + H_{piston}$$

$H_{piston} = 6.576$ m of water absolute or 3.724 m of water vacuum

***(c) At the end of suction stroke,*** $\theta = 180°$

$$H_{as} = -6.036 \text{ m}, H_{fs} = 0$$

$$H_{atm} = H_s - H_{as} + H_{piston}$$

$$10.3 = -6.036 + H_{piston}$$

$H_{piston}$ = 13.336 m of water absolute or 3.036 m of water gauge.

**Example 10.9** The piston diameter of a single acting reciprocating pump is 160 mm and stroke length 400 mm. The centre of the pump is 3 m above the water surface in the sump and 20 m below the delivery water level. The pipe diameters of suction and delivery pipes is 100 mm and are 5 m and 25 m long respectively. If the pump is working at 30 rpm, calculate:

(a) The pressure heads on the piston at the beginning, middle and end of suction and delivery strokes.

(b) The power required to drive the pump.

Take atmospheric pressure as 10.3 m of water.

**Solution.**

Given: $D = 0.16$ m, $L = 0.4$ m or r = 0.2 m, $H_s = 3$ m, $H_d = 20$ m, $d_s = d_d = 0.1$ m, $L_s = 5$ m, $L_d = 25$ m, $N = 30$ rpm, $H_{atm} = 10.3$ m of water.

***(a) Suction stroke***

$$H_s = \frac{L_s}{g} \cdot \frac{A}{a_s} \omega^2 r \cos\theta$$

(i) At θ = 0°,

$$H_{as} = \frac{L_s}{g} \cdot \left(\frac{D}{d_s}\right)^2 \cdot \left(\frac{2\pi N}{60}\right)^2 \times r$$

$$= \frac{5}{9.81} \times \left(\frac{0.16}{0.1}\right)^2 \times \left(\frac{2\pi \times 30}{60}\right)^2 \times 0.2$$

$$= 2.576 \text{ m of water}$$

Pressure head on piston,

$$H_{atm} = H_s + H_{as} + H_{piston}$$

$$10.3 = 3 + 2.576 + H_{piston}$$

$H_{piston}$ = 10.3 – 5.576 = 4.724 m of water absolute or 5.576 m of water vacuum.

(*ii*) At θ = 90°,

$$H_{as} = 0$$

$$H_{atm} = H_s + H_{piston}$$

$$10.3 = 3 + H_{piston}$$

$H_{piston}$ = 7.3 m of water absolute or 3 m of water vacuum.

(*iii*) At θ = 180°,

$$H_{as} = -0.576 \text{ m of water}$$

$$H_{atm} = H_s + H_{as} + H_{piston}$$

$$10.3 = 3 - 2.576 + H_{piston}$$

$$H_{piston} = 10.3 - 0.424 = 9.876 \text{ m of water absolute or } 0.424 \text{ m of water vacuum}$$

***Delivery stroke***

$$H_{ad} = \frac{L_d}{g}\left(\frac{D}{d_d}\right)^2 \omega^2 r \cos\theta$$

(*i*) At $\theta = 0°$

$$H_{ad} = \frac{25}{9.81} \times \left(\frac{0.16}{0.1}\right)^2 \times \left(\frac{2\pi \times 30}{60}\right)^2 \times 0.2$$

$$= 12.88 \text{ m of water}$$

$$10.3 = -(20 + 12.88) + H_{piston}$$

$$H_{piston} = 45.18 \text{ m of water absolute or } 32.88 \text{ m of water gauge.}$$

(*ii*) At $\theta = 90°$, $H_{ad} = 0$

$$H_{piston} = 10.3 + 20 = 30.3 \text{ m of water absolute}$$

or 20 m of water gauge

(*iii*) At $\theta = 180°$, $H_{ad} = 12.88$ m of water

$$10.3 = -(20 + 12.88) + H_{piston}$$

$$H_{piston} = 17.42 \text{ m of water absolute or } 7.12 \text{ m of water gauge.}$$

(b)

$$Q = \frac{ALN}{60} = \frac{\pi}{4}(0.16)^2 \times 0.4 \times \frac{30}{60}$$

$$= 4.02 \times 10^{-3} \text{ m}^3/\text{s}$$

Maximum pressure head, $H_{max} = H_s + H_{as} + H_d + H_{ad}$

$$= 3 + 2.576 + 20 + 12.88$$

$$= 38.456 \text{ m of water}$$

$$P = \frac{\gamma Q H}{10^3}$$

$$= \frac{9.81 \times 10^3 \times 4.02 \times 10^{-3} \times 38.456}{10^3} = 1.51 \text{ kW}$$

**Example 10.10** The cross-sectional area of piston of a reciprocating pump is 1.5 times that of delivery pipe. The delivery pipe is 60 m long and it rises upward at a slope of 1 in 6. If the piston has an acceleration of 2.25 m/s$^2$ at the end of stroke and separation pressure is 2.5 m of water, determine whether separation will occur and, if so, at which section of the pipe. Assume simple harmonic motion and take atmospheric pressure = 10.3 m of water.

**Solution.**

Given: $A = 1.5\ a_d$, $L_d = 60$ m, slope $= \frac{1}{6}$, $f_p = \omega^2 r = 2.25$ m/s$^2$, $H_{sep} = 2.5$ m of water, $H_{atm} = 10.3$ m of water.

$$H_d = \text{slope} \times L_d = \frac{1}{6} \times 60 = 10 \text{ m}$$

The possibility of separation during delivery stroke is only at the end of stroke, where $\theta = 180°$.

$$H_{ad} = \frac{L_d}{g} \cdot \frac{\#}{\#} \omega^2 r \ \cos\theta$$

$$= -\frac{60}{9.81} \times 1.5 \times 2.25 = -\ 20.64 \text{ m}$$

Pressure head on the piston at the end of delivery stroke

$$H_{piston} = H_d + H_{ad} = 10 - 20.64$$

$$= -\ 10.64 \text{ m of water gauge}$$

$$= H_{atm} + (H_d + H_{ad}) = 10.3 - 10.64$$

$$= -\ 0.34 \text{ m of water absolute}$$

Since $H_{piston} < H_{sep}$, therefore separation shall occur

Let $L$ = length of the pipe upto the section where the separation occurs.

Then $$H_d = \frac{L}{6} \text{ m}$$

$$H_{ad} = -\frac{L}{g} \times 1.5 \times 2.25$$

$$= -\frac{L}{9.81} \times 1.5 \times 2.25 = -\ 0.344 \text{ L m}$$

$$H_{atm} + (H_d + H_{ad}) = H_{sep}$$

$$10.3 + \left(\frac{L}{6} - 0.344\ L\right) = 2.5$$

$$0.1773\ L = 7.8$$

$$L = 43.985 \text{ m}$$

**Example 10.11** The plunger diameter and stroke length of a single acting reciprocating pump are 75 mm and 150 mm respectively. It takes its supply of water from a sump 3 m below the centre line of plunger through a pipe 5 m long and 40 mm in diameter. It delivers water to a storage tank 12 m above the pump through a pipe 30 mm in diameter and 20 m long. If separation occurs at 75 kN/m$^2$ below the atmospheric pressure, find the maximum speed at which pump may be operated without separation, Assume that the plunger has simple harmonic motion and atmospheric pressure = 10.3 m of water.

**Solution.**

Given: $D = 75$ mm, $L = 150$ mm or $r = 75$ mm, $H_s = 3$ m, $H_d = 12$ m, $d_s = 40$ mm, $d_d = 30$ mm, $L_s = 5$ m, $L_d = 20$ m, $p_{sep} = -75$ kN/m², $H_{atm} = 10.3$ m of water

***Suction stroke***

The possibility of separation during suction stroke is at the beginning when $\theta = 0°$.

$$H_{as} = \frac{L_s}{g} \cdot \frac{A}{a_s} \omega^2 r$$

$$= \frac{5}{9.81} \times \left(\frac{75}{40}\right)^2 \times \left(\frac{2\pi N}{60}\right)^2 \times 75 \times 10^{-3}$$

$$= 1.4737 \times 10^{-3} \times N^2 \;\; m$$

$$H_{atm} - (H_s + H_{as}) = H_{sep}$$

$$10.3 - \left(3 + 1.4737 \times 10^{-3} \times N^2\right) = 2.5$$

$$1.4737 \times 10^{-3} \times N^2 = 4.8$$

$$N = 57 \text{ rpm}$$

***Delivery stroke***

The possibility of separation during stroke is at the end when $\theta = 180°$.

$$H_{ad} = -\frac{L_d}{g}\left(\frac{D}{d_d}\right)^2 \omega^2 r$$

$$= -\frac{20}{9.81} \times \left(\frac{75}{30}\right)^2 \times \left(\frac{2\pi N}{60}\right)^2 \times 75 \times 10^{-3}$$

$$= -0.01048\, N^2 m$$

$$H_{atm} + (H_d + H_{ad}) = H_{sep}$$

$$10.3 + \left(12 - 0.01048\, N^2\right) = 2.5$$

$$0.01048\, N^2 = 19.8$$

$$N = 43.5 \text{ rpm}$$

Maximum permissible speed = 43.5 rpm

**Example 10.12** The bore and stroke of a reciprocating pump are 200 mm and 450 mm respectively. The pump delivers water through a 80 mm pipe to a tank location at 10 m above it and 20 m horizontally from it. If separation occurs at a 22 kN/m² absolute, find the safe speed at which pump should run for the following arrangements of delivery pipe:

(a) The delivery pipe is first horizontal and then vertical, and

(b) The delivery pipe is first vertical and then horizontal.

Take $H_{atm} = 10.3$ m of water and $n = \frac{\ell}{r} = 5$

**Solution.**

**Given:** $D = 200$ mm, $L = 450$ mm or $r = 225$ mm, $L_{ver} = 10$ m, $L_{hor} = 20$ mm, $d_d = 80$ mm, $H_d = 10$ m, $p_{sep} = 22$ kN/m² abs., $H_{atm} = 10.3$ m of water, $n = 5$

$$L_d = L_{hor} + L_{ver} = 20 + 10 = 30 \text{ m}$$

$$H_{sep} = \frac{p_{sep}}{\gamma} \frac{22\times10^3}{9.81\times10^3} = 2.242 \text{ m of water}$$

At the end of delivery stroke, $\theta = 180°$, and

$$H_{ad} = -\frac{Ld}{g}\cdot\frac{A}{a_d}\omega^2 r\left(1-\frac{1}{n}\right)$$

$$= -\frac{30}{9.81}\left(\frac{200}{80}\right)^2 \times \left(\frac{2\pi N}{60}\right)^2 \times 0.225 \times \left(1-\frac{1}{5}\right)$$

$$= -0.03773 \text{ N}^2\text{m}$$

***(a) The delivery pipe is first horizontal and then vertical***

The possibility of separation is at the bend where the entire delivery head is available.

$$H_{atm} + (H_d + H_{ad}) = H_{sep}$$

$$10.3 + (10 - 0.03773\, N^2) = 2.242$$

or $$0.03773\, N^2 = 18.058$$

$$\text{N} = 21.88 \text{ rpm}$$

***(b) The delivery pipe is first vertical and then horizontal***

$$H_d = 0 \text{ at the band.}$$

$$H_{atm} + (H_d + H_{ad}) = H_{sep}$$

$$10.3 + (10 - 0.03773\, N^2) = 2.242$$

$$0.03773\, N^2 = 8.058$$

$$N = 14.6 \text{ rpm}$$

The pump can run safely without separation with arrangement (a).

**Example 10.13** A single-acting reciprocating pump has a stroke length of 150 mm and suction pipe is 7 m long. The ratio of suction pipe diameter to piston is 0.75. The water level in the sump is 2.5 m below the axis of the pump cylinder. The suction pipe diameter is 75 mm. If the crank is running at 80 rpm, determine the pressure head on the piston at the beginning, middle and end of suction stroke. Take friction coefficient, $f = 0.015$.

**Solution.**

Given: $L$ = 150 mm or $r$ = 75 mm, $L_s$ = 7 m, $\frac{d_s}{D}$ = 0.75, $H_s$ = 2.5 m, $d_s$ = 75 mm, $N$ = 80 rpm, $f$ = 0.015

$$H_{as} = \frac{L_s}{g}\cdot\frac{A}{a_s}\omega^2 r\cos\theta$$

$$= \frac{7}{9.81}\times\left(\frac{1}{0.75}\right)^2\times\left(\frac{2\pi\times 80}{60}\right)^2\times 0.075\times\cos\theta$$

$$= 6.6774\cos\theta \text{ m}$$

$$H_{fs} = \frac{4fL_s}{2gd_s}\left(\frac{A}{a_s}\omega r\sin\theta\right)^2$$

$$= \frac{4\times 0.015\times 7}{2\times 9.81\times 0.075}\left[\left(\frac{1}{0.75}\right)^2\times\frac{2\pi\times 80}{60}\times 0.075\times\sin\theta\right]^2$$

$$= 0.356\sin^2\theta$$

***(a) At the beginning of suction stroke,*** $\theta = 0°$

$$H_{as} = 6.6774 \text{ m}$$

$$H_{fs} = 0$$

$$H_{piston} = H_{atm} - (H_s + H_{as}) = 10.3 - (2.5 + 6.6774)$$

$$= 1.1226 \text{ m of water absolute}$$

or 9.1774 m of water vacuum

***(b) At the middle of suction stroke,*** $\theta = 90°$

$$H_{as} = 0$$

$$H_{fs} = 0.356 \text{ m of water}$$

$$H_{piston} = H_{atm} - (H_s + H_{fs}) = 10.3 - (2.5 + 0.356)$$

$$= 7.444 \text{ m of water absolute}$$

or 2.856 m of water vacuum.

***(c) At the end of suction stroke,*** $0 = 180°$

$$H_{as} = -6.6774 \text{ m}$$

$$H_{fs} = 0$$

$$H_{piston} = H_{atm} - (H_s + H_{as}) = 10.3 - (2.5 - 6.6774)$$

$$= 14.4774 \text{ m of water absolute}$$

or 4.1774 m of water vacuum.

**Example 10.14** The bore and stroke of a double-acting single-cylinder reciprocating pump, running at 30 rpm, are 200 mm and 400 mm respectively. The pump draws water from a sump 1.5 m below the pump through a suction pipe 100 mm in diameter and 3 m long. The water is delivered to a tank 30 m above the pump through a delivery pipe 100 mm diameter and 40 m long. Assuming the motion of the piston to be simple harmonic, determine the net force due to fluid pressure on the piston when it has moved through a distance of 100 mm from the inner dead centre.

Take friction factor for both the suction and delivery pipes as 0.01. Neglect the size of piston rod.

**Solution.**

Given: $D = 200$ mm, $L = 400$ mm or $r = 200$ mm, $H_s = 1.5$ m, $d_s = 100$ mm, $L_s =$ 3 m, $N = 30$ rpm, $H_d = 30$ m, $d_d = 100$ mm, $L_d = 40$ m, $x = 100$ mm, $f = 0.01$

***Suction Side***

$$x = r(1 - \cos\theta)$$

$$100 = 200(1 - \cos\theta)$$

$$\cos\theta = 0.5$$

$$\theta = 60°$$

$$H_{as} = \frac{L_s}{g}\cdot\frac{A}{a_s}\omega^2 r\cos\theta$$

$$= \frac{3}{9.81}\times\left(\frac{200}{100}\right)^2\times\left(\frac{2\pi\times 30}{60}\right)^2\times 0.2\times 0.5$$

$$= 1.207 \text{ m}$$

$$H_{fs} = \frac{4fL_s}{2gd_s}\left(\frac{A}{a_s}\omega r\sin\theta\right)^2$$

$$= \frac{4\times 0.01\times 3}{2\times 9.81\times 0.1}\times\left[\left(\frac{200}{100}\right)^2\times\frac{2\pi\times 30}{60}\times 0.2\times\sin 60°\right]$$

$$= 0.2897 \text{ m}$$

$$H_{piston} = H_s + H_{as} + H_{fs}$$

$$= 1.5 + 1.207 + 0.2897 = 2.9967 \text{ m of water vacuum}$$

or $\quad 10.3 - 2.9167 = 7.3033$ m of water absolute

Force on the piston from suction side $= \gamma H A_{piston}$

$$= 9.81\times 10^3\times 7.3033\times\frac{\pi}{4}(0.2)^2 = 1250.8 \text{ N}$$

***Delivery Side***

Angular displacement from O.D.C. for delivery stroke = 180° – 0 from I.D.C. = 180° – 60°= 120°

$$H_{ad} = \frac{L_d}{g}\cdot\frac{A}{a_d}\omega^2 r\cos\theta$$

$$= \frac{40}{9.81} \times \left(\frac{200}{100}\right)^2 \times \frac{(2\pi \times 30)^2}{60} \times 0.2 \times \cos 120^\circ$$

$$= -16.097 \text{ m}$$

$$H_{fd} = \frac{4fL_d}{2gd_d}\left(\frac{A}{a_d}\omega r \sin 0\right)^2$$

$$= \frac{4 \times 0.01 \times 40}{2 \times 9.81 \times 0.1}\left[\left(\frac{200}{100}\right)^2 \times \frac{2\pi \times 30}{60} \times 0.2 \times \sin 120^\circ\right]^2$$

$$= 3.863 \text{ m}$$

$$H_{piston} = H_d + H_{ad} + H_{fd} = 30 - 16.097 + 3.863$$

$$= 17.766 \text{ m of water vacuum}$$

or $10.3 + 17.766 = 28.066$ m of water absolute

$$\text{Force on the piston from delivery side} = 9.81 \times 10^3 \times 28.066 \times \frac{\pi}{4}(0.2)^2$$

$$= 8649.7 \text{ N}$$

$$\text{Net force on piston} = 8649.7 - 2250.8 = 6398.9 \text{ N}$$

## 10.13 Air Vessel

Air vessel is a closed chamber fitted on the suction as well as on the delivery side, near the pump cylinder, to reduce the acceleration head. The functions of the air vessel are:

### *On the suction side*

1. Reduces the possibility of separation.
2. Pump can be run at a higher speed.
3. Length of the suction pipe below the air vessel can be increased.

### *On the delivery side*

1. A large amount of power consumed in supplying acceleration head can be saved.
2. Constant rate of discharge can be ensured.

### *Working of Air Vessel*

The air vessel in a reciprocating pump acts like a flywheel of an engine. The top of the vessel contains compressed air which can contract or expand to absorb most of the pressure fluctuations. When the pressure in the air rises, water in excess of the mean discharge is forced into the air vessel, thereby compressing the air in the air vessel. When the water pressure in the pipe falls, the compressed air forces the excess water out into the pipe.

The air vessel acts like an intermediate reservoir. On the suction side, the water first accumulates into the air vessel and is then transferred to the pump cylinder. On the delivery side, the water first goes to the vessel and then flows into the delivery pipe with a uniform velocity. The column of water, which now fluctuates, is only between the pump cylinder and the air vessels. This column of water is very small due to the vessels being fitted as near to the pump cylinder as possible.

### 10.13.1 Suction in Pump with Air Vessel

Two air vessels fitted on suction side and delivery side of a single-acting reciprocating pump are shown in Fig. 10.16. The suction takes place in two steps. First the water flows through the suction pipe into the air vessel and then it is raised from the air vessel to the cylinder.

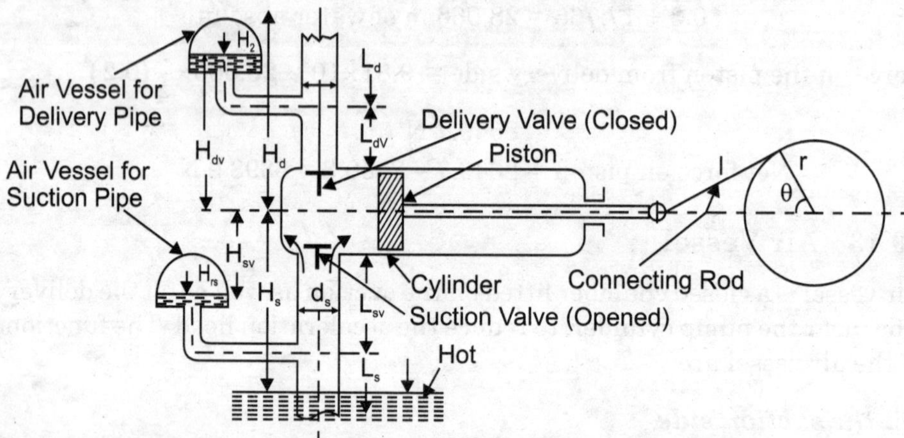


**Fig. 10.17 Air Vessel on Suction and Delivery Sides.**

Let $H_{sv}$ = static head between the cylinder centre line and water level in air vessel.

$L_{sv}$ = length of suction pipe from cylinder to centre line of inlet/outlet pipe to air vessel.

$H_{v1}$ = pressure head acting on water surface in air vessel

Head in air vessel = static head + acceleration head + frictional head + valve raising head + valve frictional head + vapour pressure + kinetic head

$$H_{v1} = H_{sv} + H_{asv} + H_{fsv} + H_{vs} + H_{vfs} + H_{vap} + \frac{\upsilon_{sv}^2}{2g} \quad \text{...(10.28)}$$

(*i*) At the beginning of suction stroke, $\theta = 0°$ and $H_{as\upsilon}$ is maximum, whereas $H_{fsv} = 0$, because $\upsilon_{sv} = 0$.

$$\therefore \quad H_{v1} = H_{sv} + (H_{asv})_{max} + 0 + H_{vs} + H_{vfs} + H_{vap}$$

$$= H_{sv} + \frac{L_{sv}}{g} \cdot \frac{A}{a_s} \omega^2 r + H_{vs} + H_{vfs} + H_{vap} \qquad \text{...(10.29)}$$

For the portion of pipe between the sump and the air vessel, there is no fluctuating head as the flow is uniform.

Atmospheric head = static head + Head in air vessel
+ frictional head due to uniform flow
+ kinetic head due to uniform velocity

$$H_{atm} = (H_s + H_{sv}) + H_{v1} + \frac{4fL_s}{d_s} \cdot \frac{v_s^2}{2g} + \frac{v_s^2}{2g} \qquad \text{...(10.30)}$$

Adding eqs. (10.29) and (10.30), we have

$$H_{atm} = H_s + \frac{4fL_s}{d_s} \cdot \frac{v_s^2}{2g} + \frac{L_{sv}}{g} \cdot \frac{A}{a_s} \omega^2 r + H_{vs} + H_{vfs} + \frac{v_s^2}{2g} \qquad \text{...(10.31)}$$

From Eq. (10.31) it may be observed that acceleration head had reduced drastically as $L_{sv}$ is much smaller than $(L_{sv} + L_s)$. Thus, the pump can be placed at a greater height above the available water surface.

Neglecting valve losses, pressure head on the piston at the beginning of suction stroke becomes,

$$H_{piston} = H_s + \frac{4fL_s}{d_s} \cdot \frac{v_s^2}{2g} + \frac{L_{sv}}{g} \cdot \frac{A}{a_s} \omega^2 r + \frac{v_s^2}{2g} \qquad \text{...(10.32)}$$

(*ii*) At the middle of suction stroke, $\theta = 90°$

and $\qquad H_{asv} = 0$

but $H_{fsv} \neq 0$, $\frac{v_s^2}{2g} \neq 0$.

$$H_{piston} = H_s + H_{fs} + H_{vfs} + (H_{KE})_{sv} + (H_{KE})_s$$

$$= H_s + \frac{4fL_s}{d_s} \cdot \frac{v_s^2}{2g} + \frac{4fL_s}{2gd_s}\left(\frac{A}{a_s}\omega r\right)^2 + \frac{v_{sv}^2}{2g} + \frac{v_s^2}{2g} \qquad \text{...(10.33)}$$

(*iii*) At the end of suction stroke, $\theta = 180°$, and $H_{as}$ is negative.

$$H_{piston} = H_s + H_{fs} - (H_{asv})_{max} + H_{KE}$$

$$= H_s + \frac{4fL_s}{d_s} \cdot \frac{v_s^2}{2g} - \frac{L_{sv}}{g} \cdot \frac{A}{a_s} \omega^2 r + \frac{v_s^2}{2g} \qquad \text{...(10.34)}$$

**Example 10.15** The diameter of plunger in a reciprocating pump is 250 mm and stroke 450 mm. The suction pipe is 125 mm in diameter and 12 m long. The suction lift is 3 m. The pump is fitted with an air vessel on the suction side such that length of suction pipe from the cylinder to the air vessel is 1.5 m and the mean water level in the air vessel is 0.6 m below the centre line of cylinder. Find the speed at which the pump can operate without separation at the beginning of suction stroke. Assume $f = 0.01$.

**Solution.**

Given: $D = 250$ mm, $L = 450$ mm or $r = 225$ mm, $d_s = 125$ mm, $L_{sv} = 1.5$ m, $L_s = 12 - 1.5 = 10.5$ m, $H_s = 3$ m, $H_{sv} = 0.6$ m, $f = 0.01$

$$v_s = \frac{Q}{a_s} = \frac{ALN/60}{\frac{\pi}{4}d_s^2}$$

$$= \frac{\frac{\pi}{4}(0.25)^2 \times 0.45 \times N/60}{\frac{\pi}{4}\times(0.125)^2} = 0.03\text{ N m/s}$$

$$H_{atm} = H_s + (H_{asv})_{\max} + H_{sep} + H_{vfs} + \frac{v_s^2}{2g}$$

$$= H_s + \frac{L_{sv}}{g}\cdot\left(\frac{A}{a_s}\right)\omega^2 r + H_{sep} + \frac{4fL_s}{d_s}\times\frac{v_s^2}{2g} + \frac{v_s^2}{2g}$$

$$10.3 = 3 + \frac{1.5}{9.81}\times\left(\frac{250}{125}\right)^2 \times 0.225 + 2.5$$

$$+ \frac{4\times0.01\times10.5}{0.125}\times\frac{(0.03N)^2}{2\times9.81} + \frac{(0.03N)^2}{2\times9.81}$$

$$= 3 + 2.5 + (0.001509 + 0.000154 + 0.0000459)N^2$$

$$N^2 = \frac{4.8}{0.0017089} = 2808.8$$

$$N = 53\text{ rpm}$$

**Example 10.16** The diameter and stroke of a single acting reciprocating pump are 300 mm and 600 mm respectively. The pump takes its supply of water from a sump 3.5 m below the pump axis through a pipe 8 m long and 200 mm diameter. If separation occurs at 2.5 m of water absolute, determine:

(a) the speed at which separation may take place at the beginning of suction stroke, and

(b) the speed of the pump if an air vessel is fitted on the suction side 2.5 m above the sump water level.

The atmospheric pressure head = 10.3 m of water, and friction coefficient, $f = 0.01$.

**Solution.**

Given: $D = 300$ mm, $L = 600$ mm or $r = 300$ mm, $H_s = 3.5$ m, $L_s = 8$ m, $d_s =$ 200 mm, $H_{sep} = 2.5$ m, $L_{sv} = 8 - 2.5 = 5.5$ m

(a) $$H_{atm} - (H_s + H_{as}) = H_{sep}$$

$$10.3-\left[3.5+\frac{L_s}{g}\left(\frac{D}{d_s}\right)^2\left(\frac{2\pi N}{60}\right)^2\times 0.3\right]=2.5$$

$$6.8-\frac{8}{9.81}\times\left(\frac{300}{200}\right)^2\times\left(\frac{2\pi N}{60}\right)^2\times 0.3=2.5$$

$$6.0364\times 10^{-3}\times N^2=4.3$$

$$N=26.7 \text{ rpm}$$

(b) Now $L_s = 2.5$ m and $L_{sv} = 5.5$ m

$$H_{atm}=H_s+(H_{asv})_{max}+H_{sep}+H_{vfs}+\frac{v_s^2}{2g}$$

$$v_s=\frac{ALN/60}{\frac{\pi}{4}d_s^2}=\frac{\frac{\pi}{4}(0.3)^2\,0.6\times N}{60\times\frac{\pi}{4}(0.2)^2}$$

$$=0.0225 \text{ N m/s}$$

$$H_{atm}=H_s+\frac{L_{sv}}{g}\cdot\left(\frac{D}{d_s}\right)^2\omega^2 r+H_{sep}+\frac{4fL_s}{d_s}\frac{v_s^2}{2g}+\frac{v_s^2}{2g}$$

$$10.3=3.5+\frac{5.5}{9.81}\left(\frac{300}{200}\right)^2\left(\frac{2\pi N}{60}\right)^2\times 0.3+2.5$$

$$+\frac{4\times 0.01\times 2.5}{0.2}\times\frac{(0.0225\,N)^2}{2\times 9.81}+\frac{(0.0225N)^2}{2\times 9.81}$$

$$4.3=(0.00415+0.0000129+0.0000258)N^2$$

$$=0.0041887N^2$$

$$N^2=1026.57$$

$$N=32 \text{ rpm}$$

### 10.13.2 Delivery in Pump with Air Vessel

Let $L_{dv}$ = length of delivery pipe between cylinder and air vessel

$L_d$ = length of delivery pipe beyond air vessel.

By fitting an air vessel on the delivery side fluctuation of pressure takes place between the cylinder and the air vessel only and the flow becomes uniform beyond the air vessel. Therefore, the velocity of water will be uniform in length $L_d$.

Frictional head due to uniform velocity in delivery pipe of length $L_d$,

$$H_{fd}=\frac{4fl_d}{d_d}\cdot\frac{v_d^2}{2g}$$

Where $v_d = \frac{Q}{a_d} = \frac{4ALN}{60 \times \pi d_d^2}$ for single acting pump.

Acceleration head in the delivery pipe between cylinder and air vessel,

$$H_{adv} = \frac{L_{dv}}{g} \cdot \frac{A}{a_d} \cdot \omega^2 r \cos\theta$$

(*i*) At the beginning of delivery stroke, $\theta = 0°$, and pressure head on the piston is,

$$H_{piston} = H_d + H_{fd} + H_{adv} + H_{KE}$$

$$= H_d + \frac{4fL_d}{d_d} \cdot \frac{v_d^2}{2g} + \frac{L_{dv}}{g} \cdot \frac{A}{a_d} \cdot \omega^2 r + \frac{v_d^2}{2g} \qquad ...(10.35)$$

(*ii*) At the middle of the delivery stroke, $\theta = 90°$ and $H_{adv} = 0$,

$$H_{fdv} = \frac{4fL_{dv}}{d_d \times 2g}\left(\frac{A}{a_d \omega r}\right)^2,$$

and $v_{dv} = \frac{A}{a_d} \cdot \omega r$ are maximum values.

Pressure head on the piston at the middle of stroke,

$$H_{piston} = H_d + H_{fd} + \left(H_{fdv}\right)_{\max} + \frac{v_{dv}^2}{2g} + \frac{v_d^2}{2g}$$

$$= H_d + \frac{4fL_d}{d_d} \cdot \frac{v_d^2}{2g} + \frac{4fL_{dv}}{d_d \times 2g}\left(\frac{A}{a_d}\omega r\right)^2 + \frac{v_{dv}^2}{2g} + \frac{v_d^2}{2g} \qquad ...(10.36)$$

(*iii*) At the end of delivery stroke, $\theta = 180°$ and $H_{adv} = -\frac{L_{dv}}{g} \cdot \frac{A}{a_d} \cdot \omega^2 r$, $H_{fdv} =$ 0, and $v_{dv} = 0$,

Pressure head on the piston at the end of delivery stroke,

$$H_{piston} = H_d + H_{fd} - (H_{adv})_{\max} + \frac{v_d^2}{2g}$$

$$= H_d + \frac{4fL_d}{d_d} \cdot \frac{v_d^2}{2g} - \frac{L_{dv}}{g} \cdot \frac{A}{a_d} \omega^2 r + \frac{v_d^2}{2g} \qquad ...(10.37)$$

### 10.13.3 Maximum Speed of Pump with Air Vessel

The maximum speed of a reciprocating pump is determine by finding the vapour pressure (or separation) head, the value of which should not exceed 2.5 m of water absolute. Neglecting valve losses, the vapour pressure head at the beginning of suction stroke, from Eq. (10.32) is,

$$H_{vap} = H_{atm} - \left(H_s + H_{fs} + H_{asv} + \frac{v_s^2}{2g}\right)$$

$$= H_{atm} - \left( H_s + \frac{4fL_s}{d_s} \cdot \frac{v_s^2}{2g} + \frac{L_{sv}}{g} \cdot \frac{A}{a_s} \omega^2 r + \frac{v_s^2}{2g} \right) \quad \text{...(10.38)}$$

For uniform flow, $Q = \dfrac{ALN}{60}$ m$^3$/s for single acting pump

$$v_s = \frac{Q}{a_s} = \frac{ALN}{60} \times \frac{4}{\pi d_s^2}$$

$$= \frac{\pi D^2 \times 2r \times N}{4 \times 60} \times \frac{4}{\pi d_s^2} = \left(\frac{D}{d_s}\right)^2 \times \frac{rN}{30} \quad \text{...(10.39)}$$

$$= \left(\frac{D}{d_s}\right)^2 \times \frac{rN}{15} \text{ for double acting pump} \quad \text{...(10.39a)}$$

### 10.13.4 Power Required to Drive the Pump Fitted with Air Vessels

Total head against which the pump has to work

$$H = H_s + H_d + H_{fs} + H_{fd} + H_{asv} + H_{adv} + \frac{v_s^2}{2g} + \frac{v_d^2}{2g}$$

$$= (H_s + H_d) + \frac{4fL_s}{d_s} \cdot \frac{v_s^2}{2g} + \frac{4fL_d}{d_d} \cdot \frac{v_d^2}{2g}$$

$$+ \frac{L_{sv}}{g} \cdot \frac{A}{a_s} \cdot \omega^2 r + \frac{L_{dv}}{g} \cdot \frac{A}{a_d} \cdot \omega^2 r + \frac{v_s^2}{2g} + \frac{v_d^2}{2g}$$

Power required by the pump, $P = \dfrac{\gamma Q H}{10^3 \times \eta_o}$ ...(10.40)

Where $\eta_o$ = overall efficiency of pump

**Example 10.17** The plunger diameter of a double acting reciprocating pump is 300 mm and stroke length is 400 mm. The diameter of delivery pipe is 150 mm. The length of suction and delivery pipes are 10 m and 450 m respectively. An air vessel has been fitted near the pump cylinder on the delivery side. The height of the pump above the level of water in the sump is 3 m. The plunger moves with SHM and makes 60 strokes per minute. The water is raised through a net head of 75 m. Find the diameter of the suction pipe in order to avoid the separation at the beginning of suction stroke. The separation occurs when the pressure in the cylinder falls below 2.5 m of water absolute. Calculate the power required by the pump neglecting all losses except due to friction in delivery pipe. Take $f = 0.01$.

**Solution.**

Given: $D = 300$ mm, $L = 400$ mm or $r = 200$ mm, $d_d = 150$ mm, $L_s = 10$ m, $L_d = 450$ m, $H_s = 3$ m, $H_s + H_d = 75$ m, $N = 60$ strokes/min or 30 rpm, $H_{vap} = 2.5$ m, $f = 0.01$

Neglecting kinetic head and valve losses,

$$H_{atm} = H_s + H_{fs} + H_{as} + H_{vap}$$

$$10.3 = 3 + \frac{4fL_s v_s^2}{2gd_s} + \frac{L_s}{g} \cdot \frac{A}{a_d} \omega^2 r + H_{vap}$$

$v_s = 0$ at $\theta = 0°$

$$10.3 = 3 + 0 + \frac{10}{9.81} \times \left(\frac{0.3}{d_s}\right)^2 \times \left(\frac{2\pi \times 30}{60}\right)^2 \times 0.2 + 2.5$$

$$4.8 = \frac{0.181}{d_s^2}$$

$$d_s = 194 \text{ mm}$$

$$Q = \frac{2ALN}{60} = \frac{2 \times \pi (0.3)^2 \times 0.4 \times 30}{4 \times 60}$$

$$= 0.0283 \text{ m}^3/\text{s}$$

$$v_d = \frac{Q}{a_d} = \frac{4 \times 0.0283}{\pi (0.15)^2} = 1.6 \text{ m/s}$$

$$H_{fd} = \frac{4fL_d v_d^2}{2gd_d} = \frac{4 \times 0.01 \times 450 \times (1.6)^2}{2 \times 9.81 \times 0.15}$$

$$= 15.657 \text{ m}$$

Total head,

$$H = (H_s + H_d) + (H_{fs} + H_{fd}) + (H_{as} + H_{ad})$$

$$= 75 + (0 + 15.657) + (4.8 + 0)$$

$$= 95.457 \text{ m of water}$$

$$P = \frac{\gamma QH}{10^3} = \frac{9.81 \times 10^3 \times 0.0283 \times 95.45}{10^3}$$

$$= 26.5 \text{ kW}$$

**Example 10.18** A single-acting reciprocating pump is to raise a liquid of density 1200 kg/m³ through a vertical height of 12 m from ³ m below pump axis to 9 m above it. The plunger moves with simple harmonic motion, has diameter 125 mm and stroke 225 mm. The suction and delivery pipes are of 75 mm diameter and 4 m and 15 m long respectively. There is a long air vessel placed on the delivery pipe near the pump axis. There is no air vessel on the suction pipe. If separation takes place below 0.85 bar below atmospheric pressure, determine:

(a) Maximum speed with which the pump can run without separation taking place, and

(b) Power required to drive the pump.

Take friction factor, $f = 0.02$

**Solution.**

Given: $\rho = 1200$ kg/m³, $H_s = 3$ m, $H_d = 9$ m, $D = 125$ mm, $L = 225$ mm or $r = 112.5$ mm, $d_s = d_d = 75$ mm, $L_s = 4$ m, $L_d = 15$ m, $p_{sep} = 0.85$ bar, $f = 0.02$

Since the air vessel is placed near the pump on the delivery pipe, therefore, the velocity in the delivery pipe will be uniform so that $H_{ad} = 0$.

$$H_{sep} = \frac{p_{sep}}{\gamma} = \frac{p_{sep}}{g\rho} = \frac{0.85\times10^5}{9.81\times1200}$$

= 7.22 m of water below atmospheric.

(a) The possibility of separation to take place at the beginning of suction stroke. There is no air vessel on the suction side, therefore, there will be acceleration head in the suction pipe.

$$H_{piston} = H_s + H_{as} \text{ below atmospheric}$$

$$\therefore \quad H_s + H_{as} = H_{sep}$$

$$3 + H_{as} = 7.22$$

$$H_{as} = 4.22 \text{ m}$$

or

$$\frac{L_s}{g}\cdot\frac{A}{a_s}\omega^2 r = H_{as}$$

$$\frac{4}{9.81}\times\left(\frac{125}{75}\right)^2\times\left(\frac{2\pi N}{60}\right)^2\times112.5\times10^{-3} = 4.22$$

$$N = 54.95 \text{ rpm}$$

(b)

$$Q = \frac{ALN}{60} = \frac{\pi}{4}(0.125)^2\times0.225\times\frac{54.95}{60}$$

$$= 2.528\times10^{-3} \text{ m}^3\text{/s}$$

$$v_d = \frac{Q}{a_d} = \frac{4\times2.528\times10^{-3}}{\pi\times(0.075)^2} = 0.572 \text{ m/s}$$

$$H_{fd} = \frac{4fL_d\,v_d^2}{2g\,d_d} = \frac{4\times0.02\times15\times(0.572)^2}{2\times9.81\times0.075}$$

$$= 0.267 \text{ m}$$

$H_{fs}$ is maximum at $\theta = 90°$, because $v_s = \frac{A}{a_s}\omega r\sin\theta$

$$H_{fs} = \frac{4fL_s}{2gd_s}\left(\frac{A}{a_s}\omega r\right)^2$$

$$= \frac{4\times0.02\times4}{2\times9.81\times0.075}\left[\left(\frac{125}{75}\right)^2\times\frac{2\pi\times54.95}{60}\times112.5\times10^{-3}\right]^2 = 0.703 \text{ m}$$

Total head on the piston, $H = H_s + H_d + \frac{2}{3}H_{fs} + H_{fd}$

$$= 3 + 9 + \frac{2}{3}\times 0.703 + 0.267 = 12.73 \text{ m}$$

$$P = \frac{\gamma Q H}{10^3} = \frac{9.81\times 1200\times 2.528\times 10^{-3}\times 12.79}{10^3} = 0.378 \text{ kW}$$

### 10.13.5 Work Saved by Air Vessel in Overcoming Pipe Friction

The acceleration heads in both the suction and delivery pipes are reduced with air vessels. This result in saving of some power. The energy saved on the suction side is small as compared to that on the delivery side because the length of the suction side is very small in comparison to the length of delivery pipe. Similarly the frictional head saved shall be small on the suction side, and can be neglected.

***(a) Single acting pump***

Consider a single acting pump, the work done per second in overcoming pipe friction on delivery side without air vessel

$$W_1 = \gamma Q\cdot\left(\frac{2}{3}H_{fd}\right) = \gamma Q\cdot\frac{2}{3}\cdot\frac{4fL_d}{d_d\cdot 2g}\left(\frac{A}{a_d}\omega r\right)^2 \qquad \ldots(10.44)$$

The factor $\frac{2}{3}$ comes from the fact that the motion is simple harmonic and the curve representing frictional head is a parabola, whose area under the curve is $\frac{2}{3}$rd of the area of the rectangle contained by it.

Work done per second in overcoming pipe friction on delivery side with air vessel

$$W_2 = \gamma Q\, H_{fd} = \gamma Q\cdot\frac{4fL_d}{d_d}\cdot\frac{v_d^2}{2g}$$

Where
$$v_d = \frac{ALN}{60\times a_d} = \frac{A}{a_d}\cdot 2r\cdot\frac{\omega}{2\pi} = \frac{A}{a_d}\times\frac{r\omega}{\pi}$$

$\therefore$
$$W_2 = \gamma Q\cdot\frac{4fL_d}{d_d\times 2g}\left[\frac{A}{a_d}\cdot\frac{r\omega}{\pi}\right]^2$$

Work saved by fitting air vessel on the delivery side of pump

$$W = W_1 - W_2$$

$$= \gamma Q\cdot\frac{4fL_d}{d_d\times 2g}\left(\frac{A}{a_d}\cdot\omega r\right)^2\left[\frac{2}{3} - \frac{1}{\pi^2}\right]$$

$$\text{Percentage saving in work/s} = \frac{W}{W_1}\times 100 = \left(\frac{\frac{2}{3}-\frac{1}{\pi^2}}{\frac{2}{3}}\right)\times 100$$

$$= \left(1-\frac{3}{2\pi^2}\right)\times 100 = 84.8\%$$

Similarly it may be shown that the same percentage of work can be saved by fitting an air vessel on the suction side of a single acting pump.

**Example 10.19** An air vessel is fitted on the delivery side of a single acting reciprocating pump. The piston moves with a simple harmonic motion. The diameter and stroke of the piston are 300 mm and 600 mm respectively. The delivery pipe is 150 mm in diameter and 50 m long. Determine the power saved in overcoming friction in the delivery pipe by the air vessel. The speed of pump is 120 rpm. Take $f = 0.01$.

**Solution.**

**Given:** $D = 300$ mm, $L = 600$ mm or $r = 300$ mm, $d_d = 150$ mm, $L_d = 50$ m, $N = 120$ rpm, $f = 0.01$.

Without air vessel, work done/s on delivery side, in overcoming friction,

$$W_1 = \frac{2}{3}\gamma Q\, H_{fd} = \frac{2}{3}\gamma Q \times \frac{4fL_d}{2gd_d}\left(\frac{A}{a_d}\cdot\omega r\right)^2$$

With air vessel, work done/s on delivery side, in overcoming friction,

$$W_2 = \gamma Q\, H_{fd} = \gamma Q\cdot\frac{4fL_d}{d_d}\cdot\frac{\upsilon_d^2}{2g}$$

Where $$\upsilon_d = \frac{Q}{a_d} = \frac{ALN}{60\times a_d} = \frac{A\times 2r\times\omega}{2\pi\times a_d} = \frac{A}{a_d}\cdot\frac{r\omega}{\pi}$$

$$\therefore \quad W_2 = \gamma Q\cdot\frac{4fL_d}{d_d}\cdot\frac{1}{2g}\left(\frac{A}{a_d}\cdot\frac{r\omega}{\pi}\right)^2$$

Work saved/s, $$W = W_1 - W_2 = \gamma Q\cdot\frac{4fL_d}{2g\,d_d}\left(\frac{A}{a_d}\omega r\right)^2\left(\frac{2}{3}-\frac{1}{\pi^2}\right)$$

$$= 9.81\times 10^3\times\left[\frac{\pi}{4}\times(0.3)^2\times 0.6\times\frac{120}{60}\right]$$

$$\times\frac{4\times 0.01\times 50}{2\times 9.81\times 0.15}\left[\left(\frac{300}{150}\right)^2\times\frac{2\pi\times 120}{60}\times 0.3\right]^2\left(\frac{2}{3}-\frac{1}{\pi^2}\right)$$

$$= 565.487\times 128.557 = 72697.4 \text{ N.m/s}$$

$$\text{Power saved by fitted air vessel} = \frac{72697.4}{10^3} = 72.6974 \text{ kW}$$

***(b) Double acting pump***

For a double acting pump,

$$v_d = \frac{2ALN}{60 \times a_d} = 2\frac{A}{a_d}\cdot\frac{r\omega}{\pi}$$

$$\therefore \quad W = W_1 - W_2$$

$$= \gamma Q \cdot \frac{4fL_d}{d_d \times 2g}\left(\frac{A}{a_d}\omega r\right)^2\left[\frac{2}{3} - \frac{4}{\pi^2}\right] \quad \text{...(10.44)}$$

$$\text{Percentage saving in work/s} = \frac{\frac{2}{3} - \frac{4}{\pi^2}}{\frac{2}{3}} \times 100 = 39.2\ \%$$

### 10.13.6 Discharge In an Out of Air Vessel

***(a) Suction Pipe***

The water enters the air vessel with uniform velocity and from there it is taken to the cylinder.

Discharge going in the air vessel = $Q_{in} - Q_{out}$

$$= A\omega r\left(\frac{1}{\pi} - \sin\theta\right) \text{ for a single acting pump} \quad \text{...(10.45a)}$$

$$= A\omega r\left(\frac{2}{\pi} - \sin\theta\right) \text{ for a double acting pump} \quad \text{...(10.45b)}$$

If the values of Eqs. (10.45a) and (10.45b) are negative, then the discharge shall be going out of air vessel.

***(b) Delivery pipe***

*(i) Single acting pump.*

Velocity of water in the delivery pipe connecting pump cylinder and air vessel,

$$v_{dv} = \frac{A}{a_d}\omega r \sin\theta$$

Discharge in the delivery pipe connecting the cylinder and air vessel,

$$Q_{dv} = a_d\, v_{dv} = A\omega r \sin\theta$$

Velocity of water in the delivery pipe beyond air vessel,

$$v_d = \frac{A}{a_d}\cdot\frac{\omega r}{\pi}$$



Discharge in the delivery pipe beyond air vessel,

$$Q_d = a_a v_d = A\frac{\omega r}{\pi}$$

Discharge of water going in the air vessel,

$$Q = Q_{dv} - Q_d = A\omega r \sin\theta - \frac{A\omega r}{\pi}$$

$$= A\omega r\left(\sin\theta - \frac{1}{\pi}\right) \qquad \text{...(10.46a)}$$

*(ii) Double acting pump*

$$Q_{dv} = A\omega r \sin\theta$$

$$Q_d = 2\frac{A\omega r}{\pi}$$

$$Q = A\omega r\left(\sin\theta - \frac{2}{\pi}\right) \qquad \text{...(10.46b)}$$

## 10.14 Theory of Working of Air Vessel

Consider a single-acting reciprocating pump is which the piston is moving with SHM. Neglecting the obliquity of connecting rod, the velocity of piston at any instant is given by,

$$v_p = (v_p)_{max} \sin\theta$$

Where $(v_p)_{max}$ = maximum piston velocity

$\theta$ = crank angle measured from IDC.

Mean velocity of piston, $(v_p)_{mean} = \dfrac{\int_0^{\pi} v_p d_0}{2\pi}$

$$= \frac{(v_p)_{max} \int_0^{\pi} \sin\theta . d\theta}{2\pi}$$

$$\therefore \quad \frac{(v_p)_{mean}}{(v_p)_{max}} = \frac{1 - \cos 0 \int_0^{\pi}}{2\pi}$$

$$= \frac{(-\cos\pi \times \cos 0°)}{2\pi} = \frac{1+1}{2\pi}$$

$$= \frac{1}{\pi} \text{ for single acting pump} \qquad \text{...(10.47a)}$$

$$= \frac{2}{\pi} \text{ for double acting pump} \qquad \text{...(10.47b)}$$

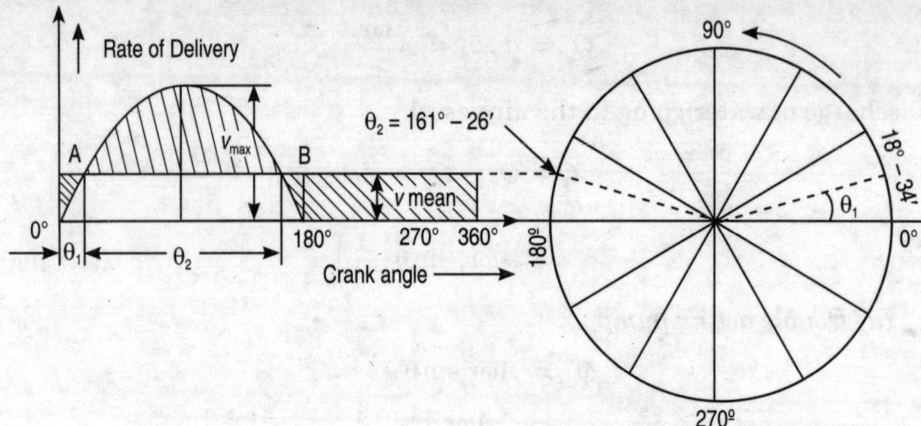


**Fig. 10.18 Delivery curve for a single acting pump.**

Fig. 10.18 shows the rate of delivery *v's* crank angle for a single acting pump. This is a sine curve. The value of mean velocity of piston is shown by a horizontal line which cuts the sine curve at points *A* and *B*, when the crank angles are $\theta_1$ and $\theta_2$ respectively.

$$v_p < (v_p)_{mean} \quad \text{for } \theta < 0 < 0 < \theta_1$$

Therefore, water is discharge out from the air vessel into the delivery pipe. Thus the volume of water in the air vessel is minimum at *A*, and the volume of air Fig. 10.17 Delivery curve for a single acting pump at *A* is maximum. From *A* to *B*, the discharge from the cylinder to the air vessel is more than its mean value, therefore at point *B*, the volume of air in the air vessel is minimum.

Let

$V_{max}$ = maximum volume of air in the air vessel.

$V_{min}$ = minimum volume of air in the air vessel.

Then fluctuating volume of air in the air vessel, $V_f = V_{max} - V_{min}$

$$V_{mean} = \frac{1}{2}(V_{max} + V_{min})$$

= mean volume in the air vessel

$$\text{Coefficient of fluctuation, } \delta_f = \frac{V_f}{V_{mean}} \qquad ...(10.48)$$

Now $\quad V_{max} - V_{min}$ = maximum volume of water

– Mean volume of water delivered by the piston

$$= \int_0^T \left[ A\{(v_p)_{max} \sin\theta\}dt - A(v_p)_{mean}\, dt \right]$$

Where $A$ = cross – sectional area of piston

$T$ = time taken by the piston to more from *A* to *B*.

$$= \int_0^T \left[ A\left\{(v_p)_{max} \sin\theta\right\}dt - \frac{A(v_p)_{max}}{\pi} dt \right]$$

Now $\theta = \omega t$

$$\therefore \quad d\theta = \omega dt \text{ or } dt = \frac{d\theta}{\omega}$$

$$\therefore \quad V_{max} - V_{min} = \frac{A(v_p)_{max}}{\omega} \int_{\theta_1}^{\theta_2} \left( \sin\theta - \frac{1}{\pi} \right) d\theta$$

$$= \frac{A(v_p)_{max}}{\omega} \left| -\cos\theta - \frac{\theta}{\pi} \right|_{\theta_1}^{\theta_2}$$

$$= \frac{-A(v_p)_{max}}{\omega} \left[ (\cos\theta_2 - \cos\theta_1) + \frac{1}{\pi}(\theta_2 - \theta_1) \right]$$

$$= \frac{A(v_p)_{max}}{\omega} \left[ \cos\theta_1 - \cos\theta_2 - \frac{\theta_2 - \theta_1}{\pi} \right]$$

For a single-acting pump,

$$(v_p)_{max} \sin\theta = (v_p)_{mean} \quad \text{at point } A \text{ and } B$$

$$= \frac{(v_p)_{max}}{\pi}$$

$$\therefore \quad \sin\theta = \frac{1}{\pi} = 0.3183$$

$$\theta_1 = 18.56° \text{ or } 18° - 34' \text{ or } 0.3239 \text{ rod}$$

$$\theta_2 = 180° - (18° - 34') = 161° - 26'$$

or rod or 161.44° or 2.8176 rod

$$\therefore \quad V_{max} - V_{min} = \frac{A(v_p)_{max}}{\omega} \left[ (\cos 18.56° - \cos 161.44°) - \frac{2.8176 - 0.3239}{\pi} \right]$$

$$= \frac{A(v_p)_{max}}{\omega} \times 1.1022 \quad \text{...(10.49)}$$

Piston volume per stroke, $V_p = AL$

$$= A(v_p)_{mean} \times \text{time for one revolution of crank.}$$

$$= \frac{A(v_p)_{max}}{\pi} \times \frac{2\pi}{\omega} = \frac{2A}{\omega}(v_p)_{max}$$

$$\therefore \quad \frac{V_{max} - V_{min}}{V_p} = \frac{\frac{A}{\omega}(v_p)_{max} \times 1.1022}{\frac{2A}{\omega}(v_p)_{max}} = 0.551 \quad \text{...(10.50)}$$

Similarly it can be shown that,

$$\frac{V_{\max} - V_{\min}}{V_p} = 0.21 \text{ for a double-acting single cylinder or two throw pump.} \quad \text{... (10.51a)}$$

$$= 0.009 \text{ for a three throw pump.} \quad \text{...(10.51b)}$$

### 10.14.1 Volume of Air Vessel

The change in volume of air in the vessel takes place according to Boyle's low.

or $pV = $ const.

$$\therefore \quad p_{\min} \times V_{\max} = p_{\max} \times V_{\min} = p_{mean} \times V_{mean} = \text{const.}$$

$$p_{\min} = \frac{p_{mean} \times V_{mean}}{V_{\max}}$$

And

$$p_{\max} = \frac{p_{mean} \times V_{mean}}{V_{\min}}$$

Coefficient of fluctuation,

$$\delta_f = \frac{p_{mean} \times p_{\min}}{p_{mean}} = \frac{p_{mean} \times V_{mean}\left(\dfrac{1}{V_{\min}} - \dfrac{1}{V_{\max}}\right)}{p_{mean}}$$

$$= \frac{V_{mean}\left(V_{\max} - V_{\min}\right)}{V_{\min} \times V_{\max}}$$

Now $V_{\min} \times V_{\max} \simeq V_{mean}^2$

$$\therefore \quad \delta_f = \frac{V_{\max} - V_{\min}}{V_{mean}}$$

Hence

$$\delta_f = \frac{p_{\max} - p_{\min}}{p_{mean}} = \frac{V_{\max} - V_{\min}}{V_{mean}} \quad \text{...(10.52)}$$

$$V_{mean} = \frac{V_{mean}}{v_{\max} - V_{\min}} \times \frac{V_{\max} - V_{\min}}{V_p} \times V_p$$

$$= \frac{1}{\delta_f} \times \frac{V_{\max} - V_{\min}}{V_p} \times V_p \quad \text{...(10.53)}$$

**Example 10.20** A double acting reciprocating pump, running at 40 rpm, is discharging 1.0 m³/min of water. The pump has a stroke of 400 mm and piston rod diameter is 200 mm. The delivery and section heads are 20 m and 5 m respectively. Find the slip of the pump and power required to drive the pump.

**Solution.**

Given: $N = 40$ rpm, $Q_a = 1.0$ m³/min, $L = 0.4$ m, $D_p = 0.2$ m, $H_d = 20$ m, $H_s = 5$ m

Theoretical discharge, $Q_{th}$ = 2 ALN

$$= 2\times\frac{\pi}{4}\times D_p^2\times 0.4\times 40$$

$$= 1.0053 \text{ m}^3\text{/min}$$

$$\text{Slip} = \frac{Q_{th}-Q_a}{Q_{th}} = \frac{1.0053-1.0}{1.0053}$$

$$= 0.00528 \text{ or } 0.528\%$$

Power required,
$$P = \frac{\gamma Q_{th}(H_d+H_s)}{10^3}$$

$$= \frac{9.81\times10^3\times1.0053\times(20+5)}{10^3\times60} = 4.109 \text{ kW}$$

## Summary

1. Discharge through the pump per second

$$Q = \frac{ALN}{60} \text{ for single acting}$$

$$= \frac{2ALN}{60} \text{ for double acting}$$

2. Work done per second = $\dfrac{\gamma ALN(H_s+H_d)}{60}$ for single acting

$$= \frac{2\gamma ALN(H_s+H_d)}{60} \text{ for double acting}$$

3. Slip = $Q_{th} - Q_{act}$

4. Coefficient of discharge, $C_d = 1-\dfrac{Q_{act}}{Q_{th}}$

5. Percentage slip = $(1-C_d)\times100$

6. Slider crank mechanism:
   Displacement from *IDC*, $x = r(1-\cos\theta)$
   Piston velocity, $v_p = \omega r\sin\theta$
   Acceleration of piston, $f_p = \omega^2 r\cos\theta$

$$= \omega^2 r\left[\cos\theta \pm \frac{\cos 2\theta}{n}\right]$$

   at the beginning (and end) of suction stroke

7. Acceleration of water in pipes:

$$f_s = \omega^2 r \cdot \frac{A}{a_s} \cos\theta, \qquad \text{in suction pipe}$$

$$f_d = \omega^2 r \cdot \frac{A}{a_d} \cos\theta \qquad \text{in delivery pipe}$$

8. Acceleration head in pipes:

Suction pipe: $H_{as} = \frac{L_s}{g} \cdot \omega^2 r \cdot \frac{A}{a_s}$

at $\theta = 0°$ i.e. beginning of stroke

$= 0$ at $\theta = 90°$ i.e. middle of stroke

$= -\frac{L_s}{g} \omega^2 r \frac{A}{a_s}$ at $\theta = 180°$ i.e. end of stroke

Delivery pipe: $(H_{ad})_{max} = \frac{L_d}{g} \cdot \omega^2 r \frac{A}{a_d}$ at $\theta = 0°$ i.e. beginning of suction stroke

$= 0$ at $\theta = 90°$ i.e. middle of delivery stroke

$= -\frac{L_s}{g} \omega^2 r \frac{A}{a_s}$ at $\theta = 180°$ i.e. end of delivery stroke.

9. For separation to take place,

$$\left[H_{atm} - (H_a + H_{as})\right] < H_{vap}$$

or $(H_s + H_{as}) < 7.8$ m of water vacuum.

10. Maximum speed of reciprocating pump

$$N_{max} = \left[\frac{g}{L_s} \times \left(\frac{d_s}{D}\right)^2 \times \left(\frac{60}{2\pi}\right)^2 \times \frac{(H_{as})_{max}}{r}\right]^{1/2}$$

11. Friction head in pipes:

$$H_f = \frac{4fL}{2gd}\left(\frac{A}{a}\omega r\right)^2 \sin^2\theta \qquad \text{(parabolic variation)}$$

$H_f = 0$ at $\theta = 0°$ and $180°$

$$(H_f)_{max} = \frac{4fL}{2gd}\left(\frac{A}{a}\omega r\right)^2 \text{ at } \theta = 90°$$

12. Air Vessel:

Neglecting valve losses, we have

Suction pipe:

$$H_{piston} = H_s + \frac{4fL_s}{d_s} \cdot \frac{v_s^2}{2g} + \frac{L_{sv}}{g} \cdot \frac{A}{a_s} \omega^2 r + \frac{v_s^2}{2g}$$

at the beginning of suction stroke.

$$= H_s + \frac{4fL_s}{d_s} \cdot \frac{v_s^2}{2g} + \frac{4fL_{sv}}{2gds}\left(\frac{A}{a_s}\omega r\right)^2 + \frac{v_{sv}^2}{2g} + \frac{v_s^2}{2g}$$

at the middle of suction stroke

$$= H_s + \frac{4fL_s}{d_s} \cdot \frac{v_s^2}{2g} + \frac{L_{sv}}{g} \cdot \frac{A}{a_s} \cdot \omega^2 r + \frac{v_s^2}{2g}$$

at the end of suction stroke.

Delivery pipe:

$$H_{piston} = H_d + \frac{4fL_d}{d_d} \cdot \frac{v_d^2}{2g} + \frac{L_{dv}}{g} \cdot \frac{A}{d_s} \omega^2 r + \frac{v_d^2}{2g}$$

at the beginning of delivery stroke.

$$= H_d + \frac{4fL_d}{d_d} \cdot \frac{v_d^2}{2g} + \frac{4fL_{dv}}{d_d \times 2g}\left(\frac{A}{a_d}\omega r\right)^2 + \frac{v_{dv}^2}{2g} + \frac{v_d^2}{2g}$$

at the middle of delivery stroke.

$$= H_d + \frac{4fL_d}{d_d} \cdot \frac{v_d^2}{2g} + \frac{L_{dv}}{g} \cdot \frac{A}{a_d} \cdot \omega^2 r + \frac{v_d^2}{2g}$$

at the end of delivery stroke.

13. Power saved by air vessel.

Air vessel fitted on delivery side:

$$\text{Work saved, } W = \gamma Q \times \frac{4fL_d}{d_d \times 2g}\left(\frac{A}{a_d}\omega r\right)^2 \left[\frac{2}{3} - \frac{1}{\pi^2}\right]$$

with single acting pump % work saved = 84.8% with single acting pump

$$W = \gamma Q \times \frac{4fL_d}{d_d \times 2g}\left(\frac{A}{a_d}\omega r\right)^2 \left[\frac{2}{3} - \frac{4}{\pi^2}\right]$$

for double acting pump

% work saved = 39.2%

## Multi-Choice Questions

1. Reciprocating pump is started with the delivery valve
   (a) closed (b) open
   (c) 50% open (d) partially open

2. Slip for a reciprocating pump is defined as
   (a) $\frac{Q_a - Q_{th}}{Q_{th}}$ (b) $\frac{Q_{th} - Q_a}{Q_a}$
   (c) $\frac{Q_{th} - Q_a}{Q_{th}}$ (d) $\frac{Q_a - Q_{th}}{Q_a}$

3. The pump suitable for small discharge at high pressure head is
   (a) axial flow (b) centrifugal
   (c) mixed flow (d) reciprocating

4. Air vessel in reciprocating pump is used to
   (a) smoothen out flow (b) increase pump efficiency
   (c) increase pump head (d) reduce acceleration head

5. Discharge through a double-acting reciprocating pump is given by
   (a) $\frac{ALN}{30}$ (b) $\frac{ALN}{60}$
   (c) $\frac{ALN}{90}$ (d) $\frac{ALN}{120}$

6. The acceleration head in the suction pipe is given by
   (a) $\frac{L_s}{g} \cdot \frac{A}{a_s} \omega r \cos\theta$ (b) $\frac{L_s}{g} \cdot \frac{A}{a_s} \omega r \sin\theta$
   (c) $\frac{L_s}{g} \cdot \frac{A}{a_s} \omega^2 r \cos\theta$ (d) $\frac{L_s}{g} \cdot \frac{A}{a_s} \omega^2 r \sin\theta$

7. The function (s) of air vessels in reciprocating pump is (are):
   (a) To obtain continuous supply of liquid at uniform rate.
   (b) To reduce the acceleration head.
   (c) To save the power required to drive the pump.
   (d) All of the above.

8. The maximum saving in energy by fitting an air vessel on the delivery side of a single acting reciprocating pump is
   (a) 39.2% (b) 42.4%
   (c) 78.4% (d) 84.8%

9. The ratio of fluctuation of volume of water in air vessel to piston stroke volume in a single acting reciprocating pump is
   (a) 0.21 (b) 0.42
   (c) 0.55 (d) 0.65
10. The separation head in a reciprocating pump is
   (a) 10.3 m of water abs (b) 2.5 m of water abs
   (c) 7.8 m of water abs (d) 5.3 m of water abs

---

**Answers**

1. (d) 2. (c) 3. (d) 4. (d) 5. (a) 6. (c) 7. (d) 8. (d) 9. (c) 10. (b)

---

## Review Questions

1. Define a pump. How a liquid is raised by a pump?
2. Explain the working principle of a reciprocating pump.
3. How do you classify pumps ?
4. What is a plunger pump? Where it is used?
5. Define slip and differentiate between positive and negative slip.
6. Define coefficient of discharge of a reciprocating pump.
7. What is the relationship between slip and coefficient of discharge?
8. Define separation in a reciprocating pump.
9. What preventive measures are usually taken to reduce separation in a reciprocating pump?
10. What is an air vessel?
11. What are the functions of an air vessel?
12. Draw the indicator diagram for a reciprocating pump incorporating the effects of acceleration heads and frictional heads.
13. Draw the indicator diagram for a reciprocating pump with air vessels fitted on suction and delivery sides.
14. How is the discharge going in or coming out of air vessel calculated?
15. What percentage of work is saved by fitting an air vessel on the delivery side in case of single acting and double acting pumps?

## Numerical Problems

1. A double acting piston pump runs at 90 rpm. Its bore and stroke are 150 mm and 450 mm respectively. Determine the theoretical discharge in $m^3/h$. If the total head to be lifted by the pump is 60 m, determine the power required to drive the pump if its efficiency is 80 %.

[**Ans.** 85.88 $m^3/h$, 17.552kW]

2. A double acting reciprocating pump has a piston diameter of 160 mm and stroke length 350 mm. The suction and delivery heads are 5 m and 18 m respectively. The pump speed is 70 rpm and percentage slip is 3 %. Calculate (a) the actual discharge per minute, (b) coefficient of discharge, and (c) power required to drive the pump. Assume efficiency of pump to be 8 %.

[**Ans.** 0.955 $m^3$/mm, 0.97, 4.224kW]

3. A single acting reciprocating pump delivers 45 $m^3$/h of water against a total head of 25 m. The pump speed is 90 rpm. If slip is 5 %, calculate (a) volume of cylinder in litres, and (b) power consumption if overall efficiency is 80 %.

[**Ans.** 8.77 liters, 4.034 kW]

4. A single acting reciprocating pump has its piston diameter as 150 mm and stroke 250 mm. The piston makes 50 double strokes per minute. The suction and delivery heads are 5 m and 15 m respectively. Find the force required to work the piston during the suction and delivery strokes. Assume the efficiency of suction and delivery strokes as 60 % and 75 % respectively. Determine the power required by the pump.

[**Ans.** 1442 N, 3669 N 1.065 kW]

5. A double acting reciprocating pump has a piston of 250 mm diameter and piston rod of 50 mm diameter, which is on one side of the piston only. Length of piston stroke is 380 mm and the speed of the crank moving the piston is 60 rpm. The suction and delivery heads are 5 m and 20 m respectively. Find the force required to work the piston during the inwards and outwards strokes. Determine the quantity of water in litres per minute raised by the pump and power required.

[**Ans.** 11632.7 N, 11921 N, 2195 litres/min, 8.97kW]

6. In a bucket pump, the diameters of bucket and pump rod are 300 mm and 50 mm respectively. The stroke is 1 m. The suction and delivery heads are 6 m and 12 m respectively. Determine the pull on the rod during the up stroke and the volume of water discharged during one upstroke and down stroke.

7. A single acting reciprocating pump has a plunger diameter of 200 mm and a stroke of 300 mm. It is placed with its centre line 4 m above the water level in the sump. The suction pipe is 75 mm in diameter and 5 m long. If separation occurs when the absolute pressure head is 2.5 m of water, find the maximum speed of pump to avoid separation at the commencement of suction stroke. Assume SHM and effective height of barometer as 10 m of water.

8. A single acting horizontal reciprocating plunger pump has a suction pipe 60 mm diameter and 6 m long. The water sump is 2 m below the cylinder axis, and the pump plunger is 150 mm diameter and of 250 mm stroke.

If the pump is working at 30 rpm, find the absolute pressure head in bar at the beginning and at the end of suction stroke. If separation occurs at 0.8 bar below atmospheric pressure, state whether the separation would occur at the beginning or at the end of suction stroke. Take friction factor, $f = 0.01$ and barometric pressure = 10.3 m of water.

9. A single acting reciprocating pump runs at 60 rpm. The diameter of piston is 125 mm and a stroke of 300 mm. The centre of the pump is 3.5 m above the water surface of sump. The suction pipe diameter is 100 mm and its length is 4 m. Find the pressure heads at the commencement and at the end of suction stroke. What conditions on the delivery side will cause negative slip?
10. The diameter of plunger of a reciprocating pump is 200 mm and stroke 300 mm. The pump is driven with SHM at 60 rpm and it draws water from a sump 4 m below pump centre line. Find the least diameter of the suction pipe which is 5 m long in order to avoid separation at this speed.
11. A Single acting reciprocation pump has cylinder bore of 200 mm and stroke 300 mm. The delivery pipe has 150 mm diameter and a large air vessel fitted on the delivery side. The crank revolves at 30 rpm and the piston has SHM. Find the discharge going in and coming out of the air vesselwhen the crank makes 60°, 120° and 180° with the inner dead centre. Calculate the crank angle at which there is no flow to or from the air vessel.
12. A single – acting reciprocating pump has plunger 400 mm diameter and stroke 600 mm. The delivery pipe is 150 mm diameter and 90 m long, the frictional coefficient being 0.008. Find the power lost in friction in the delivery pipe when the pump runs at 50 rpm,
    (a) When a large air vessel is fitted near the pump outlet,
    (b) Assuming SHM of plunger and no air vessel.
    (c) Find the power saved by installing the air vessel.

CHAPTER 11

# Centrifugal Pump

## 11.1 Introduction

A centrifugal pump is a device to raise liquids from a lower to a higher level by creating the required pressure with the help of centrifugal action. In general, it can be defined as a machine which increases the pressure energy of a fluid. Whirling motion is imparted to the liquid by means of backward curved blades mounted on a wheel known as the impeller. Liquid enters the impeller at the eye of the pump and discharges into the casing surrounding the impeller, as shown in Fig. 11.1 the pressure head developed by centrifugal action is entirely due to the velocity imparted to the liquid by the rotating impeller.

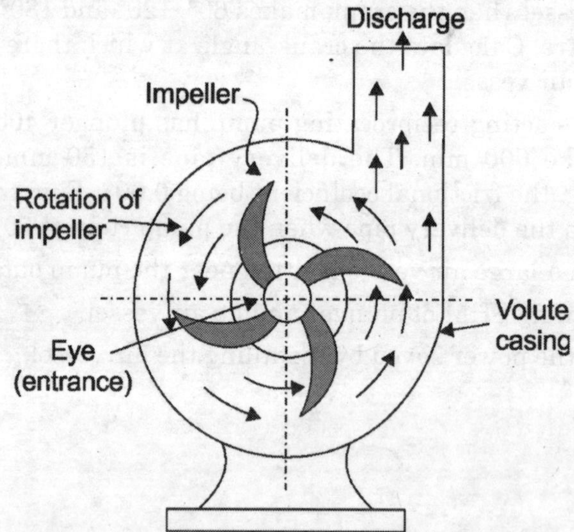


**Fig. 11.1 Principle of centrifugal pump.**

## 11.2 Comparison of Cantrifugal and Reciprocating Pumps

The advantages of centrifugal pumps over reciprocation pumps are given in Table 11.1.

**Table 11.1 Comparison of centrifugal and reciprocating pumps**

| Centrifugal pump | Reciprocating |
|---|---|
| (i) Smooth and even flow. | Pulsating flow |
| (ii) Low initial cost | Initial cost as high as four times that of centrifugal pump. |
| (iii) Compact. Occupies less floor space. Vertical type requires even lesser space | Occupies 6 to 8 times the space required for horizontal centrifugal pump. |
| (iv) Gross weight is small. | Gross weight is considerable. |
| (v) Installation is easy. | Installation is more difficult then centrifugal pump. |
| (vi) Efficiency of low head pumps is high. | Efficiency of low head pumps may be as low as 40% chiefly because of relatively higher energy losses viz. frictional, valve losses etc. |
| (vii) Construction is simplified by elimination of many parts such as non-return valves, glands, air vessel etc. therefore less number of spare parts are required. | Complicated construction. Therefore a number of spare parts are necessary. |
| (viii) Low maintenance cost. Periodical check up is sufficient. | Maintenance charges r high because parts like valves require constant attention. |
| (ix) High speed. Can be couple directly through flanged coupling to electric motors of steam turbine. | Low speed due to separation difficulties belt drive indispensable. |
| (x) Uniform torque. | Torque not uniform. |
| (xi) Easy handling of highly viscous fluid such as oils. Muddy and sewage water, paper pulp, sugar molasses, chemicals, etc | Valves and glands cause trouble when requires to transmit viscous fluids. |

## 11.3 Principle of Operation

First of all priming of the pump is cried out. The power source is switched on and the revolution of the pump impeller inside the casing full of water produces a forced vortex which is responsible for imparting a centrifugal head to water. Rotation of impeller effects a reduction of pressure at the centre. This causes the water in the suction pipe to rush into the eye. The speed of the pump should be high enough to produce centrifugal head sufficient to initiate discharge against the delivery head.

Now $$v = \sqrt{2gH}$$

Or $$H = \frac{v^2}{2g}$$

Where $v$ = outlet velocity of water in a pump

$H$ = head developed by centrifugal action.

## 11.4 Classifications of Centrifugal Pumps

The centrifugal pumps may be classified based on the following characteristic features:

1. Working head.
2. Type of casing.
3. Number of impellers per shaft.
4. Relative direction of flow through impeller.
5. Number of entrances to impeller.
6. Disposition of shaft.
7. Liquid handled.
8. Specific speed.

**1. Working head**

Low lift : up to 15 m.

Medium lift : up to 40 m.

High lift : above 40 m.

**2. Type of casing**

The casing is of two types: volute and turbine or diffusion type.

**(a) Volute pumps.** A volute casing type of pump is shown in Fig. 11.2 (a). It has a volute casing into which the impeller discharges water at a high velocity. Volute is of a spiral form and the cross-sectional area of the moving stream gradually increases from the tongue towards the delivery pipe. The cross-sectional area at any point is therefore proportional to the quantity of water flowing across that section and therefore the mean velocity remains constant. This eliminates the possibility of loss of kinetic head.

The functions of volute casing are:

(i) To collect water from the periphery of impeller and to transmit it to the delivery pipe at a constant velocity.

(ii) To eliminate the loss of velocity head.

(iii) To increase the efficiency of the pump by eliminating loss of velocity head.

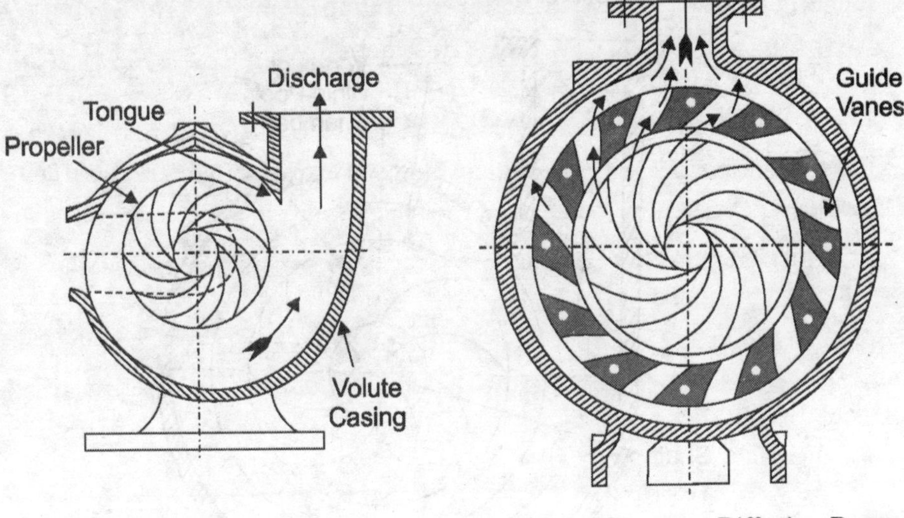


(a) Volute Pump (b) Turbine Pump or Diffusion Pump

**Fig. 11.2**

**(b) Turbine or Diffusion Pump.** This pump as shown in Fig. 11.2 (b) consists of an impeller surrounded by a guide wheel consisting of a number of stationary vanes or diffusers. The diffusers provide outlets with cross-section gradually enlarging towards the periphery. Water emerging from the impeller flows past the guide vanes and as the area across flow increases, velocity decreases, and pressure increases.

Diffusion pumps may be either horizontal or vertical shaft type. They are used in narrow wells and mines etc.

**(c) Vortex Casing.** The vortex or whirlpool chamber type casing is shown in Fig. 11.2 (c). This is an improved type of volute chamber where an annular space is provided between the volute and impeller forming a combination of spiral and circular chamber. This arrangement reduces vortices to a considerable extent and improves the performance of the pump.

**3. Number of Impellers**

(a) Single stage – for low lift.

(b) Multiple stage – for high lift.

**4. Relative Direction of Flow through Impeller**

(a) Radial flow – small discharge at medium head.

(b) Axial flow – for large discharge at low head.

(c) Mixed flow – Radial + axial flow, large discharge and low head, used for irrigation purposes.

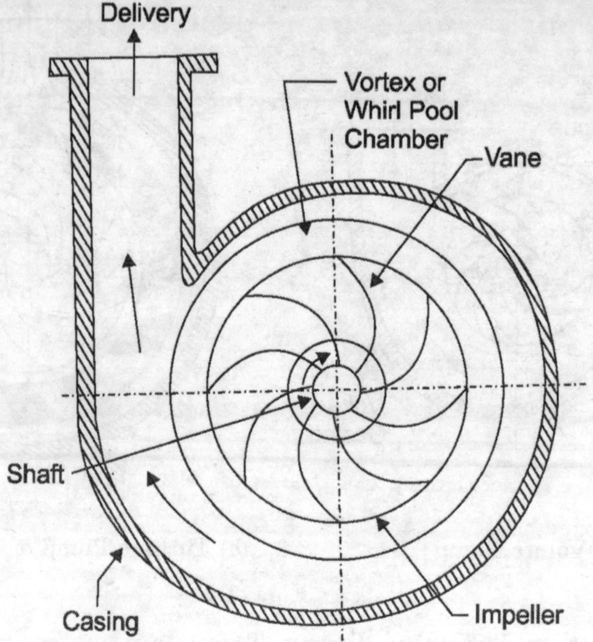


(c) Pump with Vortex Chamber

**Fig. 11.2 (a) (b) and (c) Types of Casing.**

5. **Number of Entrances**

   (a) Single entrance – suction from one side of impeller.

   (b) Double entrance – suction from both sides of impeller.

6. **Disposition of Shaft**

   (a) Horizontal – generally used.

   (b) Vertical – rarely used.

7. **Liquid Handled**

   (a) Closed impeller – for non-viscous liquids such as ordinary water, hot water, hot oil and acids etc.

   (b) Open impeller – for handling mixtures of water, sand, pebbles and clay.

8. **Specific Speed**

It is defined as the speed of geometrically similar pump when delivering one m³/s against a head of one metre.

$$\text{Specific speed, } N_s = \frac{N\sqrt{Q}}{H^{3/4}}$$

Where $N$ = actual speed of pump in rpm

$Q$ = Quantity of liquid flowing, $m^3/s$

$H$ = delivery head, m ...(11.1)

## 11.5 Accessories and Layout of Centrifugal Pump

The main accessories are:

1. Strainer and foot valve.
2. Suction pipe with fittings.
3. Delivery or regulating valve.
4. Delivery pipe with fittings.

1. **Strainer.** It keeps the floating bodies in water, such as leaves, wooden pieces and other kind of rubbish, away from the pump and avoids choking.

   **Foot valve.** It is a non-return or one way type valve, opening upwards. It does not allow the water to move downward.

2. **Suction pipe.** It leads water from the sump to the pump. It should be laid first vertical and then horizontal.

3. **Delivery or regulating valve.** It should be closed while starting the pump to allow built-up of pressure. It should be closed before stoping the pump.

4. **Delivery pipe.** It is used to deliver water from the pump to the reservoir.

The layout of a centrifugal pump is shown in Fig. 11.3

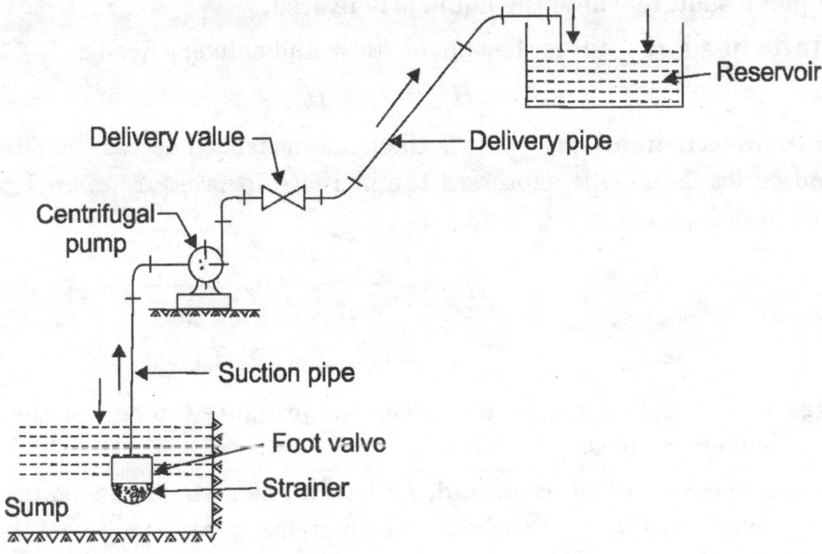


**Fig. 11.3 Layout of a centrifugal pump.**

## 11.6 Starting of a Centrifugal Pump

A centrifugal pump works on the principle that when a certain mass of fluid is rotated by an external source, it is thrown away from the central axis of rotation and a centrifugal head is impressed which enables it to rise to a higher level.

The working of centrifugal pump is as follows:

1. The delivery valve is closed and the pump is primed.
2. Keeping the delivery valve closed the electric motor is started to rotate the impeller which causes suction at the eye of casing.
3. The speed of impeller is gradually increased till the impeller rotates at its normal speed and develops normal energy required for pumping the liquid.
4. Now the delivery valve is opened and water is sucked up from the sump through the foot valve and suction pipe to the pump casing.
5. The water is then passed on through the delivery pipe to the desired place.
6. To stop the pump the delivery valve is first closed to check the back flow from the reservoir.

## 11.7 Important Terms and Formulae in Centrifugal Pumps

**Suction head, $H_s$ :** It is the vertical height of the centre line of pump shaft above the surface of liquid from which it is being raised.

**Delivery head, $H_d$ :** It is the vertical height measured from the centre line of pump shaft to where the liquid is delivered.

**Static head, $H_{stat}$ :** It is the sum of static and delivery heads.

$$H_{stat} = H_s + H_d \qquad ...(11.2)$$

**Manometric head, $H_{mano}$ :** It is the head measured across the pump inlet and outlet flanges. It expresses the increase in pressure energy per unit weight of liquid handled by impeller.

$$H_{mano} = \frac{p_d - p_s}{\gamma} + h_g = \frac{u_2 v_{u2}}{g}$$

$$= (H_{mano})_d - (H_{mano})_s + h_g \qquad ...(11.3)$$

Where $h_g$ = vertical distance between the pressure tapings for the suction and delivery gauges.

**Total, gross or effective head, *H*:** This is the actual head against which the pump has to work. It expresses the increase in total energy of the liquid between inlet and outlet.

$$H = H_{mano} + \frac{v_d^2 - v_s^2}{2g} \qquad ...(11.4)$$

Where $v_d$, $v_s$ = velocity of liquid in delivery and suction pipes respectively.

**Power required to drive the pump.**

$$P = \frac{\gamma Q H_{mano}}{10^3 \times \eta_{overall}} kW \qquad ...(11.5)$$

Where $\gamma$ = specific weight of liquid handled, N/m³

$Q$ = discharge, m³/s

$H_{mano}$ = manometric head, m

$\eta_{overall}$ = overall efficiency of pump

**Overall Efficiency:** it is the ratio of power supplied by the pump to the power delivered to the pump shaft.

$$\eta_{overall} = \frac{output}{input}$$

$$= \frac{\gamma Q \, H_{mano}}{10^3 \times (BP - power \text{ lost in coupling mechanism})} \qquad ...(11.6)$$

Where $BP$ = brake power of electric motor or prime mover which drives the pump.

*Static Efficiency* It accounts for all hydraulic losses in the pipelines.

$$\eta_{stat} = \frac{\gamma Q H_{stat}}{1.^3 \times BP} \qquad ...(11.7)$$

## 11.8 Loss of Head in Pipe Lines and Pipe Fittings

Total head loss in pipes and fittings

$$H_\ell = H_{fs} + H_{fd} + \Sigma H_{\ell f} + \frac{v_e{}^2}{2g} \qquad ...(11.8)$$

Where $H_{fs}$ = frictional head loss in suction pipe = $\frac{4fL_s}{d_s} \cdot \frac{v^2{}_s}{2g}$

$H_{fd}$ = frictional head loss in delivery pipe = $\frac{4fL_d}{d_d} \cdot \frac{v_d^2}{2g}$

$\Sigma H_{lf}$ = sum of head losses in pipe fittings

$\frac{v_e^2}{2g}$ = loss of kinetic head due to velocity of water at exit.

### 11.8.1 Determination of head losses

**1. Friction head loss in pipes:**

$$H_f = \frac{4fL}{d} \cdot \frac{v^2}{2g} \qquad ...(11.9)$$

Where $f$ = frictional factor

**2. Sudden enlargement of pipe cross-section:**

$$H_\ell = \frac{k(v_1 - v_2)^2}{2g} \quad \text{...(11.10)}$$

Where $k$ = a factor 0.2-1.1

**3. Sudden contraction of pipe line:**

$$H_\ell = k\frac{v_2^2}{2g} \quad \text{...(11.11)}$$

$$k = f\left(\frac{d_1}{d_2}\right) = 0.52\text{-}1.0 \text{ for } \frac{d_1}{d_2} = 0 - 1$$

**4. 90° bend:**

$$H_\ell = k\frac{v^2}{2g} \quad \text{...(11.12)}$$

**5. Pipe fittings:**

$$H_{\ell f} = k\frac{v^2}{2g} \quad \text{...(11.13)}$$

**Example 11.1** A centrifugal pump draws water through a 15m long 150mm diameter suction pipe which is fitted with 90° bend. The maximum capacity of the pump is 75 lps. Determine the static suction head. Assume the maximum vacuum as 5.5 m of water. Loss of head in the bend may by taken as 10 cm if water. Take $f$ = 0.006.

**Solution.**

Given: $L_s$ = 15 m, $d_s$ = 150 mm, $Q$ = 75 lps, $H_{vac}$ = 5.5 m, $f$ = 0.006

$$v_s = \frac{Q}{a_s} = \frac{4Q}{\pi d_s^2} = \frac{4 \times 75 \times 10^{-3}}{\pi(0.15)^2} = 4.244 \text{ m/s}$$

$$H_{fs} = \frac{4fL_s}{d_s} \cdot \frac{v_s^2}{2g} = \frac{4 \times 0.006 \times 15}{0.150} \times \frac{(4.244)^2}{2 \times 9.81}$$

$$= 2.20 \text{ m}$$

Loss of head due to bend, $H_{lf} = 0.1\text{m}$

$$\Sigma H_{lf} = 0.1m$$

$$H_{ke} = \frac{v_s^2}{2g} = \frac{(4.244)^2}{2 \times 9.81} = 0.918m$$

$$H_\ell = H_s + H_{fs} + \Sigma H_{lf} + \frac{v_s^2}{2g} = H_{vac}$$

$$5.5 = H_s + 2.20 + 0.1 + 0.918$$

$$H_s = 5.5 - 3.218 = 2.282m$$

## 11.9 Fundamental Equation of Centrifugal Pump

Let $i$, $d$ = point on the liquid path at suction inlet and casing outlet respectively.

1, 2 = points at impeller inlet and outlet respectively.

Applying Bernoulli's theorem between points $i$ and 1 shown in Fig.11.4, we have

$$\frac{p_i}{\gamma}+\frac{v_i}{2g}+z_i = \frac{p_1}{\gamma}+\frac{v_1^{\;2}}{2g}+z_1+H_{l(i-1)} \qquad ...(1)$$

For flow of liquid from points 1 to 2, we have

$$\frac{p_1}{\gamma}+\frac{v_{r1}^{\;2}}{2g}-\frac{v_1^{\;2}}{2g}+z_1 = \frac{p_2}{\gamma}+\frac{v_{r2}^{\;2}}{2g}-\frac{u_2^2}{2g}+z_2+H_l(1-2) \qquad ...(2)$$

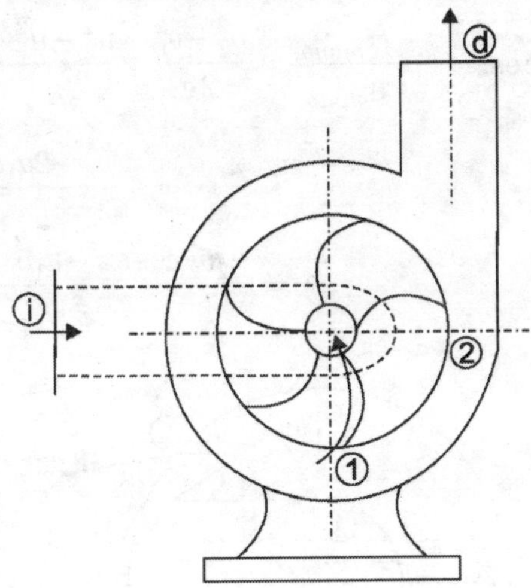


**Fig. 11.4 Liquid flow through a centrifugal pump.**

For flow of liquid from points 2 to $d$, we have

$$\frac{p_2}{\gamma}+\frac{v^2_{\;2}}{2g}+z_2 = \frac{p_d}{\gamma}+\frac{v_d^2}{2g}+z_d+H_l(2-d) \qquad ...(3)$$

Adding Eqs. (1), (2) and (3), we get

$$\frac{v_2^2-v_1^{\;2}}{2g}+\frac{v_{r1}^2-v_{r2}^2}{2g}+\frac{u_2^2-u_1^2}{2g} = \left[\left(\frac{p_d}{\gamma}+\frac{v_d^2}{2g}+z_d\right)-\left(\frac{p_i}{\gamma}+\frac{v_i^2}{2g}+z_i\right)\right]$$

$$+\left[H_{\ell(i-1)}+H_{\ell(1-2}+H_{\ell(2-d)}\right]$$

$$= \mathrm{H}_{\mathrm{mano}}+\Delta H_{mano} \qquad ...(11.14)$$

Where, $$H_{mans} = H_{static}+H_f+\frac{v_d^2}{2g}$$

Eq. (11.14) is known as the fundamental equation of centrifugal pump.

Manometric efficiency, $\eta_{mano} = \dfrac{H_{mano}}{H_{mano} + \Delta H_{mano}}$ ...(11.15)

Hence Eq. (11.14) becomes,

$$\frac{v_2^2 - v_1^2}{2g} + \frac{v_{r1}^2 - v_{r2}^2}{2g} + \frac{u_2^2 - u_1^2}{2g} = \frac{H_{mano}}{\eta_{mano}} \quad ...(11.16)$$

The velocity triangles are shown in Fig. 11.5.

$$v_{r1}^2 = u_1^2 + v_1^2 - 2u_1 v_1 \cos\alpha_1$$

$$v_{r2}^2 = u_2^2 + v_2^2 - 2u_2 v_2 \cos\alpha_2$$

$$\therefore \quad v_{r1}^2 - u_{r2}^2 = (u_1^2 - u_2^2) + (v_1^2 - v_2^2)$$

$$-2u_1 v_1 \cos\alpha_1 + 2u_2 v_2 \cos\alpha_2$$

Eq. (11.16) becomes,

$$\frac{H_{mano}}{\eta_{mano}} = \frac{v_2^2 - v_1^2}{2g} + \frac{u_2^2 - u_1^2}{2g} + \frac{u_1^2 - u_2^2}{2g} + \frac{v_1^2 - v_2^2}{2g}$$

$$+ \frac{-2u_1 v_1 \cos\alpha_1 + 2u_2 v_2 \cos\alpha_2}{2g}$$

$$= \frac{u_2 v_2 \cos\alpha_2 - u_1 v_1 \cos\alpha_1}{g}$$

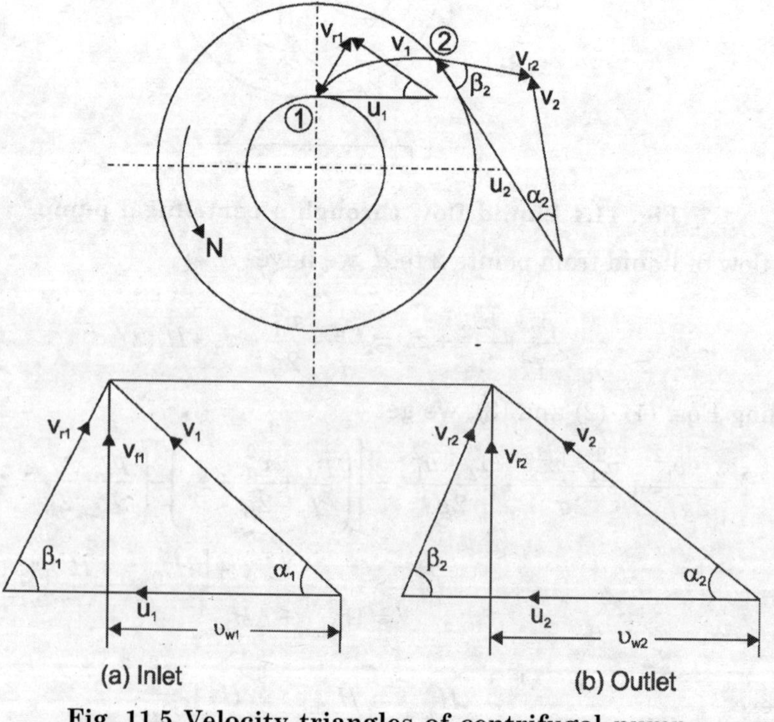


**Fig. 11.5 Velocity triangles of centrifugal pump.**

$$\frac{H_{mano}}{\eta_{mano}} = \frac{u_2 - v_{w2} - u_1 v_{w1}}{g} \qquad \text{...(11.17a)}$$

For $\alpha_1 = 90°$, $v_{w1} = 0$

$$\therefore \quad \frac{H_{mano}}{\eta_{mano}} = \frac{u_2 v_{w2}}{g} \qquad \text{...(11.17b)}$$

## 11.10 Work done on the Liquid

The velocity triangles for one vane of a centrifugal pump are shown in Fig.11.6. The liquid enters the impeller at its centre and leaves at its outer periphery. The following assumptions are made:

1. The liquid enters the impeller eye in radial direction, i.e., $\alpha_1 = 90°$.
2. Energy loss in the impeller due to friction and eddy formation is neglected.
3. No loss due to shock at entry.
4. There is uniform velocity distribution in the narrow passages formed between two adjacent vanes.

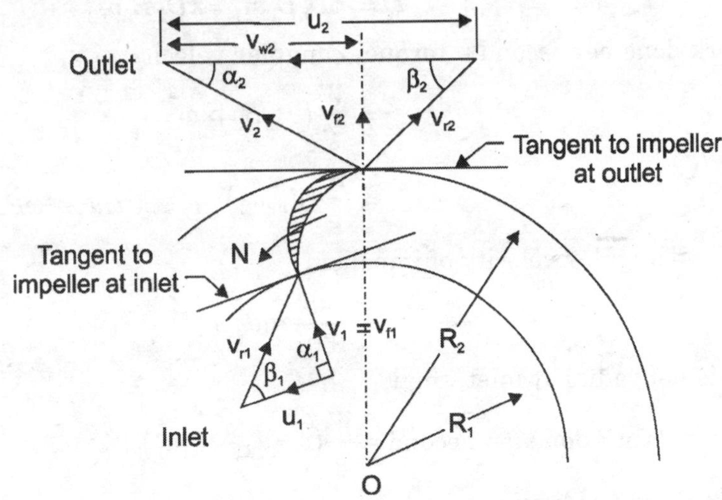


**Fig. 11.6 Velocity triangles for an impeller vane.**

Let $D_1 = 2R_1$ = diameter of impeller at inlet, m

$D_2 = 2R_2$ = diameter of impeller at outlet, m

$N$ = speed of impeller, rpm

$B_1, B_2$ = width of impeller at inlet and outlet respectively

$u_1, u_2$ = tangential velocity of impeller at inlet and outlet respectively

$v_1, v_2$ = absolute velocity of liquid at inlet and outlet respectively

$v_{w1}, v_{w2}$ = velocity of whirl of liquid at inlet and outlet respectively

$v_{r1}, v_{r2}$ = relative velocity of liquid at inlet and outlet respectively

$v_{f1}, v_{f2}$ = velocity of flow of liquid at inlet and outlet respectively

$\alpha_1, \alpha_2$ = angles made by absolute velocity of liquid at inlet and outlet respectively with the direction of motion of vane

$\beta_1, \beta_2$ = angles made by relative velocity of liquid at inlet and outlet respectively with the direction of motion of vane.

Torque on the impeller = Rate of change of moment of momentum

Since $v_{w1} = 0$, therefore,

Moment of momentum at inlet = 0

$$\text{Moment of momentum at outlet} = \frac{W}{g}(v_{w2}R_2)$$

$$\therefore \quad \text{Torque, } T = \frac{W}{g}(v_{w2}R_2)$$

Where, $W$ = weight of liquid flowing through the impeller

$= \gamma Q$

$\gamma$ = specific weight of fluid = $\rho g$

$\rho$ = density if fluid

$Q = \pi D_1 B_1 v_{f1} = \pi D_2 B_2 v_{f2}$

Work done per second = torque × angular velocity

$$= \frac{W}{g}(v_{\omega 2}R_2) \times \omega$$

$$= \frac{W}{g}(v_{w2}u_2) \qquad (\because u_2 = \omega R_2) \quad \text{...(11.18)}$$

Where $\omega$ = angular velocity of impeller

$$= \frac{2\pi N}{60} rad/s$$

If flow is not radial at inlet, then

$$\text{Work done per second} = \frac{W}{g}(v_{w2}u_2 - v_{w1}u_1) \quad \text{...(11.19)}$$

For unit weight of liquid,

Work done per second per unit weight of liquid

$$= \frac{1}{g}(v_{w2}u_2 - v_{w1}u_1) \quad \text{...(11.20)}$$

This equation is known as the *Euler momentum equation* for a centrifugal pump.

Now from outlet velocity triangle, we have

$$v_{r2}^2 = v_{f2}^2 + (u_2 - v_{w2})^2$$

or

$$v_{f2}^2 = v_{r2}^2 - (u_2 - v_{w2})^2 \quad \text{...(1)}$$

Also $$v_{f2}^2 = v_2^2 - v_{w2}^2 \qquad ...(2)$$

Comparing Eqs. (1) and (2), we have

$$v_2^2 - v_{w2}^2 = v_{r2}^2 - (u_2 - v_{w2})^2 = v_{r2}^2 - u_2^2 - v_{w2}^2 + 2u_2 v_{w2}$$

or $$v_2^2 = v_{r2}^2 - u_2^2 + 2u_2 v_{w2}$$

or $$u_2 v_{w2} = \frac{1}{2}\left(v_2^2 - v_{r2}^2 + u_2^2\right)$$

Similarly from inlet triangle, we can obtain

$$u_1 v_{w1} = \frac{1}{2}\left(v_1^2 - v_{r1}^2 + u_1^2\right)$$

Substituting these values in Euler's equation, we have

Work done per second per unit weight of liquid

$$= \frac{1}{2g}\left[\left(v_2^2 + u_2^2 - v_{r2}^2\right) - \left(v_1^2 + u_1^2 - v_{r1}^2\right)\right]$$

$$= \frac{v_2^2 - v_1^2}{2g} + \frac{u_2^2 - u_1^2}{2g} + \frac{v_{r1}^2 - v_{r2}^2}{2g} \qquad ...(11.21)$$

= increase in kinetic energy + increase in static pressure + change in kinetic energy due to retardation of flow relative to the impeller

This equation is known as the *fundamental equation of centrifugal pump*.

## 11.11 Losses in Centrifugal Pump

*1. Hydraulic losses:*

(i) Shock or eddy losses at the entrance to and exit from the impeller.

(ii) Losses due to friction in the impeller.

(iii) Friction and eddy losses in the guide vanes/diffuser and casing.

(iv) Friction losses in suction and delivery pipes.

*2. Mechanical losses:*

(i) Losses due to disc friction between impeller and liquid filling the space between impeller and casing

(ii) Losses due to friction in main bearing and glands.

*3. Leakage loss:*

The loss of energy due to leakage of liquid is known as leakage loss.

## 11.12 Pressure Rise in Pump Impeller

Applying Bernoulli's theorem between inlet and outlet edge if impeller, we have

Energy at inlet = Energy at outlet – useful work done by impeller

$$\frac{p_1}{\gamma}+\frac{v_1^2}{2g}=\frac{p_2}{\gamma}+\frac{v_2^2}{2g}-\frac{u_2 v_{w2}}{g}$$

Pressure rise between outlet and inlet edges of impeller,

$$\frac{p_2-p_1}{\gamma}=\frac{v_1^2}{2g}-\frac{v_2^2}{2g}+\frac{u_2 v_{w2}}{g}$$

Now $$v_{w2}=u_2-v_{f2}\cot\beta_2$$

and $$v_2^2=v_{f2}^2+v_{w2}^2=v_{f2}^2+\left(u_2-v_{f2}\cot\beta_2\right)^2$$

$$=v_{f2}^2+u_2^2+v_{f2}^2\cot_2\beta_2-2u_2v_{f2}\cot\beta_2$$

$$=v_{f2}^2(1+\cot^2\beta_2)+u_2^2-2u_2v_{f2}\cot\beta_2$$

$$\therefore \quad \frac{p_2-p_1}{\gamma}=\frac{v_1^2}{2g}-\frac{1}{2g}\left[v_{f2}^2\left(1+\cot^2\beta_2\right)+u_2^2-2u_2v_{f2}\cot\beta_2\right]+\frac{u_2}{g}\left(u_2-v_{f2}\cot\beta_2\right)$$

$$=\frac{1}{2g}\left[v_1^2-v_{f2}^2\operatorname{cosec}^2\beta_2-u_2^2+2u_2v_{f2}\cot\beta_2+2u_2^2-2u_2v_{f2}\beta_2\right]$$

$$=\frac{1}{2g}\left[v_1^2+u_1^2-v_{f2}^2\operatorname{cosec}^2\beta_2\right] \quad \text{...(11.22)}$$

Also $$H_{mano}=\eta_{mano}\frac{u_2v_{w2}}{g}$$

$$=\eta_{mano}\frac{u_2\left(u_2-v_{f2}\cot\beta_2\right)}{g} \quad \text{...(11.23)}$$

## 11.13 Efficiencies of a Centrifugal Pump

1. **Manometric efficiency, $\eta_{mano}$ :** the ratio of the manometric head developed by the pump to the head imparted by the impeller to the liquid is called the manometric efficiency.

$$\eta_{mano}=\frac{\text{Manometric head}}{\text{Head imparted by impeller to liquid}}$$

Head imparted by impeller to liquid

$$=\frac{H_{mano}}{\frac{u_2v_{w2}}{g}}=\frac{gH_{mano}}{u_2v_{w2}} \quad \text{...(11.24)}$$

2. **Volumetric efficiency, $\eta_{vol}$ :** it is defined as the ratio of quantity of liquid discharged per second ($Q$) from the pump to quantity passing per second through the impeller, ($Q_a$).

$$\eta_{vol} = \frac{Q}{Q_a = Q + Q_l} \quad ...(11.25)$$

Where $Q_l$ = leakage of liquid per second from the impeller

3. **Mechanical efficiency, $\eta_{mech}$** : It is defined as the ratio of power delivered by the impeller to the liquid to the power input to the pump shaft.

$$\eta_{mech} = \frac{\gamma Q_a u_2 v_{w2}}{gP} = \frac{P - P_{mechloss}}{P} \quad ...(11.26)$$

4. **Overall efficiency, $\eta_o$** : it is the ratio of power outlet of pump to the power input to the pump shaft.

$$\eta_o = \frac{P_{output}}{P_{inout}} = \frac{\gamma Q H_{mano}}{P} \quad ...(11.27)$$

Also
$$\eta_o = \eta_{mano} \times \eta_{vol} \times \eta_{mech} \quad ...(11.28)$$

$$= \frac{H_{mano} \times g}{u_2 v_{w2}} \times \frac{Q}{Q + Q_l} \times \frac{\gamma(Q + Q_l) u_2 v_{w2}}{gP}$$

Where $P$ = shaft power = $sp$ = $BP$ of driving unit – power lost in coupling

$$= \frac{u_2\, v_{w2}}{g \times 10^3}$$

$$WP = \text{water power} = \frac{\gamma Q H_{mano}}{10^3}$$

$$P_{hyd} = \frac{\gamma Q \,.\, \Delta H_{mano}}{10^3} \quad ...(11.29)$$

Power required to overcome head losses due to the following:

(i) Circulatory or secondary flow.

(ii) Friction of volute and impeller.

(iii) Turbulence.

**Example 11.2** Water is raised to a height of 40 m by means of a centrifugal pump. If the quantity of water raised is 350 m³/h, determine the power of electric motor required to drive the pump. Take efficiency of pump to be 85 percent and that of electric motor 95%

**Solution.**

Given: $H$ = 40 m, $Q$ = 350 m³/h, $\eta_p$ =0.95, $\eta_m$ = 0.95

Power of motor required,
$$P = \frac{\gamma Q H}{10^3 \times \eta_p \times \eta_m}$$

$$= \frac{9.81 \times 10^3 \times 350 \times 40}{3600 \times 10^3 \times 0.85 \times 0.95} = 47.245 \text{ kW}$$

**Example 11.3** A centrifugal pump lifts water against a static α head of 30 m of which suction lift is 3m. The loss of head in suction pipe is 1 m and 5 m in delivery pipe. Speed of pump is 1440 rpm.

The combined efficiency of motor and transmission system is 0.85. The flow rate is 60 litres/s. Calculate the power consumed by pump.

**Solution.**

Given: $H_{stat} = H_s + H_d = 30m, H_s = 3m, N = 1440\ rpm, \eta = 0.85, Q = 60$ litres/s or 0.06 m³/s, $h_{fs} = 1m$, $h_{fd} = 5m$

$$H_{mano} = H_{stat} + h_{fs} + h_{fd}$$

$$= 30 + 1 + 5 = 36\text{ m}$$

Power required to drive the pump

$$= \frac{\gamma Q H_{mano}}{10^3 \times \eta}$$

$$= \frac{9.81 \times 10^3 \times 0.06 \times 36}{10^3 \times 0.85}$$

$$= 24.929\text{ kW}$$

**Example 11.4** A centrifugal pump is to discharge 0.118 m³/s at a speed of 1450 rpm against a head of 25m. the impeller diameter at outlet is 0.25m, its width at outlet is 50 mm and manometric efficiency is 75 percent. Determine the vane angle at the outer periphery of the impeller.

**Solution.**

Given: $Q = 0.118$ m³/s, $N = 1450$ rpm, $H_{mano} = 25$ m, $D_2 = 0.25$m, $B_2 = 0.050$ m, $\eta_{mano} = 0.75$

The velocity triangles are shown is Fig. 11.7

$$u_2 = \frac{\pi D_2 N}{60} = \frac{\pi \times 0.25 \times 1450}{60}$$

$$= 18.98\text{ m/s}$$

$$Q = \pi D_2 B_2 v_{f2}$$

$$0.118 = \pi \times 0.25 \times 0.050 \times v_{f2}$$

$$v_{f2} = 3\text{ m/s}$$

$$\eta_{mano} = \frac{gH_{mano}}{u_2 v_{w2}}$$

$$0.75 = \frac{9.81 \times 25}{18.98 \times v_{w2}}$$

$$v_{w2} = 17.23\text{ m/s}$$

$$\tan \beta_2 = \frac{v_{f2}}{u_2 - v_{w2}}$$

$$= \frac{3}{18.98 - 17.23} = 1.7143$$

$$\beta_2 = 59.74^\circ$$

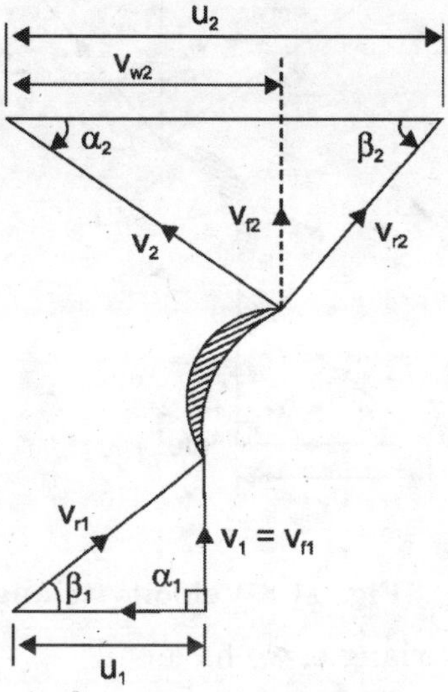

**Fig. 11.7 Velocity triangles.**

**Example 11.5** The impeller of a centrifugal pump is of 0.35m diameter and 60 mm width at the periphery. The tip angle of blades is inclimed backwards $\alpha$ at 60° from the radius. The pump delivers 20 m³/min of water. The impeller rotates at 1200 rpm. For radial inlet, calculate:

(a) Speed and direction of water as it leaves the impeller,

(b) Torque exerted by the impeller on water,

(c) Shaft power required, and

(d) Lift of the pump.

Take mechanical efficiency = 94% and hydraulic efficiency = 76%

**Solution.**

Given: $D_2 = 0.35$m, $B_2 = 0.06$ m, $\beta_2 = 60°$, $\alpha_1 = 90°$, $Q = 20$m³/min, $N = 1200$ rpm, $\eta_{mech} = 0.94$, $\eta_{hyd} = 0.76$

(a) Tangential velocity of impeller at outlet,

$$u_2 = \frac{\pi D_2 N}{60} = \frac{\pi \times 0.35 \times 1200}{60} = 22 m/s$$

Now

$$Q = \pi D_2 B_2 v_{f2}$$

$$\frac{20}{60} = \pi \times 0.35 \times 0.06 \times v_{f2}$$

$$v_{f2} = 5.05 m/s$$

The velocity diagrams are shown in Fig. 11.8

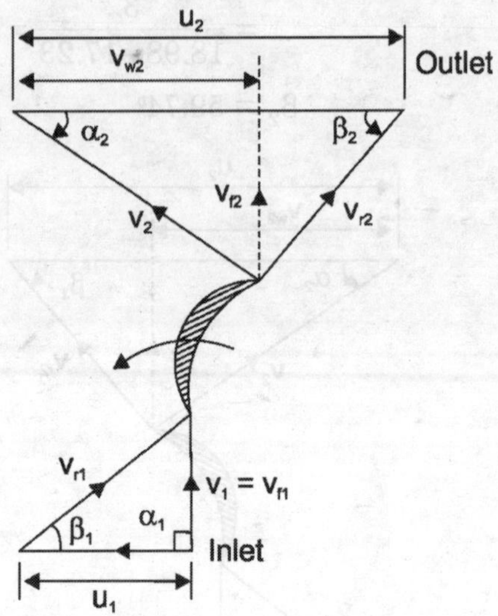


**Fig. 11.8 Velocity triangles.**

From outlet velocity triangle, we have

$$\tan \beta_2 = \frac{v_{f2}}{u_2 - v_{w2}}$$

$$\tan 60^\circ = \frac{5.05}{22 - v_{w2}}$$

$$22 - v_{w2} = 2.91$$

$$v_{w2} = 19.09 m/s$$

$$v_2 = \sqrt{v_{w2}^2 + v_{f2}^2} = \sqrt{(19.09)^2 + (5.05)^2} = 19.75 m/s$$

$$\tan \alpha_2 = \frac{v_{f2}}{v_{w2}} = \frac{5.05}{19.09} = 0.2645$$

$$\alpha_2 = 14.82^\circ$$

(b) Torque exerted by the impeller on water,

$$T = \frac{\gamma Q}{g}(v_{w2} R_2) = \rho Q (v_{w2} R_2)$$

$$= 1000 \times \frac{20}{60} \times 19.09 \times \frac{0.35}{2}$$

$$= 1113.58 \text{ N.m}$$

(c) Shaft power required, $SP = \dfrac{2\pi NT}{60 \times \eta_{mech}}$

$$= \frac{\pi \times 1200 \times 1113.58}{60 \times 0.94} = 148.87 kW$$

(d) $$P = \frac{\gamma QH}{\eta_{mech} \times \eta_{hyd}}$$

Pump lift, $H = \dfrac{148.87 \times 0.94 \times 0.76 \times 60}{9.81 \times 10^3 \times 20} = 32.52 m$

**Example 11.6** The following data relate to a centrifugal pump:

Impeller diameter at inlet = 200 mm

Impeller diameter at outlet = 400 mm

Width if impeller at inlet = 15 mm

Width of impeller at outlet = 7.5 mm

Flow rate = 18 lps

Impeller speed = 1500rpm

Vane angle of outlet = $45^0$

Find the pressure rise in the impeller for radial inlet

**Solution.**

Given: $D_1$ = 200 mm, $D_2$ = 400 mm, $B_1$ = 150 mm, $B_2$ = 75 mm, $Q$ = 18 lps, $N$ = 1500 rpm, $\beta_2$ = 45°, $\alpha_1$ = 90°

$$v_{f1} = \frac{Q}{\pi D_1 B_1} = \frac{18 \times 10^{-3}}{\pi \times 0.2 \times 0.015} = 1.91 m/s$$

$$v_{f2} = \frac{Q}{\pi D_2 B_2} = \frac{18 \times 10^{-3}}{\pi \times 0.4 \times 0.075} = 1.91 m/s$$

$$u_2 = \frac{\pi D_2 N}{60} = \frac{\pi \times 0.4 \times 1500}{60} = 31.416 m/s$$

Pressure rise in impeller $= \dfrac{1}{2g}\left[v_{f1}^2 + u_2^2 \operatorname{cosec}^2 \beta_2\right]$

$$= \frac{1}{2 \times 9.81}\left[(1.91)^2 + (31.416)^2 - (1.91)^2 \operatorname{cosec}^2 45^0\right] = 50.12 m$$

**Example 11.7** A centrifugal pump is to discharge 0.2 m³/s at a speed of 1400 rpm against a head of 24 m. The impeller diameter is 0.3 m, width at outlet is 0.05 m and the manometric efficiency is 80 per cent.

Determine the vane angle at the outlet periphery of the impeller and label the angle by sketching the impeller.

**Solution.**

Given: $Q = 0.2\ m^3/s$, $N = 1400$ rpm, $H_{mano} = 25m$, $D_2 = 0.3m$, $B_2 = 0.05$, $\eta_{mano} = 0.8$

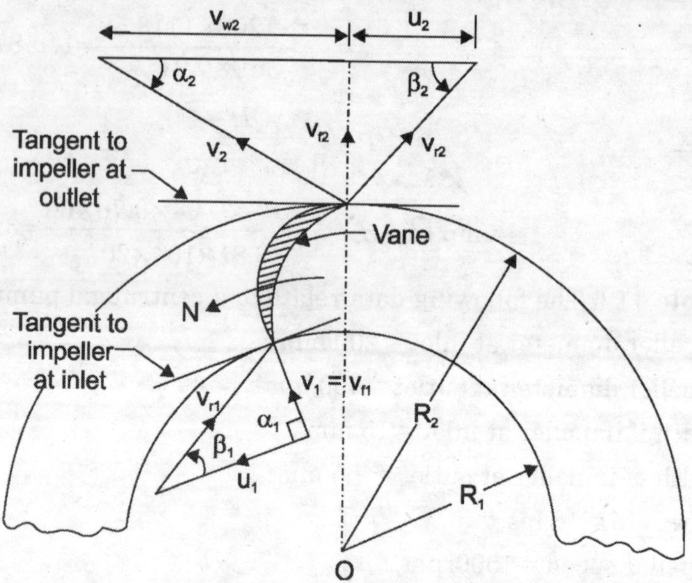


**Fig. 11.9 Velocity triangles.**

$$u_2 = \frac{\pi D_2 N}{60} = \frac{\pi \times 0.3 \times 1400}{60} = 22\,m/s$$

The velocity triangles are shown in Fig. 11.9

Manometric efficiency,
$$\eta_{mano} = \frac{gH_{mano}}{u_2 v_{w2}}$$

$$0.8 = \frac{9.81 \times 25}{22 \times v_{w2}}$$

$$v_{w2} = \frac{9.81 \times 25}{22 \times 0.8} = 13.93\,m/s$$

$$v_{f2} = \frac{Q}{\pi D_2 B_2} = \frac{0.2}{\pi \times 0.3 \times 0.05} = 4.244\,m/s$$

$$\tan \beta_2 = \frac{v_{f2}}{u_2 - v_{w2}} = \frac{4.244}{22 - 13.93} = 0.5259$$

$$\beta_2 = 27.74°$$

**Example 11.8** A centrifugal pump has an impeller of 80 cm in diameter and it delivers 1 m³/s against a head of 80 m. The impeller runs at 1000rpm and the width at outlet is 8 cm. If the leakage loss is 3 per cent of the discharge, external mechanical loss is 10 k W and the manometric efficiency is 80 per cent, calculate the blade angle at outlet, the power required and over all efficiency of the pump.

**Solution.**

Given: $D_2$ = 0.80m, $Q_a$ = 1m³/s , $H_{mano}$ = 80*m*, $B_2$ = 0.08*m*, $Q_{leak}$ = 0.03 *Q*, $P_{mechioss}$ = 10 kW, $\eta_{mano}$ = 0.8, *N* = 1000*rpm*

Theoretical discharge, $Q_{th} = Q_a + Q_{leak} = 1 + 0.03 \times 1 = 1.03\ m^3/s$

$$u_2 = \frac{\pi D_2 N}{60} = \frac{\pi \times 0.8 \times 1000}{60} = 41.89 m/s$$

$$Q_{th} = \pi D_2 B_2 v_{f2}$$

$$1.03 = \pi \times 0.8 \times 0.08 \times v_{f2}$$

$$v_{f2} = 5.12 m/s$$

$$\eta_{mano} = \frac{gH_{mano}}{u_2 v_{w2}}$$

$$v_{\omega 2} = \frac{9.81 \times 80}{41.89 \times 0.8} = 23.42\ m/s$$

$$\tan\beta_2 = \frac{v_{f2}}{u_2 - v_{w2}} = \frac{5.12}{41.89 - 23.42} = 0.2772$$

$$\beta_2 = 15.49^\circ$$

Blade angle at outlet = 15.49°

$$\text{Power required} = \frac{\gamma Q_{th} \times u_2 v_{w2}}{g \times 10^3} + \alpha 10 kW$$

$$= 1.03 \times 41.89 \times 23.42 + 10$$

$$= 1010.5 + 10 = 1020.5\ kW$$

$$\text{Mechanical efficiency} = \frac{1010.5}{1020.5} = 0.99 \text{ or } 99\%$$

Over all efficiency,
$$\eta_0 = \eta_{mano} \times \eta_{mech}$$

$$= 0.8 \times 0.99 = 0.792 \text{ or } 79.2\%$$

**Example 11.9** A centrifugal pump is required to discharge 600 litres of water per second and develop a head of 15m when the impeller rotates at 750 rpm. The manometric efficiency is 80 per cent. The loss of head in the pump due to fluid resistance is being assumed to be 0.27 $v^2$ where $v$ is the absolute velocity with which water leaves the impeller. Water enters the impeller without shock or whirl and the velocity of flow is 3.2m/s. Determines:

(a) The impeller diameter and outlet area, and

(b) Blade angle at the outlet edge.

**Solution.**

Given: $Q = 600$ lps or $0.6\ m^3/s$, $H_{mano} = 15$ m, $N = 750$ rpm, $\eta_{mano} = 0.8$, $h_f = 0.027\ v^2$, $v_f = 3.2$m/s

(a)
$$\eta_{mano} = \frac{gH_{mano}}{u_2 v_{w2}}$$

$$H_{mano} = \frac{u_2 v_{w2}}{g} - h_f = \frac{H_{mano}}{\eta_{mano}} - h_f$$

$$15 = \frac{15}{0.8} - 0.027 v_2^2$$

$$v_2^2 = 15\left(\frac{1}{0.8} - 1\right) \times \frac{1}{0.027} = 138.89$$

$$v_2 = 11.785 \text{ m/s}$$

Now
$$v_{f1} = v_{f2} = v_f = 3.2 m/s$$

$$\text{Discharge, } Q = A_f \times v_f$$

$$\text{Area of flow, } A_f = \frac{0.6}{3.2} = 0.1875 m^2$$

Tangential velocity at outlet,
$$u_2 = \frac{\pi D_2 N}{60}$$

$$= \frac{\pi \times D_2 \times 750}{60} = 39.27\, D_2\, m/s$$

Now
$$u_2 v_{w2} = \frac{gH_{mano}}{\eta_{mano}} = \frac{9.81 \times 15}{0.8} = 183.94$$

$$\therefore \quad v_{w2} = \frac{183.94}{39.27 \times D_2} = \frac{4.684}{D_2}$$

The velocity triangles for the centrifugal pump are shown in Fig. 11.10 From outlet velocity triangle, we have

$$v_{f2}^2 = v_2^2 - v_{w2}^2$$

$$(3.2) = (11.785)^2 - v_{w2}^2$$

$$v_{w2} = 11.34\ m/s = \frac{4.684}{D_2}$$

$$D_2 = 0.413 m \text{ or } 413\ mm$$

Outlet area across flow,
$$A_{f2} = \pi D_2\, B_2 v_{f2} = \frac{Q}{v_{f2}} = \frac{0.6}{3.2} = 0.1875 m^2$$

$$u_2 = 39.27 \times 0.0413 = 16.22 m/s$$

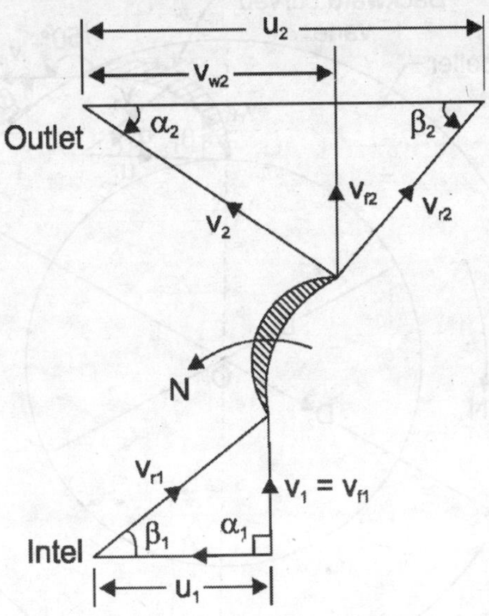


**Fig. 11.10 Velocity triangles.**

(b) From outlet velocity triangle, we have

$$\tan \beta_2 = \frac{v_{f2}}{u_2 - v_{w2}}$$

$$= \frac{3.2}{16.22 - 11.34}$$

$$= 0.66573$$

$$\beta_2 = 33.25°$$

**Example 11.10** A centrifugal pump is required to lift 9100 litres per minute of water against a head of 6m while running at 500 rpm. The velocity of flow through the wheel is 2m/s and the manometric efficiency is 60% the angle of vane tip with the direction of motion is 150°. Work out the following:

(a) Sketch an impeller of centrifugal pump and draw the velocity triangles.

(b) The diameter of impeller.

(c) The width of impeller.

**Solution.**

Given: $Q = 9100\ lpm = \dfrac{9100 \times 10^3}{10^6 \times 60} = 0.1517\, m^3/s,\ H_{mano} = 6m,\ N = 500 rpm,$

$v_f = 2m/s,\ H_{mano} = 0.6\text{m},\ \beta_2 = 180° - 150° = 30°$

(a) The sketch of pump impeller and velocity triangles are shown in Fig.11.11.

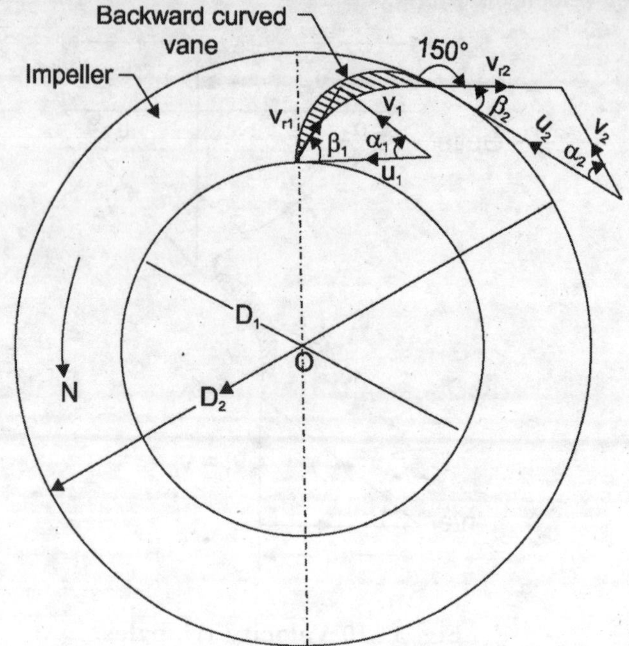


(a) Sketch of impeller.

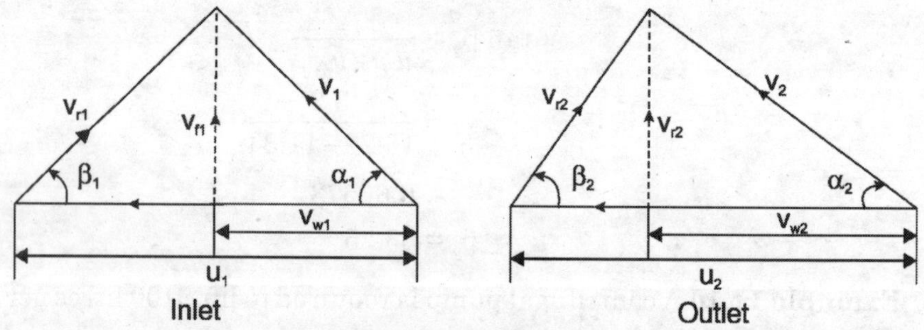


(b) Velocity triangles.

**Fig. 11.11**

(b)
$$\eta_{mano} = \frac{H_{mano}}{\left(\dfrac{u_2 v_{w2}}{g}\right)}$$

$$u_2 v_{w2} = \frac{9.81 \times 6}{0.6} = 98.1$$

$$u_2 = \frac{\pi D_2 N}{60} = \frac{3.14 \times D_2 \times 500}{60} = 26.18 D_2 \ m/s$$

$$v_{f1} = v_{f2} = 2\text{m/s}$$

$$v_{w2} = \frac{98.1}{26.18 D_2} = \frac{3.747}{D_2}$$

From outlet velocity triangle, we have

$$\tan \beta_2 = \frac{v_{f2}}{u_2 - v_{w2}}$$

$$\tan 30^\circ = \frac{2}{26.18D_2 - v_{w2}}$$

$$0.57735 = \frac{2}{26.18D_2 - \frac{3.747}{D_2}}$$

Or $$26.18.D_2^2 - 3.747 = 3.464D_2$$

Or $26.18.D_2^2 - 0.13232D_2 - 0.14312 = 0$

$$D_2 = \frac{1}{2}\left[0.13232 \pm \sqrt{(0.13232)^2 + 4 \times 0.14312}\right]$$

$$= 0.45 m \text{ or } 450 mm$$

(c) $$Q = \pi D_2 B_2 v_{f2}$$

$$B_2 \frac{0.1517}{\pi \times 0.45 \times 2} = 0.054 \text{ m or } 54 \text{ mm}$$

**Example 11.11** A centrifugal pump having internal and external diameters of 250 mm and 500 mm respectively is running at 1200 rpm. The rate of flow through the pump is 0.06 m³/s and velocity of flow is constant and equal to 2.5 m/s. The suction and delivery pipes are 200 mm and 150 mm diameters respectively. The suction and delivery heads are 6 m and 30 m of water absolute respectively. If the power required to drive the pump is 25 kW and the outlet vane angle is $45^0$, determine:

(a) Inlet vane angle,

(b) Overall efficiency of the pump, and,

(c) Manometric efficiency of the pump.

**Solution.**

Given: $D_1 = 250$ mm, $D_2 - 500$ mm, $N = 1200$ rpm, $Q = 0.06$ m³/s, $v_{f1} - v_{f2} = 2.5$ m/s, $d_d$ 200 mm, $d_d = 150$ mm, $H_s = 6$ m, $H_d = 30$ m, $p = 25$ kW, $\beta_2 = 45°$ $z_1 = z_2$.

(a) The velocity triangle are shown in Fig. 11.12.

$$u_1 = \frac{\pi D_1 N}{60} = \frac{\pi \times 0.25 \times 1200}{60} = 15.71 m/s$$

From inlet velocity triangle, we have

$$\tan \beta_1 = \frac{v_{f1}}{u_1} = \frac{2.5}{15.71} = 0.15915$$

$$\beta_1 = 9.04°$$

Inlet vane angle = 9.04°

(b) Manometric head,
$$H_{mano} = \frac{p_2 - p_1}{\gamma} + \frac{v_2^2 - v_1^2}{2g} + (z_2 - z_1)$$

$$v_1 = v_s = \frac{4Q}{\pi d_s^2} = \frac{4 \times 0.06}{\pi (0.2)^2} = 1.91 m/s$$

$$v_2 = v_d = \frac{4Q}{\pi d_s^2} = \frac{4 \times 0.06}{\pi (0.15)^2} = 3.395 m/s$$

$$H_{mano} = (30 - 60) + \frac{(3.395)^2 - (1.91)^2}{2 \times 9.81} = 24.4 m$$

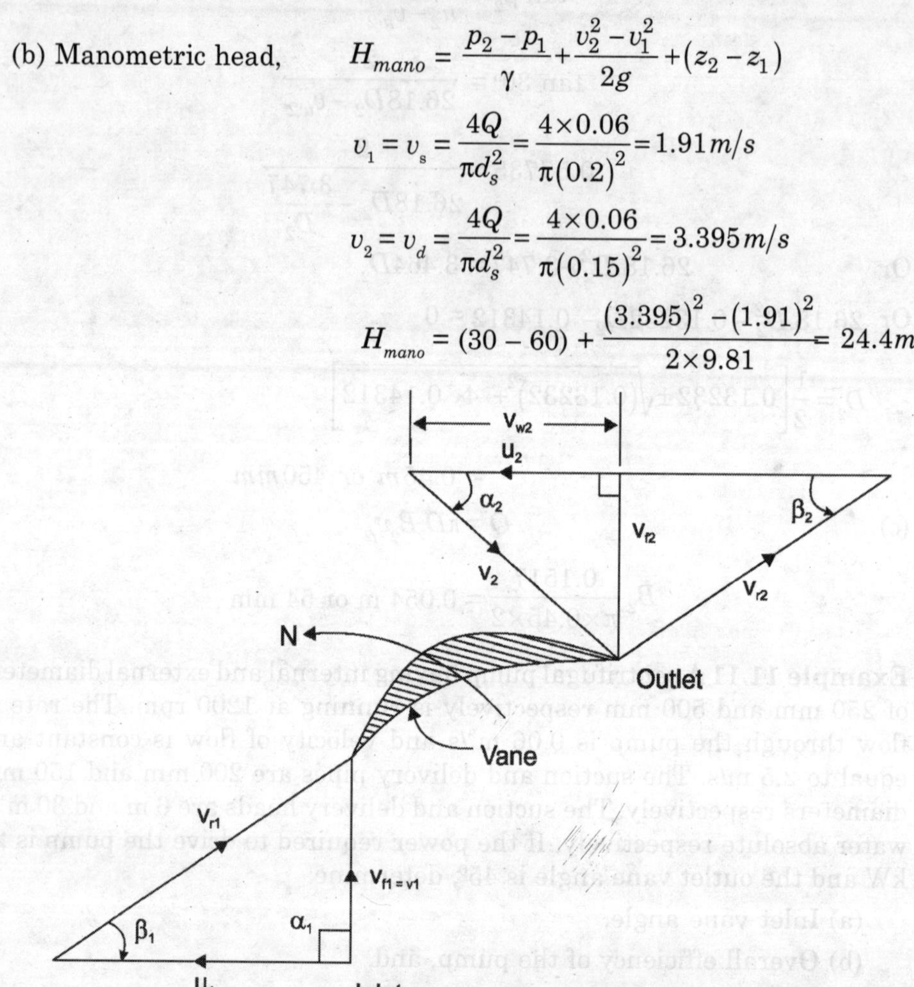


**Fig. 11.12 Velocity triangles**

Overall efficiency,
$$\eta_o = \frac{\gamma Q H_{mano}}{P}$$

$$= \frac{9.81 \times 10^3 \times 0.06 \times 24.4}{25 \times 10^3}$$

$$= 0.5745 \text{ or } 57.45\%$$

(c)
$$\eta_{mano} = \frac{g H_{mano}}{u_2 v_{w2}}$$

$$u_2 = \frac{\pi D_2 N}{60} = \frac{\pi \times 0.5 \times 1200}{60} = 31.416 m/s$$

$$v_{w2} = u_2 - v_{f2} \cot \beta_2 = 31.416 - 2.5 \cot 45° = 28.916 \text{ m/s}$$

$$\eta_{mano} = \frac{9.18 \times 24.4}{31.416 \times 28.916} = 0.2635 \text{ or } 26.35\%$$

**Example 11.12** A centrifugal pump impeller has outlet diameter of 350 mm and width 50 mm. the vanes are curved backwards at 40° to the tangent at outer periphery. The thickness of vanes occupies 20 percent of the peripheral area and the velocity of flow is constant from inlet to outlet. The impeller rotates at 750 rpm. If the flow rate through the pump is 0.15 m³/s determine:

(a) The pressure rise in the impeller, and

(b) The percentage of total work converted to outlet kinetic energy.

**Solution.**

Given: $D_2 = 350$ mm, $B_2 = 50$ mm, $k_t = 0.8$, $\beta_2 = 40°$, $N = 750$ rpm, $Q = 0.15$ m³/s

Refer to Fig. 11.12

(a)

$$u_2 = \frac{\pi D_2 N}{60} = \frac{\pi \times 0.35 \times 750}{60} = 13.74\, m/s$$

$$0.15 = 0.8 \times \pi \times 0.35 \times 0.05 \times v_{f2}$$

$$0.15 = 0.8 \times \pi \times 0.35 \times 0.05 \times v_{f2}$$

$$v_{f2} = 3.41 \text{ m/s}$$

$$v_{f1} = v_{f2} = 3.41 \text{ m/s} \qquad \textit{(given)}$$

$$\tan \beta_2 = \frac{v_{f2}}{u_2 - v_{w2}}$$

$$\tan 40° = \frac{3.41}{13.74 - v_{w2}}$$

$$v_{w2} = 9.67 \text{ m/s}$$

$$\text{Pressure rise} = \frac{1}{2g}\left(v_{f1}^2 + u_2^2 - v_{f2}^2 \operatorname{cosec}^2 \beta_2\right)$$

$$= \frac{1}{2 \times 9.81}\left[(3.41)^2 + (13.74)^2 - (3.41)^2 \operatorname{cosec}^2 40\right]$$

$= 8.78 m$ of water

(b)

$$v_2 = \sqrt{v_{f2}^2 + v_{w2}^2}$$

$$= \sqrt{(3.41)^2 + (9.67)^2} = 10.25 \text{ m/s}$$

Kinetic energy in exit velocity from impeller

$$= \frac{v_2^2}{2g} = \frac{(10.25)^2}{2 \times 9.81}$$

= 5.35 N.m per unit weight of water

Work done per unit weight of water = $\frac{u_2 v_{w2}}{g} = \frac{13.74 \times 9.67}{9.81}$ 13.54 N.m

Percentage of work converted to kinetic energy = $\frac{5.35}{13.54} \times 100 = 39.51\%$

**Example 11.13** A centrifugal pump lifts water under a static head of 40m of water, out of which 5m is suction lift. The suction and delivery pipes are both 150 mm in diameter. The head loss in suction pipe is 2m and in delivery pipe 6m. The impeller is 400 mm in diameter and 25 mm wide at the mouth. It revolves at 1200rpm. Its exit blade angle is 35°. If the manomrtric efficiency of the pump is 85 percent, determine:

(a) The discharge through the pump and

(b) The pressure at suction and delivery branches of the pump.

**Solution.**

Given: $H_{static} = 40m$, $H_s = 5m$, $d_s = d_d = 150mm$, $H_{fs} = 2m$, $H_{fd} = 6m$, $D_2$ = 400mm, $B_2 = 25mm$, $N = 1200\ rpm$, $\beta_2 = 35°$, $\eta_{mano} = 0.85$

(a) $H_{mano} = H_{static} + H_{fs} + H_{fd} = 40 + 2 + 6 = 48m$

$$u_2 = \frac{\pi D_2 N}{60} = \frac{\pi \times 0.4 \times 1200}{60} = 25.13 m/s$$

Let $\alpha_1 = 90°$

Work done per unit weight of water $= \frac{u_2 v_{w2}}{g}$

$$\eta_{mano} = \frac{g H_{mano}}{u_2 v_{w2}}$$

$$v_{w2} = \frac{9.81 \times 48}{25.13 \times 0.85} = 22.04 m/s$$

$$\tan \beta_2 = \frac{v_{f2}}{u_2 - v_{w2}}$$

$$\tan 35° = \frac{v_{f2}}{25.13 - 22.04} = \frac{v_{f2}}{3.09}$$

$$v_{f2} = 2.16 m/s$$

$$Q = \pi D_2 B_2 v_{f2}$$

$$= \pi \times 0.4 \times 0.025 \times 2.16 = 0.06786 m^3/s$$

(b) $$v_s = v_d = \frac{4Q}{\pi d_s^2} = \frac{4 \times 0.06786}{\pi \times (0.15)^2} = 3.84 m/s$$



$$\text{Velocity head} = \frac{v_s^2}{2g} = \frac{v_d^2}{2g} = \frac{(3.84)^2}{2 \times 9.81} = 0.75m$$

Total effective pressure on delivery side

$$= H_d + H_{fd} + \frac{v_d^2}{2g}$$

$$= (40 - 5) + 6 + 0.75 = 41.75 \text{ m of water}$$

$$= 9.81 \times 10^3 \times 41.75 = 409.56 \text{ kN/m}^2$$

Total effective pressure on suction side

$$= H_s + H_{fs} + \frac{v_s^2}{2g}$$

$$= 5 + 2 + 0.75 = 7.75 \text{ m of water (vacuum)}$$

or $$10.3 - 7.75 = 2.55 \text{ m of water abs.}$$

$$= 9.81 \times 10^3 \times 2.55 = 25.01 \text{ kN/m}^2$$

**Example 11.14** A centrifugal pump is required to deliver 0.25 m³/s of water against a head of 25m when the impeller rotates at 1500rpm. The manometric efficiency is 80 percent. The head loss due to fluid resistance is 0.03 $v_2^2$ metres, where $v_2$ is the velocity of water leaving the impeller in m/s. The area of impeller outlet surface is 1.2 $D_2^2$ $m^2$, where $D_2$ is impeller outlet diameter in metres. Determine:

(a) The impeller diameter and

(b) The outlet vane angle.

Assume that water enters the impeller without whirl.

**Solution.**

Given: $Q = 0.25\ m^3/s$, $H_{mano} = 25m$, $N = 1500\ rpm$, $\eta_{mano} = 0.8$, $H_l = 0.03 v_2^2$, $A_{f2} = 1.2\ D^2_2$, $\alpha_1 = 90°$

(a) $$\eta_{mano} = \frac{gH_{mano}}{u_2 v_{w2}}$$

$$0.8 = \frac{9.81 \times 25}{u_2 v_{w2}}$$

Or $$u_2\, v_{w2} = 306.5625$$

Also $$H_{mano} = \frac{u_2 v_{w2}}{g} - H_l$$

$$H_l = \frac{306.5625}{9.81} - 25 = 6.25m$$

∴ $$0.03 v_2^2 = 6.25$$

$$v_2 = 14.4\ m/s$$

$$v_{f2} = \frac{Q}{A_{f2}} = \frac{0.25}{1.2D_2^2} = \frac{0.2083}{D_2^2}\ m/s$$

$$v_2 = \frac{\pi D_2 N}{60} = \frac{\pi D_2 \times 1500}{60} = 78.54\, D_2\ m/s$$

$$v_{w2} = \frac{306.5625}{78.54 D_2} = \frac{3.09}{D_2}\ m/s$$

$$v_{f2} = \sqrt{v_2^2 - v_{w2}^2}$$

$$\frac{0.2083}{D_2^2} = \sqrt{(14.4)^2 - \left(\frac{3.90}{D_2}\right)2}$$

$$\frac{0.0434}{D_2^4} = 207.36 - \frac{15.21}{D_2^2}$$

or $207.36\, D_2^4 - 15.21 D_2^2 - 0.0434 = 0$

$$D_2^{\ 2} = \frac{1}{2}\left[0.0733 \pm \sqrt{(0.0733)^2 + 4\times 1\times 2.09\times 10^{-4}}\right]$$

$$= \frac{1}{2}[0.0733 \pm 0.0786] = 0.0759$$

$$D_2 = 0.275m \text{ or } 275mm$$

(b)

$$\tan\beta_2 = \frac{v_{f2}}{u_2 - v_{w2}} = \frac{\dfrac{0.2083}{(0.275)^2}}{78.54\times 0.275 - \dfrac{3.90}{0.275}}$$

$$= \frac{2.7544}{7.4167} = 0.37/38$$

$$\beta_2 = 20.4°$$

**Example 11.15** The following data refer to a radial single stage, double suction, centrifugal pump:

$Q_2 = 100\ lps$, $D_1 = 100\ mm$, $D_2 = 300\ mm$, $H_{mano} = 42\ m$, $N = 1800\ rpm$, $B_1 = 25\ mm/side$, $B_2 = 23\ mm\ total$, $\eta_o = 60\%$, $\Delta Q = 2.5 lps$, $P_{mech} = 1.5$ kW, $K_t = 0.87$, $\beta_2 = 30°$, $\alpha_1 = 90°$

Determine:

*(a)* $\beta_1$ *(b)* $\alpha_2$ (c) *speed ratio* (d) $v_2$ (e) $\eta_{mano}$ *(f)* $\eta_{vol}$ *(g)* $\eta_{mech}$.

**Solution.**

(a) $Q_t = Q_2 + \Delta Q = 100 + 2.5 = 102.5\ lps$

or $= 0.1025\ m^3/s$

$$Q_{perside} = \frac{Q_t}{2} = \frac{0.1025}{2} = 0.05125\, m^3/s$$

$$u_1 = \frac{\pi D_1 N}{60} = \frac{\pi \times 0.1 \times 1800}{60} = 9.425\ m/s$$

$$Q = K_t \times \pi D_1 B_1 v_{f1}$$

$$0.05125 = 0.87 \times \pi \times 0.1 \times 0.025 \times v_{f1}$$

$$v_{f1} = 7.5\ m/s$$

$$\tan \beta_1 = \frac{v_{f1}}{u_1} = \frac{7.5}{9.425} = 0.79575$$

$$\beta_1 = 38.5°$$

(b)

$$Q = K_t \times \pi D_2 B_2 v_{f2}$$

$$B_2 = \frac{0.023}{2} = 0.0115\, m \text{ for one side}$$

$$v_{f2} = \frac{0.05125}{0.87 \times \pi \times 0.3 \times 0.0115} = 5.435\, m/s$$

$$v_2 = \frac{\pi D_2 N}{60} = \frac{\pi \times 0.3 \times 1800}{60} = 28.27\, m/s$$

$$\tan \beta_2 = \frac{v_{f2}}{u_2 - v_{w2}}$$

$$\tan 30° = \frac{5.435}{28.27 - v_{w2}}$$

$$v_{f2} = 18.85\ m/s$$

$$\tan \alpha_2 = \frac{v_{f2}}{v_{w2}} = \frac{5.435}{18.85} = 0.28833$$

$$\alpha_2 = 16.08°$$

(c) Speed ratio,

$$K_{u2} = \frac{u_2}{\sqrt{2gH_{mano}}} = \frac{28.27}{\sqrt{2 \times 9.81 \times 42}} - 0.985$$

(d)

$$v_2 = \frac{v_{w2}}{\cos \alpha_2} = \frac{18.85}{\cos 16.08} = 19.6\, m/s$$

(e)

$$\eta_{mano} = \frac{gH_{mano}}{u_2 v_{w2}}$$

$$= \frac{9.81 \times 42}{28.27 \times 18.85} = 0.7732 \text{ or } 77.32\%$$

(f)

$$\eta_{vol} = \frac{Q_2}{Q_2 + \Delta Q} = \frac{100}{100 + 2.5}$$

$$= \frac{100}{102.5} = 0.9756 \text{ or } 97.56\,\%$$

(g)
$$\eta_{mech} = \frac{\text{shaft power} - \text{mechanical losses}}{\text{shaft power}}$$

$$\text{Shaft power} = \frac{\gamma Q_2}{2} \times \frac{H_{mano}}{\eta_0}$$

$$= 9.81 \times \frac{0.1}{2} \times \frac{42}{0.6} = 34.335\, kW$$

$$\eta_{mech} = \frac{34.335 - 1.5}{34.335} = 0.9563 \text{ or } 95.63\,\%$$

**Example 11.16** The impeller of a centrifugal pump has an outer diameter of 250 mm and an effective are a of 0.017 m$^2$ the blades are bent backwards so that the direction of outlet relative velocity makes an angle of 148° with the tangent drawn in the direction of impeller rotation. The diameters of suction and delivery pipes are 150 mm and 100 mm respectively. The pump delivers 0.031 m$^3$/s at 1450 rpm. When the gauge points on the suction and delivery pipes are close to the pump, show heads of 4.6 m below and 18 m above atmospheric respectively. The head losses in the suction and delivery pipes are 2 m and 2.9 m respectively. The motor driving the pump delivers 8.67 kW. Assuming that water enters the pump without shock and whirl, determine:

(a) The manometric efficiency, and

(b) The overall efficiency of pump.

**Solution.**

Given: $D_2 = 250\ mm$, $A_{f2} = 0.017\ m^2$, $\beta_2 = 180° - 148° = 32°$, $d_s = 150\ mm$, $d_d = 100 mm$, $Q = 0.031\ m^3/s$, $N = 1450\ rpm$, $\frac{p_s}{\gamma} = -4.6\ m$, $\frac{p_d}{\gamma} = 18\ m$, $H_{fs} = 2\ m$, $H_{fd} = 2.9\ m$, $P = 8.67\ kW$, $\alpha_1 = 90°$

(a)
$$v_s = \frac{4Q}{\pi d_s^2} = \frac{4 \times 0.031}{\pi (0.15)^2} = 1.754\ m/s$$

Applying Bernoulli's equation between water surface in sump and pump inlet where suction pressure gauge is fitted we have

$$0 = H_s + H_{fs} + \frac{p_s}{\gamma} + \frac{v_s^2}{2g}$$

$$= H_s + 2 - 4.6 + \frac{(1.754)^2}{2 \times 9.81}$$

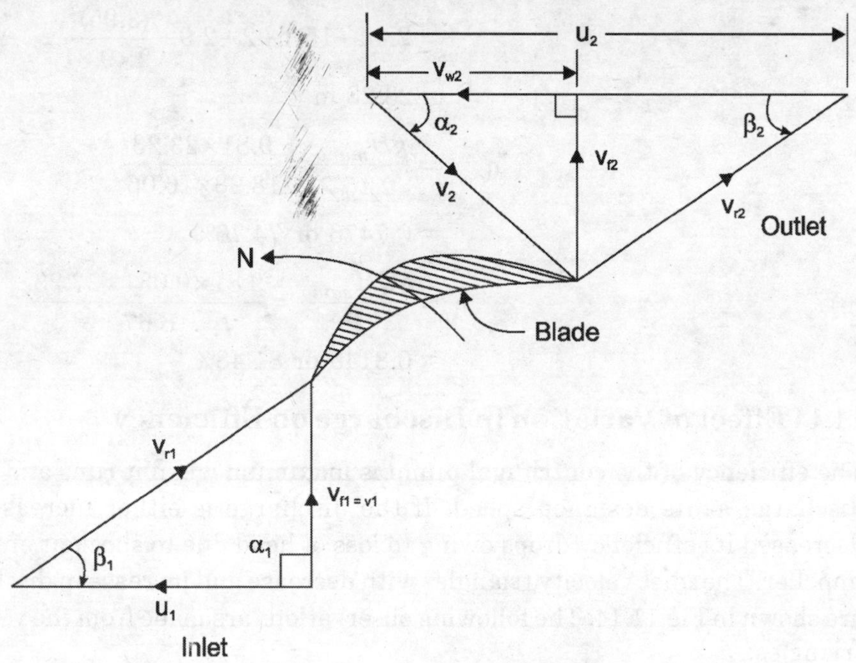


**Fig. 11.13 Velocity triangles.**

$$H_s = 2.44\ m$$

Now applying Bernoulli's equation between impeller outlet where delivery pressure gauge is fitted and outlet of delivery pipe, we have

$$\frac{p_d}{\gamma} + \frac{v_d^2}{2g} = H_d + H_{fd} + \frac{v_d^2}{2g}$$

$$H_d = \frac{p_d}{\gamma} - H_{fd} = 18 - 2.9 = 15.1 m$$

The velocity triangles are shown in Fig. 11.13

$$\tan \beta_2 = \frac{v_{f2}}{u_2 - v_{w2}}$$

Or

$$v_{w2} = v_2 - \frac{v_{f2}}{\tan\beta_2}$$

$$= \frac{\pi \times 0.25 \times 1450}{60} - \frac{0.031}{0.017} \times \frac{1}{\tan 32°}$$

$$= 18.98 - 2.92 = 16.06 \text{ m/s}$$

$$v_d = \frac{4Q}{\pi d_d^2} = \frac{4 \times 0.031}{\pi \times (0.1)^2} = 3.95 m/s$$

$$H_{mano} = H_s + H_d + H_{fs} + H_{fd} + \frac{v_d^2}{2g}$$

$$= 2.44 + 15.1 + 2 + 2.9 + \frac{(3.95)^2}{2 \times 9.81}$$

$$= 23.23 \text{ m}$$

$$\eta_{mano} = \frac{gH_{mano}}{u_2 v_{w2}} = \frac{9.81 \times 23.23}{18.98 \times 16.06}$$

$$= 0.7476 \text{ or } 74.76\%$$

(b)
$$\eta_0 = \frac{\gamma Q H_{mano}}{P} = \frac{9.81 \times 0.031 \times 23.23}{8.67}$$

$$= 0.8148 \text{ or } 81.48\%$$

## 11.14 Effect of Variation in Discharge on Efficiency

The efficiency of the centrifugal pump is maximum when it runs and gives discharge at its designed speed. If the discharge is either increased or decreased its efficiency drops owing to loss of head due to shock at entry to impeller. The inlet velocity triangles with decrease and increase in discharge are shown in Fig 11.14. The following observations are made from the velocity triangles:

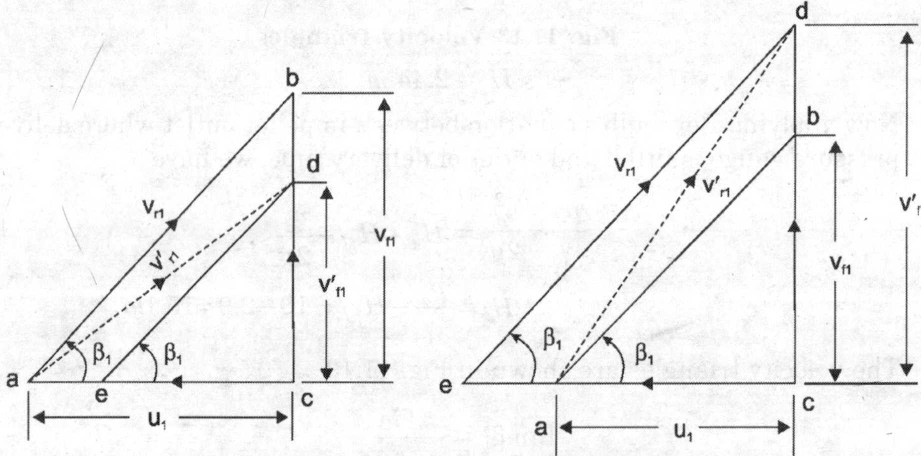


(a) Decrease in discharge (b) Inecrease in discharge.

**Fig. 11.14 Effect of variation in discharge on inlet velocity triangle.**

1. *abc* is the velocity triangle at inlet when the pump is operating under the normal conditions of discharge. The relative velocity $v_{r1}$ is along the contour of vane inlet angle $\beta_1$. The flow velocity is $v_{f1.}$
2. With the decrease or increase in discharge the flow velocity decreases or increases from $cb = v_{f1}$ to $cd = v'_{f1}$
3. With the pump running at the same speed the peripheral velocity $u_1 = ac$ remains unchanged and the effective velocity triangle becomes edc.

4. The new relative velocity $ad = v'_{r1}$ no longer remains parallel to the inlet vane angle and consequently shock occurs at entry to impeller. With fixed value of flow velocity $cd = v'_{f1}$ and flow taking place along the blade, the velocity triangle should have the form of cde, in which *ed* is parallel to *ad*. The sudden change in peripheral velocity $ac = u_1$ brings about shock, and as a consequence of which, there is a loss of head.

Loss of head at entrance to impeller,

$$H_l = \frac{(change\ of\ peripheral\ velocity)^2}{2g}$$

$$= \frac{(ae)^2}{2g}$$

(i) For decease in discharge, $H_l = \frac{(ac - ec)^2}{2g} = \frac{(u_1 - v'_{f1} \cot \beta_1)^2}{2g}$ ...(11.30)

(ii).Fir increase in discharge, $H_l = \frac{(ec - ac)^2}{2g} = \frac{(v'_{fi} \cot \beta_1 - u_1)^2}{2g}$ ...(11.31)

## 11.15 Effect of Number of Vanes of Impeller on Head and Efficiency

The ideal head, $H_{mano} = \eta_{mano} \times \frac{u_2 v_{w2}}{g}$, is developed considering infinite number of blades in the impeller through which water has to pass. In practice, the impeller must have only a finite number of blades. In such a case the velocity of whirl ($v_{w2}$) is reduced owing to the secondary or circulatory flow effects in the impeller. The new value of velocity of whirl is given by,

$$v'_{w2} = v_{w2} - \Delta v_{w2}$$

The ideal or Euler head, $(H_{th})_\infty = \frac{u_2 v_{w2}}{g}$

The head developed, taking $v'_{w2}$ into consideration is known as theoretical head considering finite number of blades or virtual head. It is denoted by $H_{th}$

Thus $$H_{th} = \frac{u_2 v'_{w2}}{g} = \frac{u_2 (v_{w2} - \Delta v_{w2})}{g}$$

Correction factor for finite number of blades in impeller,

$$\varepsilon = \frac{H_{th}}{(H_{th})} = \frac{v_{w2} - \Delta v_{w2}}{v_{w2}} = 1 - \frac{\Delta v_{w2}}{v_{w2}}$$

$\varepsilon$ is sometimes known as vane or blade efficiency or effectiveness. The correction factor $\varepsilon$ is given in Table 11.2

**Table 11.2 Correction factor $\varepsilon$ *v's* number of blades *z*.**

| z | 1 | 2 | 4 | 6 | 10 | 20 | $\infty$ |
|---|---|---|---|---|---|---|---|
| $\varepsilon$ | 0.25 | 0.40 | 0.572 | 0.666 | 0.77 | 0.87 | 1.0 |

It has been observed in practice that as the number of vanes is increased the value of $\varepsilon$ increases and approaches unity. The value of $\varepsilon$, in addition to number of vanes, depends on the shape of the vane and the outlet vane angle. For $z > 24$, $\varepsilon \to 1.0$.

The manometric head or the actual head developed by the pump is still smaller than virtual head $H_{th}$ because there are friction losses in suction passages, interior of impeller and discharge volute casing. Over and above there is head loss due to turbulence. Fig.11.15 shown the head curves against discharge.

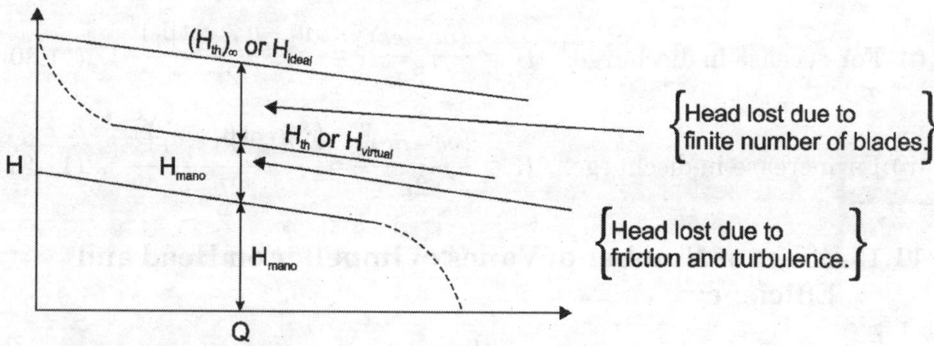


**Fig. 11.15 Ideal, virtual and manometric heads *v's* discharge.**

## 11.16 Proportions of Centrifugal Pump

### 1. Breadth of Impeller (Fig. 11.16)

Let $Q$ = discharge through the impeller, m³/s

Then $Q = \pi D_1 . B_1 . v_{f1} = \pi D_2 . B_2 . v_{f2}$

Where $D$, $B$ = diameter and breadth of impeller

$$B_1 = \frac{Q}{\pi D_1 v_{f1}} \quad ...(11.34a)$$

$$B_2 = \frac{Q}{\pi D_2 v_{f2}} \quad ...(11.34b)$$

Generally $\quad v_{f1} = v_{f2}$

$\therefore \quad B_1 D_1 = B_2 D_2$

Or $\quad BD = \text{const.} \quad ...(11.35)$

Eq. (11.35) represents a hyperbola.

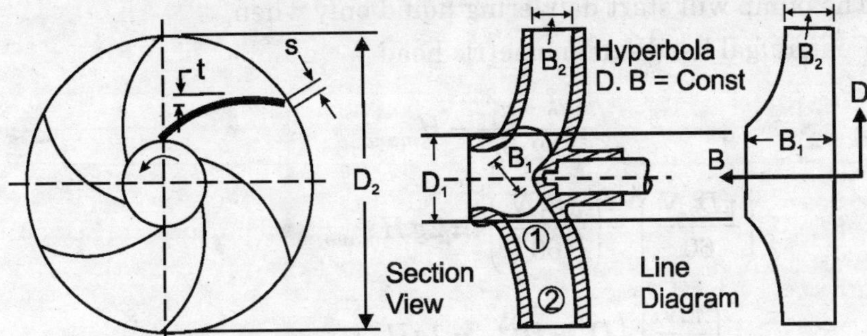


**Fig. 11.16 Breadth and diameter of impeller.**

Considering vane thickness, if $z$ be the number of blades and $s$ = blade thickness measured and the eirele, then actual metal thickness,

$$t = s \sin \beta$$

$$Q = \quad {}_1 - z.s_1).B_1.v_{f1} = (\pi D_2 - z.s_2).B_2.v_{f2}$$

$$= (\pi D - z.s).B.v_f \qquad ...(11.36)$$

**2. Speed ratio, $K_{u2}$:** It is the ratio of velocity of flow at exit to the theoretical velocity of jet corresponding to manometric head ($H_{mano}$). Thus

$$K_{u2} = \frac{u_2}{\sqrt{2gH_{mano}}} \qquad ...(11.37)$$

$K_{u2}$ Varies from 0.95 to 1.25.

**3. Flow ratio, $K_{f2}$ :** It is the ratio of velocity of flow at exit to the theoretical velocity of jet corresponding to manometric head. Thus

$$K_{f2} = \frac{v_{f2}}{\sqrt{2gH_{mano}}} \qquad ...(11.38)$$

$K_{f2}$ varies from 0.1 to 0.25.

**4. Impeller outer Diameter, $D_2$**

$$u_2 = \frac{\pi D_2 N}{60}$$

And $$u_2 = K_{u2}\sqrt{2gH_{mano}}$$

$\therefore$ $$D_2 = \frac{60K_{u2}\sqrt{2gH_{mano}}}{\pi N}$$

$$= \frac{84.6K_{u2}\sqrt{H_{mano}}}{N} \qquad ...(11.39)$$

Inner diameter if impeller $$D_1 = \left(\frac{1}{3} \text{ to } \frac{2}{3}\right) D_2 \simeq 0.5D_2$$

**5. Least Diameter of Impeller**

The pump will start delivering liquid only when,

Centrifugal head = manometric head

Or $$\frac{u_2^2 - u_1^2}{2g} = H_{mano}$$

$$\left(\frac{\pi D_2 N}{60}\right)^2 - \left(\frac{\pi D_1 N}{60}\right)^2 = 2g\, H_{mano}$$

$$\left(\frac{\pi N}{60}\right)^2 \left(D_2^2 - D_1^2\right)^2 = 2g H_{mano}$$

Let $D_1 = 0.5\, D_2$, then

$$\left(\frac{\pi N}{60}\right)^2 - \left(1 - \frac{1}{4}\right) D^2{}_2 = 2g H_{mano}$$

$$\left(\frac{\pi N}{60}\right)^2 \times 0.75\, D^2{}_2 = 2g H_{mano}$$

$$D_2 = \sqrt{\frac{2 \times 9.81 \times 3600}{\pi^2 \times 0.75} \times \frac{H_{mano}}{N_2}}$$

$$= \frac{97.68\sqrt{H_{mano}}}{N} \qquad ...(11.40)$$

**6. Diameter of Suction Pipe, $d_s$**

$$Q = \frac{\pi}{4} d_s^2\, v_s$$

$$d_s = \sqrt{\frac{4Q}{\pi v_s}} \qquad ...(11.41)$$

**7. Diameter of Delivery Pipe, $d_d$**

$$Q = \frac{\pi}{4} d_d^2\, v_d$$

$$d_d = \sqrt{\frac{4Q}{\pi v_d}} \qquad ...(11.42)$$

**8. Types of Blades (Fig. 11.17)**

(a) Blades bending backwards.

$\beta_1 < 90°$ and $\beta_2 < 90°$

(b) Radial blades.

$\beta_2 = 90°$

(c) Blades bending forwards.

$\beta_2 > 90°$

(d) For radial inlet, i.e without shock and whirl, $\alpha_1 = 90°$

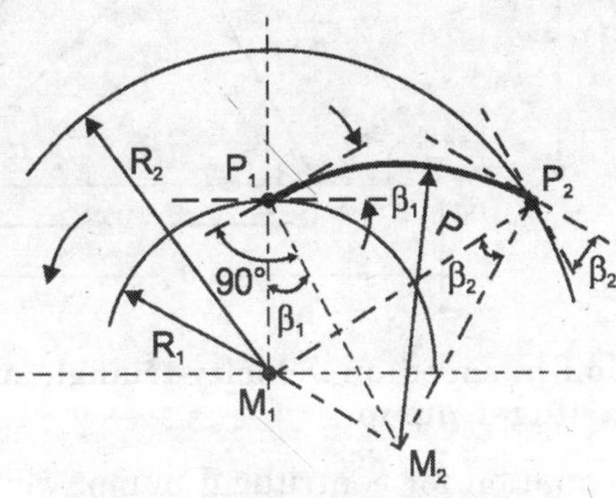


(a) Blades Bending Backwards

(b) Straight Blades

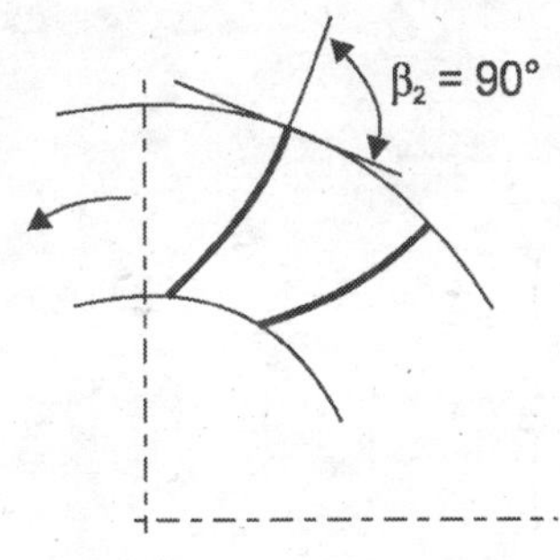


(c) Blades bending Radially

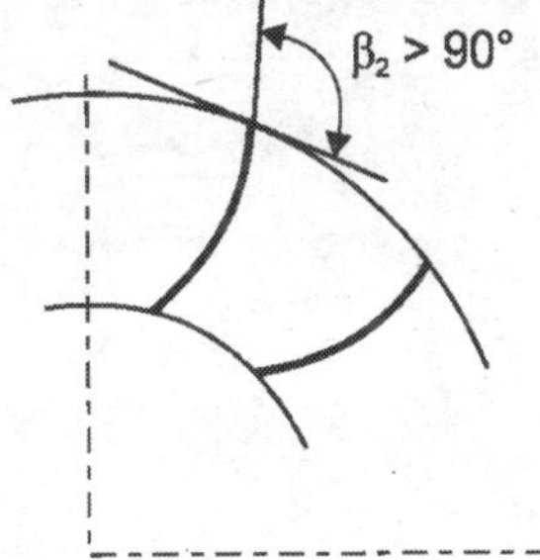


(d) Blades bending Forwards

**Fig. 11.17 Types of Blades.**

## 11.17 Variations of Speed and Diameter

### (a) Effect of Variation in Speed on Discharge, Head and Power.

Peripheral speed, $$u = \frac{\pi D N}{60}$$

$$\therefore \quad u \propto N$$

If $u$ change to $u'$ the vane angles remain constant, as is evident from Fig.11.18 then

$$v_f \propto u \propto N$$

Now, $Q$ = area across flow × $v_f$

$$\therefore \quad Q = \left(\frac{N}{N'}\right)^2$$

Thus, discharge varies directly as speed.

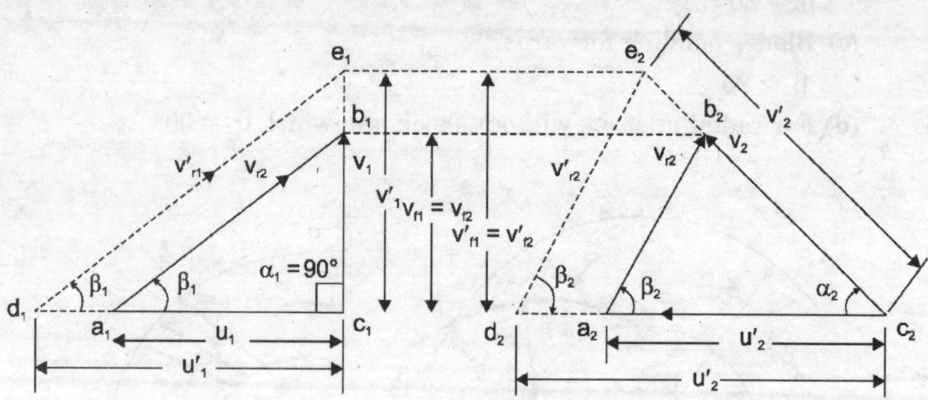


**Fig. 11.18 Effect of variation in speed on velocity triangle of a centrifugal pump.**

Also from the fundamental equation for centrifugal pump, we observe that head depends on square of velocity $u_2$, $v_2$ and $v_{r2}$. These velocities are proportional to speed $N$.

$$\therefore \qquad \frac{Q}{Q'} = \frac{N}{N'} \qquad \text{...(11.43)}$$

$$\frac{H}{H'} = \left(\frac{N}{N'}\right)^2 \qquad \text{...(11.44)}$$

And
$$\frac{P}{P'} = \left(\frac{N}{N'}\right)^3 \qquad \text{...(11.45)}$$

**(b) Effect of Alteration of Diameter on Discharge, Head and Power.**

Now $\qquad u \propto D$

$\therefore \qquad v_f \propto D$

Area across flow, $\qquad A_f \propto D$

But $\qquad Q = A_f \, v_f$

$\therefore \qquad Q \propto D^2$

Also $H$ depends on square of velocities and velocities are proportion to $D$.

$\therefore \qquad H \propto D^2$

Now power, $P \propto QH \propto D^4$

$$\frac{Q}{Q'} = \left(\frac{D}{D'}\right)^2 \qquad \text{...(11.46)}$$

$$\frac{H}{H'} = \left(\frac{D}{D'}\right)^2 \qquad \text{...(11.47)}$$

$$\frac{P}{P'} = \left(\frac{D}{D'}\right)^4 \quad \text{...(11.48)}$$

If height of the impeller is also changed. Then

$$A_f \propto D^2$$

$$Q \propto D^3$$

$$P \propto D^5$$

(c) Relation between $Q$, $H$, $P$, $N$ and $D$.

$$Q \propto D^3 . N \quad \text{...(11.49)}$$

$$H \propto D^2 . N^2 \quad \text{...(11.50)}$$

$$P \propto D^5 . N^3 \quad \text{...(11.51)}$$

## 11.18 Specific Speed of Pump - $N_s$

The specific speed of a centrifugal pump is defined as the speed of a geometrically similar pump which would deliver unit quantity against a unit head.

The expression for specific speed may be obtained as follows:

Discharge, $Q = A_f \times v_f$

$= \pi DB \times v_f$

Or $Q \propto D \times B \times v_f$

Now $B \propto D$

$\therefore$ $Q \propto D^2 \times v_f$

Also $u = \frac{\pi DN}{60}$

Or $u \propto D \times N$

And $u \propto v_f \propto \sqrt{H_{mano}}$

$\therefore$ $\sqrt{H_{mano}} \propto D \times N$

Or $D \propto \sqrt{\frac{H_{mano}}{N}}$

$\therefore$ $Q \propto \frac{H_{mano}}{N^2} \times \sqrt{H_{mano}} \propto \frac{(H_{mano})^{3/2}}{N^2}$

Or $Q = K \frac{(H_{mano})^{3/2}}{N^2}$

Where $K$ = constant of proportionality

By definition: when $H = 1$ m, $Q = 1$ m$^3$/s, then $N = N_s$, i.e.

$$1 = K \frac{(1)^{3/2}}{N_s^2}$$

Or $$K = N_s^2$$

$\therefore$ $$Q = N_s^2 \frac{(H_{mano})^{3/2}}{N^2}$$

Or $$N_s = \frac{N\sqrt{Q}}{(H_{mano})^{3/4}} \qquad ...(11.52)$$

For multi–stage pump, $$H_{mano} = \frac{\text{total head developed}}{\text{number of stages}}$$

For a double suction pump, $$Q = \frac{\text{Actual discharge}}{2}$$

The range of specific speeds for different types of pumps are given in Table 11.3.

**Table 11.3 Specific speed ranges for different types of pumps.**

| **Type of pump** | **Slow speed, Radial flow** | **Medium speed, Radial flow** | **High speed, Radial flow** | **Mixed flow (or screw pump)** | **Axial flow (or propeller pump)** |
|---|---|---|---|---|---|
| Specific speed | 10-30 | 30-50 | 50-80 | 80-160 | 160-500 |

## 11.19 Model Tasting and Geometrically Similar Pumps

The following conditions must be satisfied by the model and prototype pumps in order to know the performance of prototype.

1. **Specific speed.**

$$(N_s)_m = (N_s)_p$$

$$\left[\frac{N\sqrt{Q}}{H_{mano}^{3/4}}\right]_m = \left[\frac{N\sqrt{Q}}{H_{mano}^{3/4}}\right]_P \qquad ...(11.53)$$

2. **Peripheral tangential velocity.**

$$u = \frac{\pi DN}{60}$$

also $$u \propto \sqrt{H_{mano}}$$

$\therefore$ $$\sqrt{H_{mano}} \propto DN$$

Or $$\frac{\sqrt{H_{mano}}}{DN} = \text{const.}, \left[\frac{H_{mano}}{D^2N^2} = \text{head or lift coefficient}\right]$$

Or $$\left[\frac{\sqrt{H_{mano}}}{DN}\right]_m = \left[\frac{\sqrt{H_{mano}}}{DN}\right]_p$$

$$\left[\frac{\sqrt{H_{mano}}}{DN}\right]_m = \left[\frac{\sqrt{H_{mano}}}{DN}\right]_p \quad ...(11.54)$$

**3.** Also $Q \alpha D^2 \times v_f$

And $v_f \alpha u \alpha DN$

$\therefore$ $$Q \propto D^2 \times DN \alpha D^2N$$

or $$\frac{Q}{D^3N} = \text{const.} = \text{Flow coefficient}$$

$\therefore$ $$\left[\frac{Q}{D^3N}\right]_m = \left[\frac{Q}{D^3N}\right]_p \quad ...(11.55)$$

**4.** Power of pump, $P = \gamma Q H_{mano}$

$\therefore$ $$P \propto Q \times H_{mano}$$

Or $$P \propto D^3N \times H_{mano} \qquad [\because Q \alpha D^3 N]$$

Or $$P \propto D^3N \times D^2 N^2 \alpha D^5 N^3 \qquad [\because H_{mano} \alpha D^2 N^2]$$

Or $$\frac{P}{D^5N^3} = \text{const.} \qquad \left[\frac{P}{D^5N^3} = \text{power coefficient}\right]$$

Or $$\left[\frac{P}{D^5N^3}\right]_m = \left[\frac{P}{D^5N^3}\right]_p \quad ...(11.56)$$

**Example 11.16** A single centrifugal pump is built to give a discharge of $Q$ when working against a manometric head of 18 m. During test it was found that the head actually generated was 19.5 m for the designed discharge $Q$. If it is required to reduce the original diameter of 300 mm without reducing the speed of the impeller, compute the reduction necessary.

**Solution.**

Given: $H'_{mano} = 18$ m, $H_{mano} = 19.5\ m$, $D = 300$ mm

Head generated by pump, $H_{mano} = \dfrac{u^2}{2g} = \dfrac{(\pi DN/60)^2}{2g}$

or $H_{mano} \alpha D^2$

$\therefore$ $$\frac{H_{mano}}{H'_{mano}} = \left(\frac{D}{D'}\right)^2$$

$$D' = D\left(\frac{H'_{mano}}{H_{mano}}\right)^{1/2} = 300\left[\frac{18}{19.5}\right]^{1/2} = 288.2 \text{ mm}$$

**Example 11.17** To predict the performance of a large centrifugal pump, its model, having the following specification, was made:

$$P = 20\ kW,\ H_{mano} = 8m,\ N = 650\ rpm$$

Diameter of model pump impeller is 10 times smaller than that of the prototype. The prototype pump has to work against a head of 30m. Find its working speed and power request to drive it. Find also the rate of flow of both the pumps

**Solution.**

Given: $P_m = 20\ kW, H_m = 8\ m, N_m = 650\ rpm, \frac{(D_2)_p}{(D_2)_m} = 10, H_p = 30m,$

(a) Specific flow:

$$(Q_{11})_m = (Q_{11})_p$$

$$\frac{Q_m}{D_m^2 \cdot \sqrt{H_m}} = \frac{Q_p}{D_p^2 \cdot \sqrt{H_p}}$$

Now
$$P = \frac{\gamma Q H}{10^3}$$

$$Q = \frac{10^3 P}{\gamma H}$$

$$Q_m = \frac{10^3 \times 20}{9.81 \times 10^3 \times 8} = 0.2548 \text{ m}^3/\text{s}$$

And
$$Q_p = Q_m\left(\frac{D_p}{D_m}\right)^2 \sqrt{\frac{H_p}{H_m}}$$

$$= 0.2548 \times 10^2 \times \sqrt{\frac{30}{8}} = 49.35\ m^3/s$$

(b) Specific speed:
$$\frac{N_p\sqrt{Q_p}}{H_p^{3/4}} = \frac{N_m\sqrt{Q_m}}{H_m^{3/4}}$$

Or
$$N_p = N_m\sqrt{\frac{Q_m}{Q_p}} \cdot \left(\frac{H_p}{H_m}\right)^{3/4}$$

$$= 650\left(\frac{0.2548}{49.35}\right)^{1/2} \cdot \left(\frac{30}{8}\right)^{3/4} = 125.8 \text{ rpm}$$

(c) Specific power:

$$\frac{P_p}{D_p^2 \cdot H_p^{3/2}} = \frac{P_m}{D_m^2 \cdot H_m^{3/2}}$$

$$P_p = P_m \cdot \left(\frac{D_p}{D_m}\right)^2 \cdot \left(\frac{H_p}{H_m}\right)^{3/2}$$

$$= 20 \times 10^2 \times \left(\frac{30}{8}\right)^{1.5}$$

$$= 14523.68 \text{ kW}$$

**Example 11.18** **A centrifugal pump is delivering 0.27 m³/s of water against a head of 20 m. the impeller rotates at 600 rpm. The inner and outer diameter of impeller is 300 mm and 60 mm respectively. The area of flow is constant at 0.09 m² from inlet to outlet of impeller. If the vanes of the impeller are bent at an angle of 35° to the tangent at exit, determine:**

**(a) Manometric efficiency,**

**(b) Inlet vane angle, and**

**(c) Loss of head at inlet to impeller when the discharge is reduced by 40 percent.**

**Solution.**

Given: $Q = 0.27$ m³/s, $H_{mano} = 20$ m, $N = 600$ rpm, $D_1 = 300$ mm, $D_2 = 600$ mm, $A_f = 0.09$ m², $\beta_2 = 35°$.

(a)

$$u_1 = \frac{\pi D_1 N}{60} = \frac{\pi \times 0.3 \times 600}{60} = 9.42 \text{ m/s}$$

$$u_2 = \frac{\pi D_2 N}{60} = \frac{\pi \times 0.6 \times 600}{60} = 18.85 \text{ m/s}$$

$$v_{f1} = v_{f2} = v_f = \frac{Q}{A_f} = \frac{0.27}{0.09} = 3 \text{ m/s}$$

$$v_{w2} = u_2 - \frac{v_{f2}}{\tan \beta_2}$$

$$= 18.85 - \frac{3}{\tan 35°} = 14.56 \text{ m/s}$$

$$\eta_{mano} = \frac{gH_{mano}}{u_2 v_{w2}}$$

$$= \frac{9.81 \times 20}{18.85 \times 14.56} = 0.7148 \text{ or } 71.48 \text{ \%}$$

(b) $$\tan\beta_1 = \frac{v_{f1}}{u_1} = \frac{3}{9.42} = 0.31847$$

$$\beta_1 = 17.67°$$

(c) $$v'_{f1} = v_f(1-0.4) = 3\times 0.6 = 1.8 \text{ m/s}$$

$$\text{Loss of head at inlet} = \frac{\left(u_1 - v'_{f1}\cot\beta_1\right)^2}{2g}$$

$$= \frac{\left(9.42-1.8\cot 17.67^\circ\right)^2}{2\times 9.81}$$

$$= 0.724 \text{ m of water}$$

**Example 11.19** In order to predict the performance of a large centrifugal pump, a scale model of one eighth size was made with the following specifications:

Power, $P$ = 30 kW; Head, $H_{mano}$ = 8 m; speed, $N$ = 950 rpm,

If the prototype pump has to work under a head of 24m calculate its working speed, power required to drive it, and the ratio of flow rates handled by the two pumps.

**Solution.**

Given: $\frac{D_p}{D_m} = 8$, $P_m = 30$ kW, $(H_{mano})_m = 8$m, $N_m = 950$ rpm, $(H_{mano})_p = 24$m

(a) Using Eq. (11.54), we have

$$\left(\frac{\sqrt{H_{mano}}}{DN}\right)_m = \left(\frac{\sqrt{H_{mano}}}{DN}\right)_p$$

$$N_p = \sqrt{\frac{(H_{mano})_p}{(H_{mano})_m}} \times \frac{D_m}{D_p} \times N_m$$

$$= \sqrt{\frac{24}{2}} \times \frac{1}{8} \times 950 = 205.68 \text{ rpm}$$

(b) Using Eq. (11.56), we have

$$\left(\frac{P}{D^5N^3}\right)_m = \left(\frac{P}{D^5N^3}\right)_p$$

$$P_p = P_m \times \frac{(D_p)^5}{D_m} \times \frac{(N_p)^3}{N_m}$$

$$= 30 \times 8^5 \times \left(\frac{205.68}{950}\right)^3$$

$$= 9976.5 \text{ kW}$$

(c) Using Eq. (11.55), we have

$$\left(\frac{Q}{D^3 N}\right)_m = \left(\frac{Q}{D^3 N}\right)_p$$

$$\frac{Q_p}{Q_m} = \left(\frac{D_p}{D_m}\right)^3 \times \left(\frac{N_p}{N_m}\right)$$

$$= 8^3 \times \frac{205.68}{950} = 110.85$$

**Example 11.20** Two geometrically similar pumps are running at the same speed of 1200 rpm. One pump has an impeller diameter of 300 mm and lifts water at the rate of 0.03 m³/s against a head of 15 m. Determine the head and impeller diameter of the other pump to deliver half the discharge.

**Solution.**

Given: $N_1 = 1200\ rpm,\ D_1 = 300\ mm,\ Q_1 = 0.03\ m^3/s,\ H_1 = 15\text{m},$
$Q_2 = \frac{Q_1}{2} = 0.015\, m^3/s$

(a)
$$\left[\frac{N\sqrt{Q}}{H^{3/4}}\right]_1 = \left[\frac{N\sqrt{Q}}{H^{3/4}}\right]_2$$

$$H_2 = H_1 \left[\sqrt{\frac{Q_2}{Q_1}}\right]^{4/3} = 15 \times \left[\sqrt{\frac{1}{2}}\right]^{4/3}$$

$$= 9.45 \text{ m}$$

(b)
$$\left(\frac{\sqrt{H}}{DN}\right)_1 = \left(\frac{\sqrt{H}}{DN}\right)_2$$

$$D_2 = D_1 \sqrt{\frac{H_2}{H_1}} = 300 \sqrt{\frac{9.45}{15}} = 238\ mm$$

## 11.20 Net Positive Suction Head (NPSH)

Let $H_s$ = suction head

$v_s$ = velocity of liquid in suction pipe

$H_{fs}$ = frictional head loss in suction pipe

$p_1$ = absolute static pressure at pump inlet

$p_{atm}$ = atmospheric pressure of the liquid

$p_v$ = vapour pressure of the liquid at a given temperature

$\frac{v_s^2}{2g}$ = kinetic head in suction pipe.

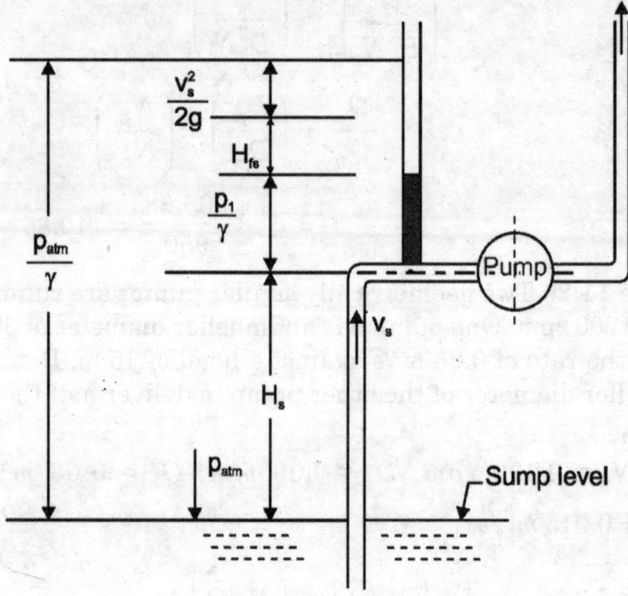


**Fig. 11.19 Pressure balance at pump section.**

$H_{sv}$ = inerease in velocity head of liquid when entering the impeller.

The various heads acting on the suction side of the centrifugal pump are shown in Fig.11.19. the pump will work without cavitation, if $\frac{p_1}{\gamma}$ is greater than $\frac{p_v}{\gamma}$ by $H_{sv}$.

$$\frac{p_1}{\gamma} - \frac{p_v}{\gamma} = H_{sv} \qquad ...(1)$$

Also

$$\frac{p_1}{\gamma} = \frac{P_{atm}}{\gamma} - \left(H_s + H_{fs} + \frac{v_s^2}{2g}\right) \qquad ...(2)$$

From Eqs. (1) and (2), we get

$$H_{sv} = \frac{P_{atm}}{\gamma} - \left(H_s + H_{fs} + \frac{v_s^2}{2g}\right) - \frac{p_v}{\gamma}$$

$$= H_{atm} - (H_s + H_{fs} + H_{ks}) - H_v$$

$$= H_{atm} - \left(H_s + H_{fs} + H_{ks}\right) - H_v$$

$$= H_{atm} - H_{st} - H_v \quad \text{...(11.57)}$$

where $H_{ks} = \dfrac{v_s^2}{2g}$ = kinetic head in suction pipe

$H_{st}$ = total suction head

$$= H_s + H_{fs} + H_{ks}$$

$$H_v = \frac{p_v}{\gamma}$$

The value of $H_{sv}$ is called the net positive suction head (NPSH).

NPSH = net inlet head – head corresponding to the vapour pressure of the liquid.

NPSH may be defined as the net head that is required to make the liquid flow through the suction pipe from the sump to the impeller.

NPSH is a parameter that can be used to check cavitation in pump. The minimum NPSH depends upon the pump design, its speed and discharge.

From Eq. (2), the minimum values of suction lift $(H_s)$ is given by,

$$H_s = \left(\frac{p_{atm} - p_v}{\gamma}\right) - H_{fs} - \frac{v_s^2}{2g} \quad \text{when } p_1 = p_v \quad \text{...(11.58a)}$$

$$= H_{atm} - \left(H_{vap} + H_{fs} + \frac{v_s^2}{2g}\right) \quad \text{...(11.58b)}$$

Suction lift is usually limited from 7 to 8 m.

## 11.21 Cavitation in Centrifugal Pump

When the liquid is flowing in the pump, it is possible that the pressure at any part of the pump may fall below the vapour pressure. At such a part the liquid will vapourise and the flow will no longer remain continuous. The vapourisation of the liquid will appear in the form of bubbles released in the low pressure region of the pump. These bubbles are carried along with water stream and when these pass though a region of high pressure, they collapse suddenly. When the bubbles collapse on a metallic surface such as tips of impeller blades, the cavities are formed. Successive bubbles collapsing at the same metallic surface produce pitting since penetration in the grain boundaries take place. Once the pitting takes place, the liquid rushes to fill the pits causing mechanical destruction and the liquid hits the blades with such a great force that it damages the impeller. The phenomenon of formation of cavities by the vapourisation of liquid is known as cavitation. A great noise is experienced due to cavitation leading to vibration of the pumping set. This trouble is mainly experienced at the impeller vane tip at inlet due to high suction lift, which must be kept within limits.

Cavitation begins to appear in centrifugal pumps when the pressure at the suction falls below the vapour pressure of the liquid. The intensity of cavitation increases with the decrease in value of NPSH. The cavitation in a pump can be noted by a sudden drop in efficiency, head and power requirement. It imposes limitations on the flow rate and speed of rotation of pump.

Thoma's cavitation factor is used to indicate the onset of cavitation. Thoma's cavitation factor for pumps is defined as:

$$\sigma = \frac{H_{atm} - H_s - H_{vap}}{H_{mano}} = \frac{H_{sv}}{H_{mano}} \quad \text{...(11.59)}$$

The cavitation will occur if the value of $\sigma$ is less than the critical value $\sigma_{crit}$ at which the cavitation just begins. The cavitation parameter $\sigma$ is a function of specific speed, efficiency of pump, and number of vanes.

**Harmful Effects of Cavitation**

1. Pitting and erosion of surface.
2. Sudden drop in head, efficiency and power delivered to the fluid.
3. Noise and vibration.

**Factors Promoting Cavitation**

1. Restricted suction.
2. High impeller speed.
3. Too high specific speed for optimum design parameters.
4. Too high temperature of flowing fluid.

**Suction Specific Speed**

Now $\quad Q \propto ND^3$

$$\text{NPSH} \propto \frac{v^2}{2g} \propto \frac{Q^2}{D^4} \propto \frac{Q^2}{\left(\frac{Q}{N}\right)^{4/3}} \propto N^{4/3} \cdot Q^{2/3} \propto \left(NQ^{1/2}\right)^{4/3}$$

Or

$$\frac{N\sqrt{Q}}{(NPSH)^{3/4}} = \text{const.} = (N_s)_{suc} \quad \text{...(11.60)}$$

Where $(N_s)_{suc}$ = suction specific speed

$$\sigma = \frac{NPSH}{H_{mano}} = \left[\frac{N_s}{(N_s)_{suc}}\right]^{4/3} \quad \text{...(11.61)}$$

**Example 11.21** Tests on a model pump indicate a cavitation parameter, $\sigma_c$ = 0.10. A geometrically similar unit is to be installed at a location where atmospheric pressure, $p_{atm}$ = 0.91 bar and vapour pressure, $p_{vap}$ = 0.035 bar absolute. The pump is to work against a water head of 25 m. What is the maximum suction head?

**Solution.**

Given: $\sigma_c = 0.10$, $p_{atm} = 0.91$ bar, $p_{vap} = 0.035$ bar (*abs.*), $H_{mano} = 25$ m.

$$H_{atm} = \frac{p_{atm}}{\gamma} = \frac{0.91\times10^5}{9.81\times10^3}$$
$$= 9.276 \text{ m of water}$$

$$H_{vap} = \frac{p_{vap}}{\gamma} = \frac{0.035\times10^5}{9.81\times10^3}$$
$$= 0.3567 \text{ m of water}$$

$$\sigma_c = \sigma = \frac{H_{atm} - H_{vap} - H_s}{H_{mano}}$$

$$0.10 = \frac{9.276 - 0.3567 - H_s}{25}$$

$$H_s = 6.419 \text{ m}$$

**Example 11.22** A single stage centrifugal pump works under a head of 150 m and has a positive suction lift of 4 m. Laboratory tests are to be conducted on a model at 115 m head. Calculate the suction lift for the laboratory tests. $H_{atm}$ = 10.2 m of water, and $H_{vap}$ = 0.32 m of water.

**Solution.**

Given: $(H_{mano})_p = 150$ m, $(H_{suc})_p = 4$ m, $(H_{mano})_m = 115$ m, $H_{atm} = 10.2$ m, $H_{vap} = 0.32$ m

$$\text{Cavitation factor for prototype, } \sigma_p = \frac{H_{atm} - H_{vap} - (H_{suc})_p}{(H_{mano})_p}$$

$$= \frac{10.2 - 0.32 - 4}{150} = 0.0392$$

For cavitation similarity, $\sigma_m = \sigma_p$

$$\sigma_m = \frac{10.2 - 0.32 - (H_{suc})_m}{115} = 0.0392$$

$$(H_{suc})_m = 5.372 \text{ m of water}$$

## 11.22 Multi – Stage Centrifugal Pumps

A multi- stage centrifugal pump is one in which two or more identical impeller are mounted on the same shaft or on different shafts. The important functions performed by a multi – stage pump are:

1. To generate heads greater than that permissible with a single impeller, keeping the discharge constant. This can be achieved by mounting the two or more identical impeller in series on the same shaft and enclosed in the same casing.
2. To generate large discharge, keeping the head constant. This can be achieved by mounting the pump impellers in parallel on separate shafts.

### 11.22.1 Pumps in Series

Two pump impellers in series are shown in Fig.11.20. The discharge from impeller-1 passes through a guided passage and enters the impeller-2.

Let $H_{mano}$ = head generated by one impeller

$n$ = number of impeller in series.

Then total manometric head generated,

$$(H_{mano})_{total} = nH_{mano}$$

However, $Q_1 = Q_2 = \ldots = Q_n$

Then total manometric is used for delivery a relatively small quantity of liquid against very high heads.

**Advantages of pump impeller in series**

1. Less frictional head loss.
2. Reduced stresses.
3.- Small slip leakage.
4. Axial thrust can be eliminated by proper arangement.
5. A higher suction lift is possible.

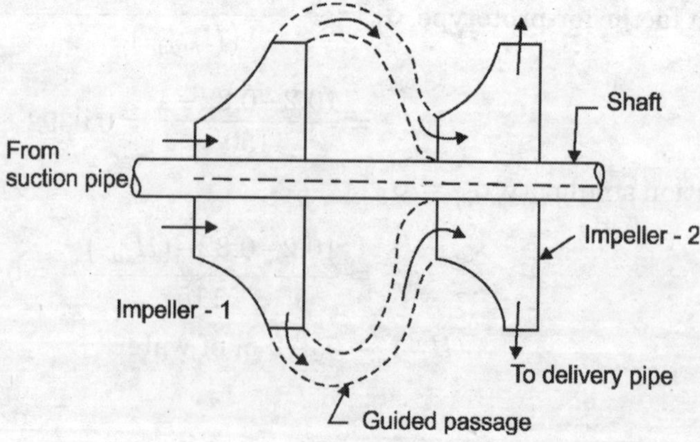


**Fig. 11.20 Pumps in series.**

### 11.22.2 Pumps in Parallel

Two pump impellers in parallel are shown in Fig.11.21. In this case, the impellers are mounted on parallel shafts. The head raised by each impeller is the same and the discharges add up.

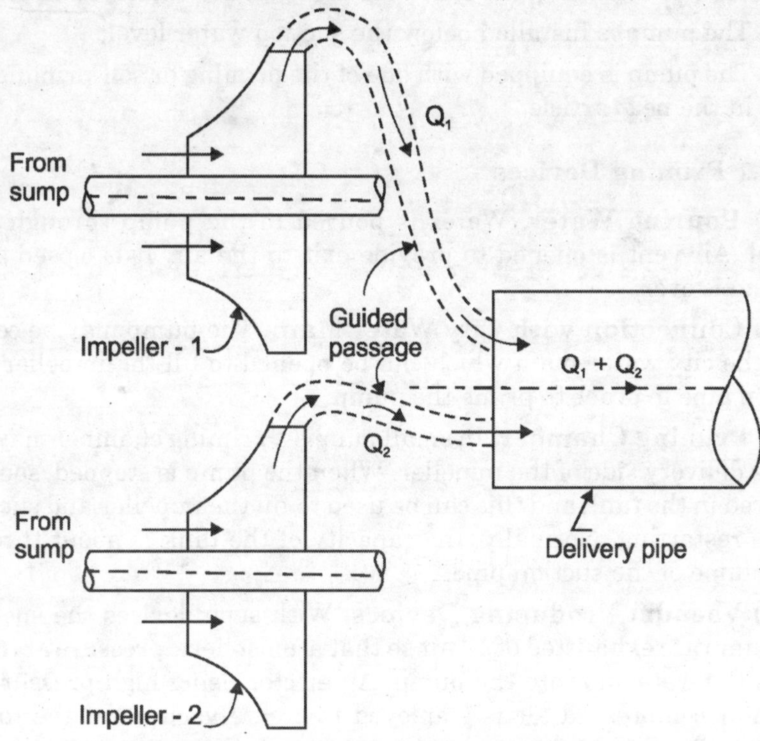


**Fig. 11.21 Pumps in parallel.**

Let $Q$ = discharge capacity of one pump

$n$ = number of identical pumps.

Then total discharge, $Q_{total} = nQ$

However $(H_{mano})_1 = (H_{mano})_2 = ..... = (H_{mano})_n$

Pumps in parallel are used to raise large quantity of liquid against a relatively small head.

## 11.23 Priming of Pump

**Priming:** Before starting a pump, its impeller and suction pipe have to be filled with water in order to remove any air, gas or vapour from the water ways of the pump. If a centrifugal pump is not primed before starting, pockets inside the impeller may give rise to vortices and cause discontinuity of flow. The wearing rings may rub and seize causing serious damage if the pump is allowed to dry. It is also essential that packing be lubricated by liquid leaking past it.

Originally, priming is done by pouring water through a funnel, displaced air being allowed to escape through air vents. When a pump is being primed or stopped, the delivery valve should be kept closed.

Necessity of priming is the main disadvantage of a centrifigual pump. To overcome this difficulty, the following methods are employed in practice.:

(a) The pump is installed below the suction water level;

(b) The pump is equipped with one of the priming or self-priming devices given in the next article.

### 11.23.1 Priming Devices

**(a) Pouring Water.** Water is poured in the pump through priming funnel. Air vent is opened to provide exit to the air. It is closed after the priming is over.

**(b) Connection with City Water Main.** The pump may be connected with the city water main which can be opened to fill the impeller and the suction pipe in order to prime the pump.

**(c) Priming Chamber.** In small pumps a priming chamber may be used on the delivery side of the impeller. When the pump is stopped, some water is stored in the tank and this can be used to fill the impeller and suction line before restarting. Normally, the capacity of the tank is about three times the volume of the suction pipe.

**(d) Vacuum Producing Devices.** With such devices the suction line and pump are exhausted of all air so that atmospheric pressure at the sump, forces the water up into the pump. An ejector using high pressure water, steam or compressed air is employed to create vacuum at the top of the casing (refer Fig. 11.22) so that water is sucked into the suction pipe and the impeller.

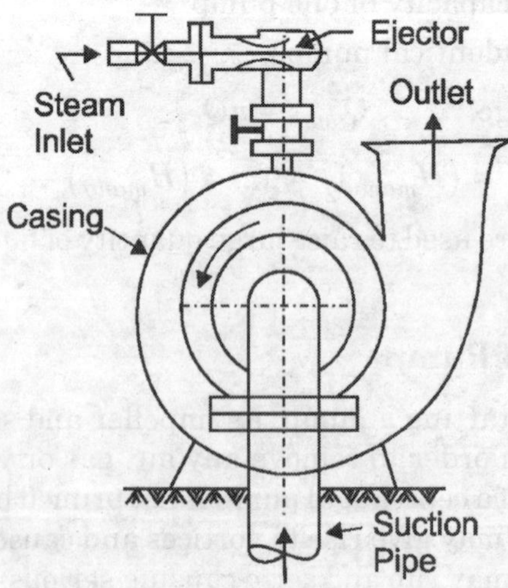


**Fig. 11.22 Priming of centrifugal pump.**

### 11.23.2 Self Priming Devices

They are of the following two types:

**(1) By Formation of Air-Water Mixture:** The pump is made to exhaust the air of the impeller and the suction pipe. The density of the air is about 800 times smaller than that of water, therefore, the head generated by the revolving impeller is 800 times less. Now if air is mixed with water, the head generated by the impeller will increase depending upon the density of the mixture.

Fig. 11.23 (a) shows an open impeller mounted in the casing such that there is a small clearance between the blades and the side walls and also between the blade tips and the circumference of the casing. In order to start for the first time, the casing is filled with water upto a level where the suction pipe begins. The suction pipe and a mixture of water and air is formed in the casing which is discharged by the impeller into a widened portion known as separator. Because of a large cross sectional area of separator, the water velocity decreases and the air is separated from the mixture and escapes through the delivery pipe which should open to atomsphere at its exit. The water free of air, comes back to the impeller from the separator through the passage made by the fixed vane as shown in Fig. 11.23 (b). The impeller goes on revolving and the water is drawn up the suction pipe gradually till whole of the air is evacuated from it by the formation of air-water mixture and the air separating out. The pump will then starts functioning normally.

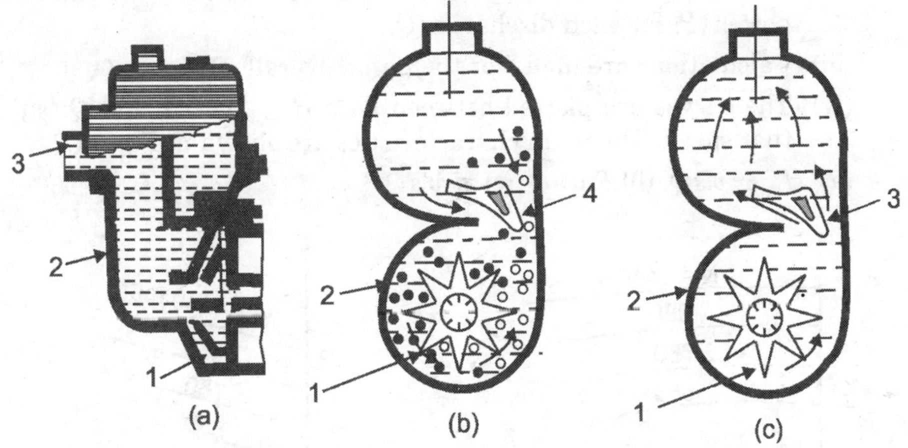


**Fig. 11.23 Principle of Operation of Self-Priming of Centrifugal Pump.**

1. Impeller 2. Casing 3. Suction Branch 4. Fixed Guide Vane

**(2) By Providing Air or Vaccum Pump.** The pump impeller and its suction pipe is primed by an air pump or vaccum pump, run by the same pump shaft and fitted on the suction side of pump [Fig. 11.23 (c)]. This air pump runs in parallel with the centrifugal pump, exhaust air from the suction pipe and discharges into the open air or in delivery pipe if it is open from its exit end. After the pump is primed, the air pump discharges the water back into the suction pipe. The air pumps are generally of two types- water ring pump and side channel pump.

## 11.24 Characteristics of Centrifugal Pumps

The characteristic of a pump is a graphical representation of its behaviour and performance under different circumstances. These curves show manometric head, brake power, and efficiency of the pump as ordinates and with its discharge as the abscissa.

Pump characteristics can be divided into four categories:

1. Main characteristics.
2. Operating characteristics.
3. Constant efficiency (or Muschel) curves
4. Constant head and constant discharge curves.

### 1. Main Characteristics

The main characteristics curves are obtained as follows:

(i) The pump is run at a constant speed and the discharge is varied over the desired range by means of the delivery valve.

(ii) Measurements are made for manometric head ($H_{mano}$) and shaft power ($P$) for each discharge ($Q$).

(iii) Calculations are made for the pump overall efficiency ($\eta_o$).

(iv) The curves are plotted between $Qv's\ H_{mano}$, $Qv'sp$, and $Qv's\eta_o$ for that speed. The main characteristics are shown in Fig. 11.24.

(a) $H_{mano}\ v's\,Q$ (b) $P\ v's\,Q$ (c) $\eta_o\ v's\,Q$

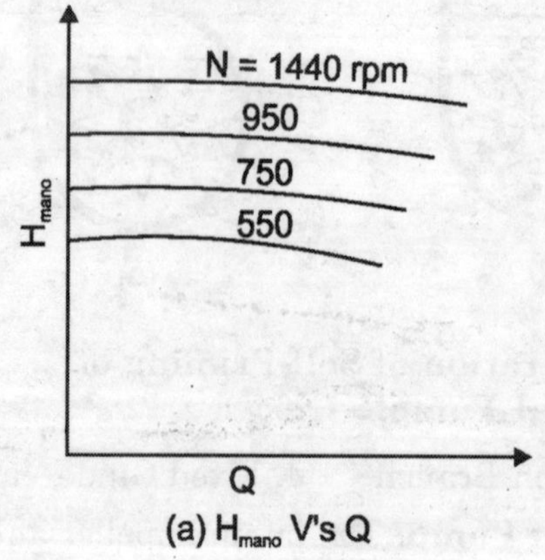


(a) $H_{mano}$ V's Q

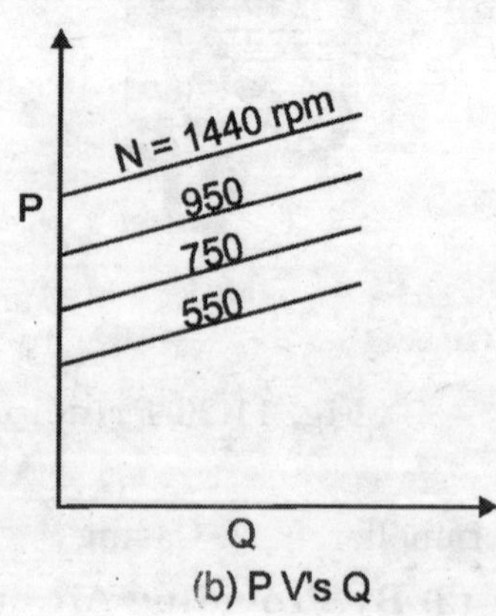


(b) P V's Q

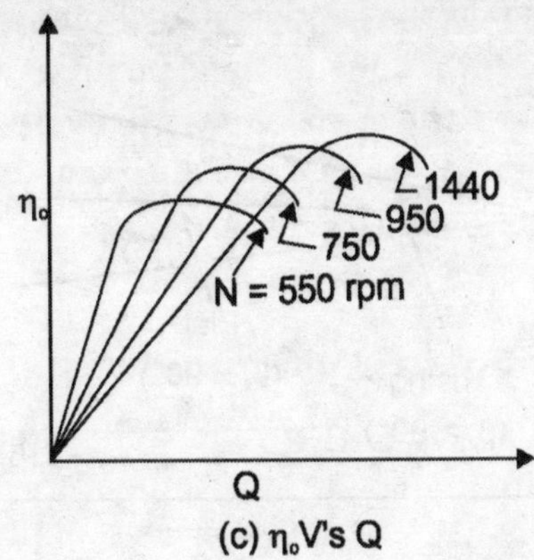

(c) $\eta_o$ V's Q

**Fig. 11.24 Main characteristics for centrifugal pump.**

## 2. Operating Characteristics

The maximum efficiency of centrifugal pump occurs at the design speed (or driving motor speed. Therefore, for optimum performance, the pump should be operated at the design speed. To obtain the operating characteristic curves, the pump is run at the design speed and discharge is varied. The values of $H_{mano}$, $P$ and $\eta_o$ are observed / calculated at various values of discharge. Then $H_{mano}$, $P$ and $\eta_o$ values are plotted against $Q$. The design discharge and head are obtained from the corresponding curve where the efficiency is maximum, The operating curves are shown in Fig.11.25.

The manometric head *v's* discharge characteristics for various types of vanes at design speed are shown in Fig. 11.26.

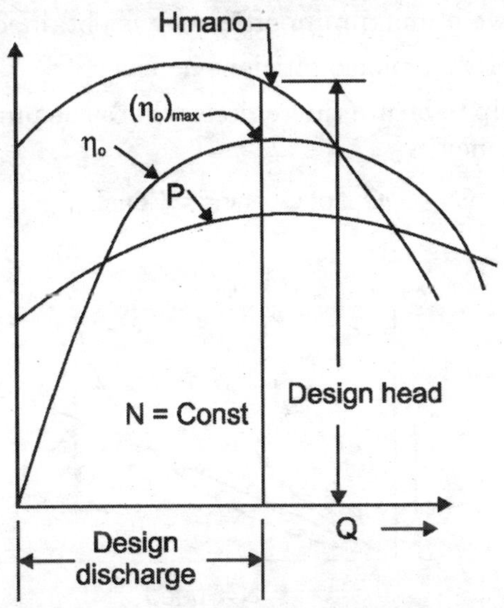


**Fig. 11.25 Operating characteristics.**

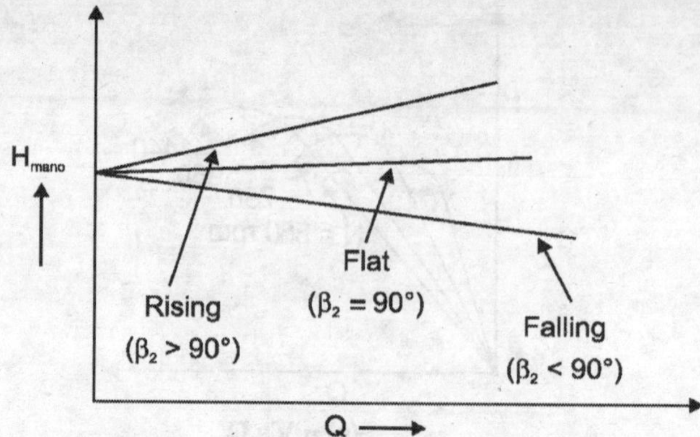


**Fig. 11.26 Rising, Flat or Falling *H* v's *Q* Curves of a Centrifugal Pump.**

### 3. Constant Efficiency (or Muschel) Curves

The constant efficiency (or iso-efficiency) curves represent the performance of a pump over its entire range of operations. These curves, shown in Fig.11.27, are obtained from main characteristics curves as follows:

(i) For a given efficiency, the values of discharges are obtained from Fig. 11.24 (c). These points are projected on the $H_{mano}$ v's $Q$ curves in Fig. 11.24 (a).

(ii) Step (i) is repeated for other valves of efficiencies.

(iii) The points corresponding to one efficiency are joined by a free hand curve.

(iv) The curve so obtained are the constant efficiency or Muschel curves.

(v) The curve of maximum efficiency is obtained by joining the peak points of various iso-efficiency curves.

These curves help to lacate the regions where the pump would operate with maximum efficiency.

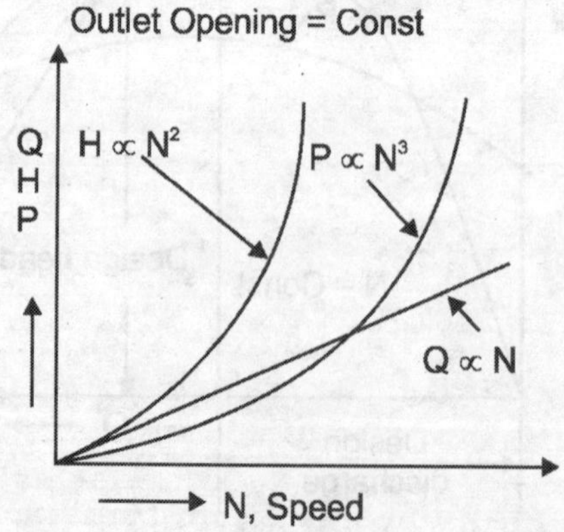


**Fig. 11.28 *Q*, *H* and *P* v's *N* curves of a centrifugal pump.**

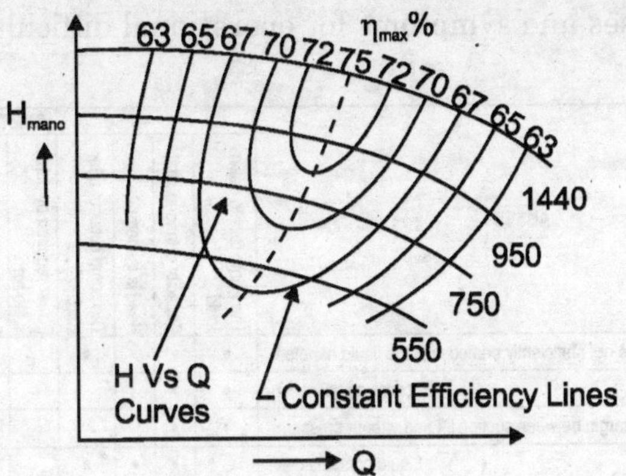


**Fig. 11.29 Constant efficiency or Muschel curves of a centrifugal pump.**

**4. Constant Head and Constant Discharge Curves**

These curves, shown in Fig. 11.28, are drawn for a variable speed pump. The curves are plotted between $Q v's N$ ($H$ = const.),

$H_{mano}$ $v's$ $N^2$ ($Q$ = const.), and $P v's$ $N^3$.

## 11.25 Selection of Pumps

The following steps may be adopted for selection of particular type of pump:

1. The main criteria for the selection of a pump is:
   (i) Required discharge, $Q$.
   (ii) Manometric head, $H_{mano}$.
   (iii) Speed, $N$ rpm.
2. From the above data, calculate the specific speed of the pump.
3. If the specific speed is low, then use multi-stage pump. The number of stages can be decided on the basis of manometric head and type of impeller to be used.
4. Select the type of impeller to be used.
   (i) Shrouded type impeller: For pumping fresh clean water.
   (ii) Unshrounded or propeller type impeller: For pumping solid-liquid mixture like sewage water, etc.
   (iii) Mixed flow impellers with diffuser vanes: For deep will or submersible pumps.
5. Very low heads (≃ 6m) and large discharge: Use axial flow pumps.
6. High heads: Use radial flow pumps.

## 11.26 Causes and Symptoms for Operational Difficulties in Centrifugal Pumps

The Operational difficulties, causes and symptoms are given in Table 11.4.

Table 11.4 Causes and symptoms for operational difficulties in centrifugal pumps.

| CAUSES | SYMTOMS: The Pump fails to deliver liquid | Capacity is inadequate | Head is inadequate | The pump cuts out after starting | Power required by pump is too high | Excessive leakage of stuffing box | Packing must be renewed too frequently | The pump vibrates or is noisy | Bearings heat up | Pump runs heavily or seizes |
|---|---|---|---|---|---|---|---|---|---|---|
| Pump and suction line are no! sufficiently primed with the liquid handled | • | • | | • | | | | • | | • |
| Excessive suction lift | • | • | | | | | | • | | |
| Insufficient margin between suction lift and vapour pressure | • | • | | | | | | • | | • |
| Liquid contains gas | | • | • | • | | | | | | |
| Air pocket in suction line | • | • | | • | | | | | | |
| Air leak in suction line | | • | | • | | | | | | |
| Air leaks into pump along stuffing box | | • | | • | | | | | | |
| Fool valve too small | | • | | | | | | | | |
| Foot valve partly clogged | | • | | | | | | • | | |
| Foot valve and suction line not fully submerged | • | • | | • | | | | • | | |
| Connection of water seal on suction stuffing box blocked | | | | • | | | • | | | |
| Lantern ring in stuffing box incorrectly fitted | | | | • | | • | • | | | |
| Speed too low | • | • | • | | | | | | | |
| Speed too high | | | | | • | | | | | |
| Incorrect direction of rotation | • | | • | | | | | | | |
| Total manometric head of the system greater than the manometric head of the pump | • | | • | | • | | | | | |
| Total manometric head of the system lower than the manometric head of the pump | | | | | • | | | | | |
| Specific gravity of liquid handled is not what it was originally supposed to be | | | | | • | | | | | |
| Viscosity of liquid handled is not what it was originally supposed to be | | • | • | | • | | | | | |
| The pump is operating at too low a capacity | | | | | | | | • | | • |
| Parallel connection is unsuitable for the specific operating conditions | • | • | • | | | | | | | • |
| Line, impeller or pump casing clogged | • | • | | | • | | | • | | |
| Pump set incorrectly aligned | | | | | • | | • | | • | • |
| Foundation not level | | | | | | | | • | | |
| Pump shaft warps | | | | | • | • | • | • | • | |
| A rotating part runs against a stationary part e.g. impeller runs against wearing rings | | | | | • | | | • | • | • |
| Bearing(s) defective | | | | | | | | • | • | • |
| Wearing rings are worn | | • | • | | | | | | | |
| Impeller is damaged | | • | • | | | | | • | | |
| Pump shaft or shaft sleeve locally at the stuffing box is worn | | | | | | • | • | | | |
| Stuffing box incorrectly packed | | | | | • | • | • | | | |
| Type of packing used unsuitable for liquid handled | | | | | | • | • | | | |
| Impeller out of balance | | | | | | • | • | • | • | • |
| Failure to apply water-cooling when handling hot liquids | | | | | | • | • | | | |
| Clearance between pump shaft and bore of pump casing at the bottom of the stuffing box too great | | | | | | | • | | | |
| Liquid for water seal contains impurities | | | | | | | • | | | |
| Gland overtightened | | | | | | | • | | | |
| Axial fit of complete pump shaft with impeller incorrect | | | | | | | | • | • | • |
| Insufficient or excessive lubrication | | | | | | | | | • | |
| Lubricant contains impurities | | | | | | | | | • | • |
| Bearings incorrectly fitted | | | | | | | | | • | • |

## Summary

1. Work done per unit weight of liquid

$$= \frac{u_2\, v_{w2}}{g} \text{ for radial inlet}$$

$$= \frac{1}{g}\left(u_2\, v_{w2} - u_1\, v_{w2}\right) \text{ for non-radial inlet}$$

2. Fundamental equation $= \dfrac{v_2^2 - v_1^2}{2g} + \dfrac{u_2^2 - u_1^2}{2g} + \dfrac{v_{r1}^2 - v_{r2}^2}{2g} = \dfrac{H_{mano}}{\eta_{mano}}$

3. Manometric efficiency, $\eta_{mano} = \dfrac{gH_{mano}}{u_2\, v_{w2}}$

4. Volumetric efficiency, $\eta_{vol} = \dfrac{Q}{Q_a = Q + Q_l}$

5. Mechanical efficiency, $\eta_{mech} = \dfrac{\gamma Q_a\, u_2\, v_{w2}}{gP} = \dfrac{P - P_{mech}\ loss}{P}$

6. Overall efficiency, $\eta_\circ = \eta_{mano} \times \eta_{vol} \times \eta_{mech} = \dfrac{\gamma Q H_{mano}}{P}$

7. Manometric head, $H_{mano} = \dfrac{u_2\, v_{w2}}{g} - H_l = \dfrac{u_2\, v_{w2}\, \eta_{mano}}{g}$

$$= \frac{\eta_{mano}}{g}\left[u_2 - \left(u_2 - v_{f2} \operatorname{cosec} \beta_2\right)\right]$$

8. Pressure rise in impeller, $\dfrac{p_2 - p_1}{\gamma} = \dfrac{1}{g}\left[v_1^2 + u_1^2 - v_{f2}^2 \operatorname{cosec}^2 \beta_2\right]$

9. Hydraulic power, $P_{hyd} = \dfrac{\gamma Q\, H_{mano}}{10^3}\, kW$

10. Loss of head at entrance to impeller,

$$H_L = \frac{\left(u_1 - v'_{f1} \cot \beta_1\right)^2}{2g}, \text{ due to decrease in discharge}$$

$$= \frac{\left(v'_{f1} \cot \beta_1 - u_1\right)^2}{2g}, \text{ due to decrease in discharge}$$

11. Correction factor for finite number of vanes in impeller,

$$\varepsilon = \frac{H_{th}}{(H_{th})_\infty} = \frac{v_{w2} - \Delta v_{w2}}{v_{w2}}$$

12. Specific speed of pump, $$N_s = \frac{N\sqrt{Q}}{H_{mano}^{3/4}}$$

13. Model testing. $$\left[\frac{N\sqrt{Q}}{H_{mano}^{3/4}}\right]_m = \left[\frac{N\sqrt{Q}}{H_{mano}^{3/4}}\right]_p$$

$$\left[\frac{\sqrt{H_{mano}}}{DN}\right]_m = \left[\frac{\sqrt{H_{mano}}}{DN}\right]_p$$

$$\left(\frac{Q}{D^3 N}\right)_m = \left(\frac{Q}{D^3 N}\right)_p$$

$$\left(\frac{P}{D^5 N^3}\right)_m = \left(\frac{P}{D^5 N^3}\right)_p$$

14. Thoma's cavitation factor,

$$\sigma = \frac{H_{atm} - H_s - H_{vap}}{H_{mano}}$$

15. Minimum suction lift, $$H_s = H_{atm} - \left(H_{vap} + H_{fs} + \frac{v_s^2}{2g}\right)$$

16. Pumps in series

$$(H_{mano})_{total} = n\, H_{mano}$$

$$Q = \text{const}$$

17. Pumps in parallel, $$Q_{total} = nQ$$

$$H_{mano} = \text{const.}$$

## Multi-Choice Questions

1. Head developed by a centrifugal pump depends upon
   (a) impeller diameter (b) speed
   (c) fluid density (d) type of casing
2. For starting a centrifugal pump, its delivery valve should be
   (a) open (b) close
   (c) partially open (d) any one of the above
3. The water in the casing of a centrifugal pump moves
   (a) radially (b) axially
   (c) tangentiallv (d) in a spiral
4. In a centrifugal pump, the liquid enters the pump at the
   (a) top (b) bottom
   (c) centre (d) sides
5. There shall be no flow of water from a centrifugal pump until the pressure rise in the impeller is greater than the
   (a) suction head (b) delivery head
   (c) static head (d) manometric head
6. The suitable pump for flood control and irrigation applications is
   (a) reciprocating (b) centrifugal
   (c) axial flow (d) mixed flow
7. A suitable pump for handling viscous fluid is
   (a) reciprocating pump (b) centrifugal pump
   (c) turbine pump (d) screw pump
8. Which one of the following is not a rotary pump?
   (a) gear pump (b) vane pump
   (c) screw pump (d) axial pump
9. Priming is required in a
   (a) gear pump (b) screw pump
   (c) reciprocating pump (d) centrifugal pump
10. For pumping molasses, it is preferable to employ
    (a) reciprocating pump (b) centrifugal pump
    (c) open impeller pump (d) rotary pump
11. For pumping mustard oil from a drum, a suitable pump is
    (a) reciprocating pump (b) vane pump
    (c) centrifugal pump (d) gear pump

12. The head of devolved by a centrifugal pump is

(a) $H = \frac{u_1 v_{w1}}{g}$ (b) $H = \frac{u_2 v_{w2}}{g}$

(c) $H = \frac{u_1 v_{w1}}{2g}$ (d) $H = \frac{u_2 v_{w2}}{2g}$

13. Which of the following is not a dimensional parameter?

(a) friction factor (b) specific speed

(c) pressure coefficient (d) Thoma's cavitation parameter

14. NPSH is given by

(a) $H_{atm} - H_{vap} - H_s - H_{fs}$

(b) $H_{atm} - H_{vap} - H_s + H_{fs}$

(c) $H_{atm} - H_{vap} + H_s - H_{fs}$

(d) $H_{atm} - H_{vap} + H_s + H_{fs}$

15. The maximum suction lift is given by

(a) $H_{atm} - H_{vap} + \frac{v_s^2}{2g} - H_{fs}$ (b) $H_{atm} - H_{vap} - \frac{v_s^2}{2g} + H_{fs}$

(c) $H_{atm} - H_{vap} - \frac{v_s^2}{2g} \ H_{fs}$ (d) $H_{atm} - H_{vap} + \frac{v_s^2}{2g} + H_{fs}$

---

**Answers**

1. (a) 2. (b) 3. (d) 4. (c) 5. (d) 6. (b) 7. (a) 8. (d) 9.(d) 10. (c)
11. (d) 12. (b) 13. (b) 14. (a) 15. (c)

---

## Review Questions

1. What are the advantages of a centrifugal pump over a reciprocating pump?
2. Explain the working of a single-stage centrifugal pump.
3. How centrifugal pumps are classified?
4. Differentiate between low lift, medium lift and high lift centrifugal pumps.
5. What are the functions of volute casing of a centrifugal pump?
6. Differentiate between single stage and multi-stage pumps.
7. Define specific speed of a centrifugal pump.
8. What values of head and discharge are used while calculating the specific speed of a multi-stage and double suction centrifugal pump respectively?
9. Define static, manometric and gross head of a centrifugal pump.
10. Define various efficiencies of a centrifugal pumps:
11. Differentiate between ideal, virtual and manometric head.
12. Explain various types of vane angles of a centrifugal pump.
13. What is the effect of number of impeller blades on the $H_{mano}$ υ's $Q$ curve?
14. Draw the effect of variation of speed on discharge, head and power.
15. What are model pumps?
16. What is priming of a pump? Describe some priming devices.
17. What is a self-priming device?
18. Describe the cavitation phenomenon in centrifugal pump.
19. What is meant by $NPSH$?
20. What is suction lift?

## Numerical Problems

1. A centrifugal pump impeller has an outer diameter of 300 mm and width at outer periphery 12.5 mm. The pump has radial inflow and delivers 0.08 $m^3/s$ of water against a total head of 40m. If the speed of pump is 1500 rpm and its manometric efficiency is 80 percent, find the blade angle at exit.

   **[Ans.** 60°]

2. A centrifugal pump is delivering 0.04 $m^3/s$ to a height of 20 m through a 150mm diameter 100m long pipeline. If the inlet losses in suction pipe are equal to 0.33 m and friction factor is 0.06 for the pipeline, find the power required to drive the pump. Assume overall efficiency of pump as 70%.

   **[Ans.** 17.4 kW]

3. A centrifugal pump runs at 750 rpm. The difference in surface levels at suction and delivery, together with the pipe friction losses is 15 m. The relative velocity of water at impeller exit is inclined backwards at 45° to the tangent. The velocity of flow through the impeller, and in suction and delivery pipes is 2 m/s. A manometric efficiency of 77 percent is expected and water enters the impeller without whirl. Find the diameter of the impeller and fraction of kinetic energy at discharge from the impeller which must be recovered as pressure if other losses are neglected.

   **[Ans.** 378 mm, 0.445]

4. A multi-stage pump running at 1500 rpm delivers 1200 lps of water against a head of 35 m. A similar multi-stage pump has a specific speed of 30 units, find the number of stages.

   **[Ans.** 15 impellers in parallel]

5. A centrifugal pump having an overall efficiency of 75 % delivers 1820 litres of water per minute to a height of 18 m through a pipe of 0.1 m diameter and 90 m length. If $f$ = 0.012, calculate the power required to drive the pump.

   **[Ans.** 7.32 kW]

6. To determine the efficiency of a centrifugal pump, the following observations were made:

   Pressure guage reading on suction side = 2.5 m

   Pressure guage reading on delivery side = 125 m of water

   Discharge = 0.25 $m^3/s$

   Input to pump = 425 kW.

   Find the efficiency of the pump.

   **[Ans.** 73.57 %]

7. A single – stage centrifugal pump has an impeller of 250 mm diameter which rotates at 1800 rpm and lifts 0.06 $m^3/s$ to 2.5 m with and efficiency of 70 %. Calculate the number of stages and diameter of each impeller of a similar multi – stage pump to lift 0.075 $m^3/s$ to 175 m at 1500 rpm.

[**Ans.** 73.57 %]

8. A centrifugal pump impeller runs at 950 rpm. Its tip and eye diameters are 500 mm and 250 mm respectively. The vanes are set back at an angle of 35° to the outer rim. If the radial velocity of water through the impeller is maintained constant at 2 m/s, find the angle of vanes at inlet, the velocity and direction, of water at outlet, and the work done by the impeller per kg of water.

[**Ans.** 9° – 9′, 22 *m/s*, 5 – 12′, 545.4 N.m ]

9. A centrifugal pump of 1.2 m diameter runs at 200 rpm and pumps 1.88 $m^3/s$, the average lift being 6 m. The angle which the vanes make at exit with the tangent to the impeller is 26°. The radial velocity of flow is 2.5 m/s. Determine the useful power and the efficiency. Find also the least speed to start pumping against a head of 6 m, the inner diameter of impeller being 0.6m.

[**Ans.** 109.8 kW, 62.7 %, 200 rpm]

10. A centrifugal pump running at 1000 rpm is installed 2 m above the water surface in the pump. The delivery head is 15 m. The friction loss in suction and delivery pipes are 0.18 $Q^2$ and 0.17 $Q^2$ respectively, where $Q$ is measured in $m^3/s$. The outside diameter of impeller is 300 mm and its width at exit is 15 mm. The blades occupy 5 % of the circumference and are curved backward at 30° to the tangent. The flow is radial at inlet and the radial component of velocity remains unchanged through the impeller. Calculate the discharge and manometric efficiency of the pump assuming that 50 % of velocity head of water leaving the impeller is converted to pressure head in the volute. Neglect all other losses.

[**Ans.** 5$m^3/s$, 62.2 %]

CHAPTER 12

# Fluid Devices and Systems

## 12.1 Gear Pump

A single start external type gear pump, shown in Fig.12.1, consists of two identical intermeshing spur wheels working with a fine clearance inside a suitably shaped casing. One gear is keyed to the driving shaft of an electric motor and the other revolves idly. Oil is entrained in the spaces between the teeth and the casing and carried round between the gears from the suction port to the discharge port. It is mainly used for pumping oils of high viscosity.

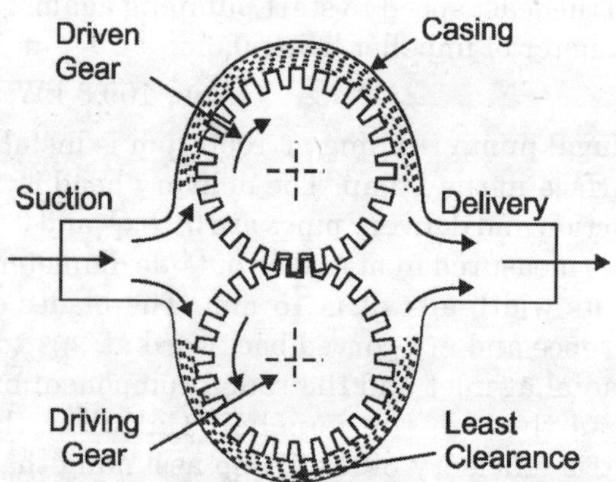


**Fig. 12.1 Line diagram of gear pump.**

Let $A$ = area enclosed between two adjacent teeth and casing.

$b$ = face width of teeth

$z$ = number of teeth on each gear

$n$ = speed in rpm.

Then, volume of liquid pumped in one revolution, $V = 2\,Abz$

Total ideal discharge, $$Q_{th} = \frac{VN}{60} = \frac{2AbzN}{60}$$

Let $\eta_{vol}$ = volumetric efficiency

Then, actual discharge, $$Q_{act} = \frac{2AbzN}{60} \times \eta_{vol} \qquad ...(12.1)$$

Empirical relation for volumetric displacement,

$$V = 0.95\,\pi\, c\,(d_0 - c)\, b \qquad \text{...(12.2)}$$

Where $c$ = centre distance between gears

$d_0$ = outside diameter of gears.

## 12.2 Screw Pump

The source of power is directly connected to the screw pump of a rotor or screw, as shown in Fig.12.2 the rotors of screw pump are extended. The screw may be single helical or double helical. The fluid is carried forward to the discharge along the rotor on pockets formed between teeth and the casing. There may be one, two or three screws. In a two-screw pump, one is a rotor and the other is idler. Screw pump is especially suitable for operation at high speed and are free from turbulence and pulsation.

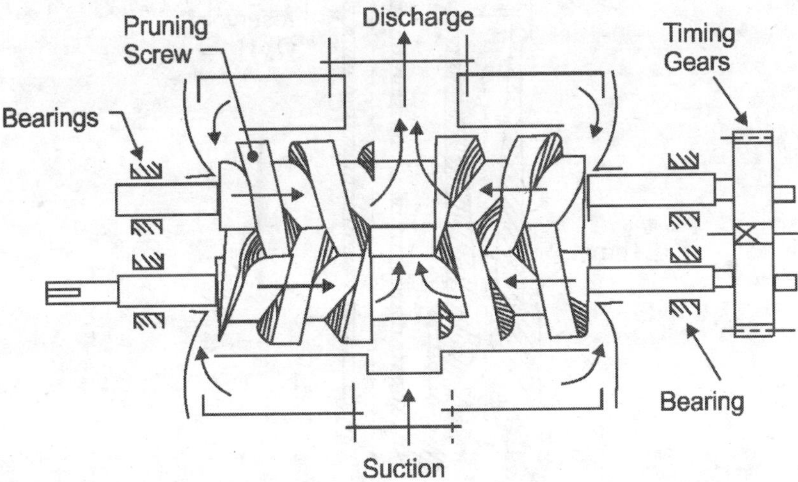


**Fig. 12.2 Screw pump.**

## 12.3 Vane Pump

A vane pump, shown in Fig.12.3, consists of a rotor disc having a number of slots into which fit sliding blades or vanes. The rotor turns in a casing which is bored eccentric to the driving shaft. The vanes are free to slide radially with the help of springs and thus form the required seal between the suction and discharge connections. When the rotor turns, the liquid flows in between its side and casing and is trapped by the vanes, which force it from suction to discharge. It is used especially for machine tools and industrial equipment.

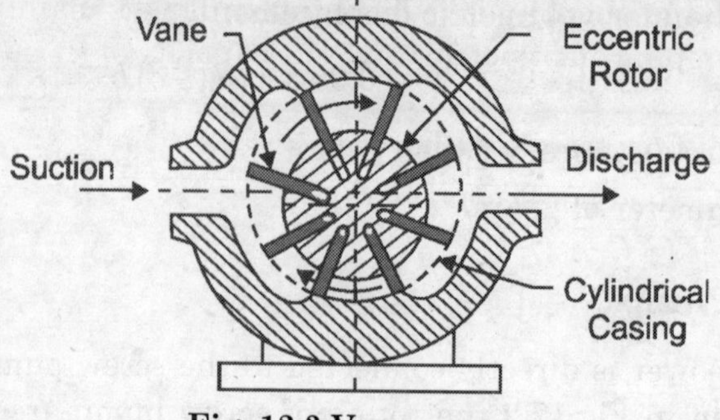


Fig. 12.3 Vane pump.

## 12.4 Jet Pump

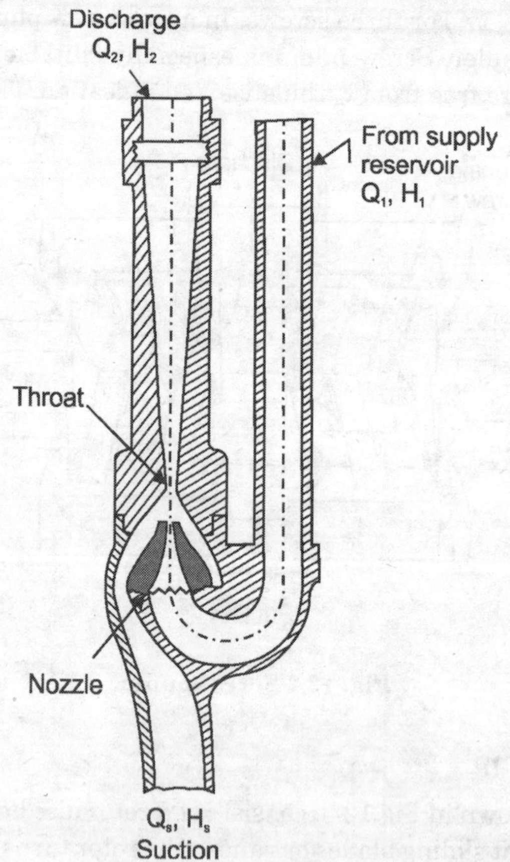


Fig. 12.4 Jet pump.

A line diagram of jet pump is shown in Fig.12.4. Water at a high pressure is forced through a fine aperture nozzle, thereby converting most of the pressure energy into kinetic energy. It results in a lowering of pressure which causes suction to take place. With water at high pressure, a more perfect vacuum can be produced so that a suction lift of 5.5 to 6 m can be obtained. The jet should be near the surface of water. Several jets may be employed if a large quantity is to be pumped. This is mainly used in mines and also for pumping oil or petroleum. The capacity of jet pump ranges upto about 0.05 $m^3/s$

Let $Q_1$ = rate of flow from supply reservoir

$Q_s$ = quantity sucked per second

Then delivery, $Q = Q_1 + Q_s$

$H_1$ = height of supply reservoir above the jet

$H_s$ = suction head

$H_d$ = delivery head

$$\text{Efficiency of jet pump} = \frac{Q_s(H_s + H_d)}{Q_1(H_1 - H_d)} \quad ...(12.3)$$

## 12.5 Air Lift Pump

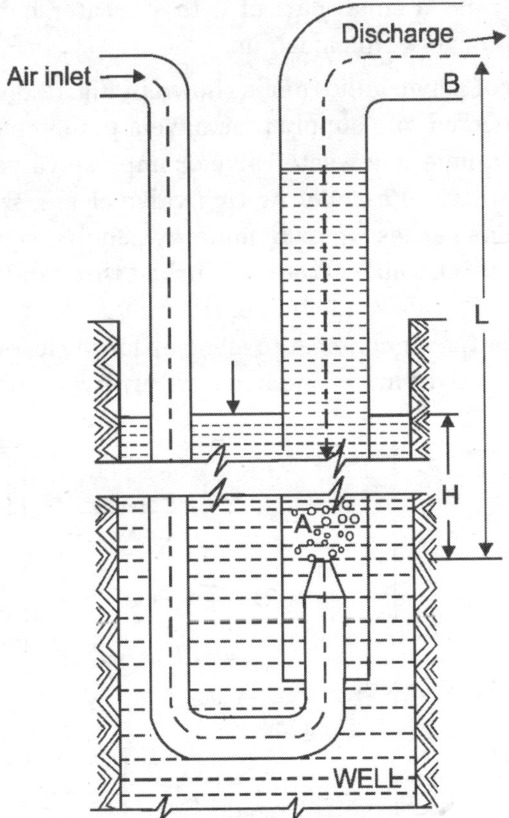


**Fig. 12.5 Air lift pump.**

It uses compressed air for raising water. Fig.12.5 shows diagrammatically the simplest method of raising water from a deep well by means of compressed air supplied by a compressor: The stream of high velocity air at

A produces a jet and entraps the water in its immediate vicinity and carries it upwards to B. The air-water mixture can rise above the water level because of its lesser density than water. The air is introduced at a considerable depth below the water surface in the well, so that the actual difference (L-H) may be sufficient to initiate flow to point B, the desired delivery level. It requires no moving parts, no valves. It is particularly suitable for draining out water from mines where air compressors are already available. It is also suitable for pumping water containing solids

## 12.6 Hydraulic Ram

It is a device to raise a part of large amount of water available at some height, to a greater height. This is employed when some natural source of water like a spring or a stream is available at some altitude, e. g., in a hilly region. Work done by a large quantity of water in falling through a small height is used to raise a small part of it to a greater height. No external power is required to work this machine.

A line diagram of a hydraulic ram is shown in Fig.12.6. A is the source of water supply connected to a supply pipe have a gate valve *G*. Water flows through the supply pipe to a waste valve or impulse valve *W*, which when open, allows the water to escape. As this valve closes, water is suddenly brought to rest. This causes water hammer which forces open the delivery valve fitted in an air chamber. Water is pushed through the delivery pipe, connected to the air chamber, into the service tank *B*. An air valve is provided which admits air for the air chamber to make up for the loss due to absorption in water. All these valves are of the non – return type.

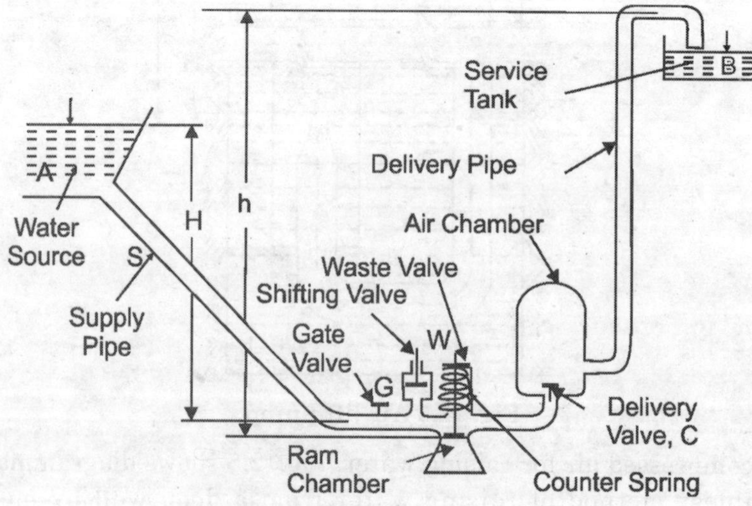


**Fig. 12.6 Hydranlic ram.**

## Principle of Working

Hydraulic ram is an impulse pump in which the impulse is developed at the expense of dynamic inertia possessed by a moving column of water. The momentum of a long column of water flowing thought the supply pipe is made to force a part of the water to a level higher than that of the supple source itself. The length of the supply pipe should be as long as possible to develop maximum impulse.

### *Cycle of Operation*

1. Delivery valve *C* and gate valve *W* are closed.
2. Vacuum is created by rebound and air is admitted through the shifting valve.
3. Waste valve opens and water starts flowing to waste through the valve.
4. Waste valve closes suddenly causing water hammer.
5. Momentarily high pressures resulting from water hammer forces open the delivery valve.
6. Water enters the air chamber through the delivery valve.
7. Pressure rises in the air chamber, slowly bringing the incoming water to rest and closing the delivery valve, thus completing the cycle

#### *Advantages*

1. No fuel or electricity is required.
2. It requires no lubrication and sealing devices.
3. Low maintenance expenses and no supervision is required
4. Gives regular water supply.
5. High efficiency and quiet operation.

#### *Calculations*

Lct $Q$ = quantity of water wasted by ram, $m^3/s$

$q$ = quantity of water pumped by ram, $m^3/s$

$h$ = height of service tank above waste valve, m

$H$ = height of source above waste valve, m

$L$ = length of supply pipe from the source of supply to the waste valve, m

$D$ = diameter of supply pipe, m

$H_{ls}$ = head loss in supply pipe, m

$H_{ld}$ = head loss in delivery pipe, m

Efficiency of ram, $$\eta = \frac{q(h+H_{ld})}{(Q+q)(H-H_{ls})} \qquad ...(12.4)$$

D' Aubisson efficiency, $$\eta_A = \frac{q \times h}{(Q+q)H} \text{ (neglecting losses)} \qquad ...(12.5)$$

Rankine efficiency, $$\eta_R = \frac{q(h-H)}{Q \times H} \qquad ...(12.6)$$

## 12.7 Hydraulic Accumulator

It is a device to accumulate liquid under pressure by a pump when it is not required by the machine. The pressure can be later supplied to the machine as and when required.

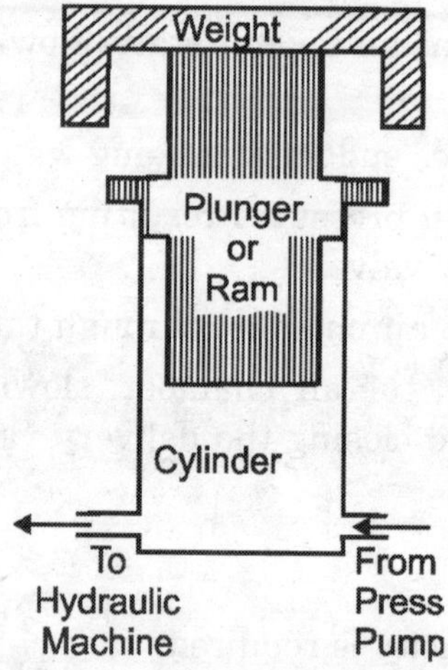


**Fig. 12.7 Hydraulic Accumulator.**

The hydraulic accumulator, shown in Fig.12.7 consists of a cylinder and a plunger, generally known as a ram. One side of the cylinder is connected to the press pump and other to the hydraulic machine. Either the cylinder or ram can be fixed. Generally the cylinder is fixed and the ram moves up and down to accommodate a variable quantity of liquid inside the cylinder. Accumulators may by dead load type or variable load (steam pressure) type.

## Capacity of Accumulator

It is the maximum amount of energy which can be stored by the accumulator. The storage capacity is equal to the potential energy of the lifted ram together with weight.

Let $D$ = diameter of ram

$L$ = stroke or ram lift

$P$ = pressure of water supplied.



Total moving weight of ram, $W = \frac{\pi}{4} D^2 p$

Capacity of accumulator = work done in lifting ram

$$= W \times L \quad ...(12.7)$$

$$= \frac{\pi}{4} D^2 L p$$

Volume of accumulator, $V = \frac{\pi}{4} D^2 L$

$$\text{Capacity of ram} = p \times V \quad ...(12.8)$$

$$= \text{capacity of accumulator}$$

## 12.8 Hydraulic Press

The hydraulic press works on the principle of Pascal's law of transmission of equal fluid pressure in all directions. The Pascal's law is depicted in Fig.12.8. It consists of two pistons of different sizes which operate inside two cylinders connecter with a pipe.

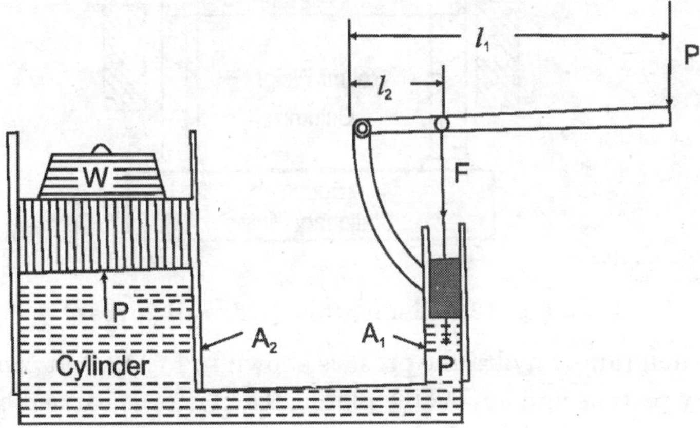


**Fig. 12.8 Principle of hydraulic press.**

Let $p$ = pressure on the pistons

$A_1$, $A_2$ = cross – sectional area of pistons.

$F_1$, $F_2$ = forces on the pistons

Then
$$\frac{F_1}{A_1} = \frac{F_2}{A_2} = p$$

$$\therefore \quad \frac{F_2}{F_1} = \frac{A_2}{A_1} \quad ...(12.9)$$

$$\text{Mechanical advantage of press} = \frac{A_2}{A_1} \quad ...(12.10)$$

If force on the smaller piston is applied by a lever which has a mechanical advantage $\frac{l_2}{l_1}$, then total mechanical advantage of machine becomes

$$= \frac{l_1}{l_2} \times \frac{A_2}{A_1} \quad \text{...(12.11)}$$

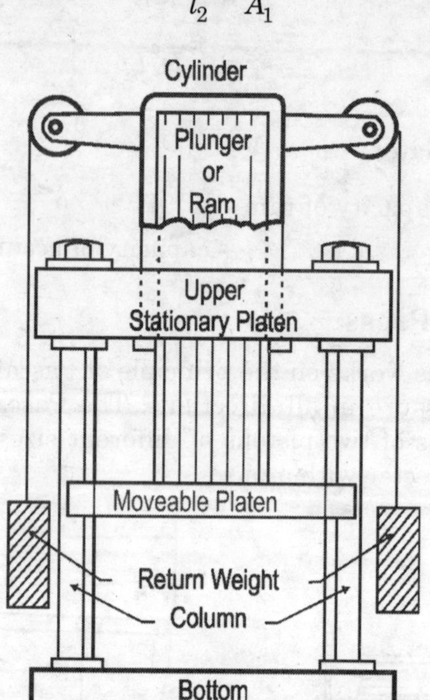


**Fig. 12.9 Elementary hydraulic press.**

A line diagram of hydraulic press is shown in Fig.12.9. It consists of two stationary platens and a movable platen. The latter is carried by a plunger or ram which passes through the upper stationary platen and moves in the cylinder. The upper and lower stationary platens are joined by columns. The ram coves down under hydraulic pressure supplied by pumps and applies tremendous pressure upon any material placed between the stationary and movable platens.

The hydraulic press is used for metal forming processes, packing, forging, moulding, etc.

## 12.9 Hydraulic Lift

A simple mechanism for hydraulic lift is shown in Fig.12.10. It consists of a cylinder and a ram. The water under pressure is forced into the cylinder, thus pushing the ram vertically upwards. A platform or cage is fitted to the top end of the ram. The platform can be made to stop at different floors of the building so that one can walk over it. The platform of cage moves between guides.

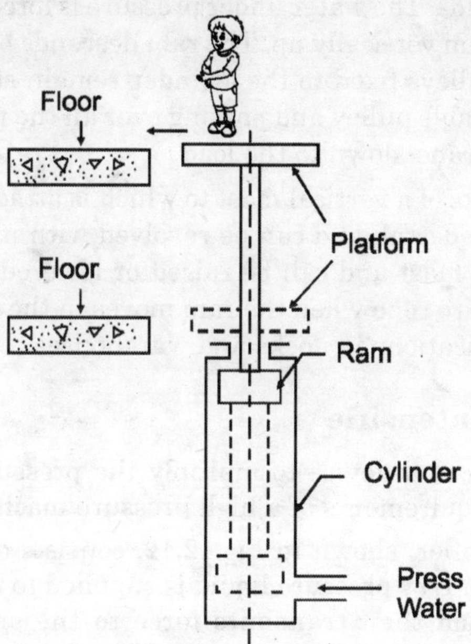


Fig. 12.10 Hydraulic lift.

## 12.10 Hydraulic Crane

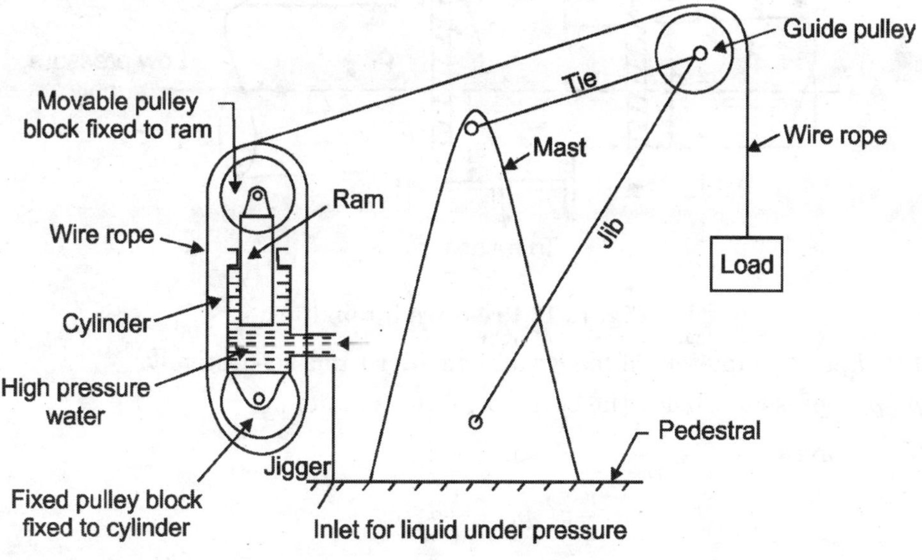


Fig. 12.11 Hydraulic crane.

A hydraulic crane shown in Fig.12.11 consists of two main parts, jigger and crane itself. The jigger is made up of a cylinder and a ram, both having a set of pulleys at their ends. The water under pressure is forced into the cylinder, which pushes the ram vertically up. The ram descends by opening an outlet valve. The set of pulleys fixed to the cylinder remain stationary while the other fixed to a movable pulley and passing over all the pulleys, is stretched over the jib of the crane, down to the load.

The crane consists of a vertical mast to which is attached a jib. The mast is supported on a pedestal, and can be revolved with its vertical axis. The jib swings with the mast and can be raised or lowered, together with the load, by means of wire rope when the ram moves in the cylinder. Hydraulic crane finds its applications in docks and warehouses.

## 12.11 Pressure Intensifie

Pressure intensifier is a device to multiply the pressure supplied by the pump to suit the requirements of a high pressure machine.

A simple intensifier, shown in Fig.12.12, consists of two coaxial rams moving in cylinders. Low pressure liquid is supplied to ram of large cross – sectional area which then transmits force to the small ram by a rod connecting the two rams $d_2$, being smaller than $d_1$, the pressure of liquid coming out will be high. The volume between the two rams must be vented.

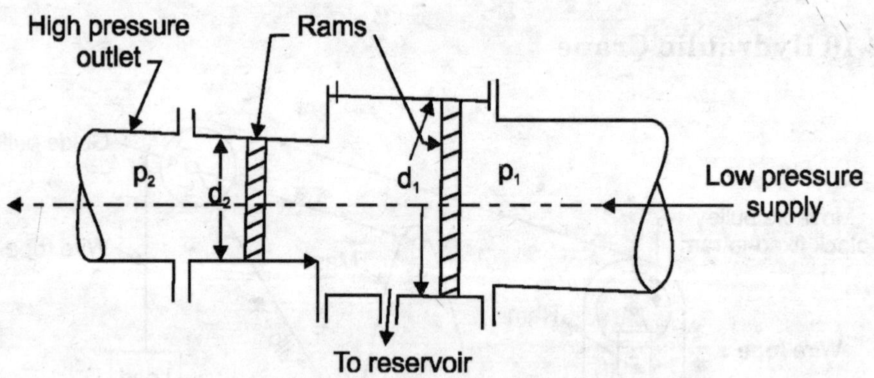


**Fig. 12.12 Pressure intensifies**

Let $d_1$, $d_2$ = diameters of bigger and smaller rams respectively.

$p_1$, $p_2$ = pressure inside the corresponding cylinders.

Then forces $$F_1 = F_2$$

$$\frac{\pi}{4} d_1^2 \, p_1 = \frac{\pi}{4} d_2^2 \, p_2$$

$$p_2 = \left(\frac{d_1}{d_2}\right)^2 p_1 \qquad \ldots (12.12)$$

### 12.11.1 Modified Pressure Intensifier

A modified form of pressure intensifier is shown in Fig. 12.13. It consists of two coaxial rams indie a cylinder. The cylinder and the inner ram are fixed. A, B and C are the valves provided for fixed cylinder, sliding ram and fixed ram respectively. To start with, the hollow sliding ram is full with liquid and valves A and C are open. The liquid having low pressure $p_1$ enters through valve A from the main supply and forces the sliding ram upwards. The liquid inside the sliding ram goes out through valve C at a greater pressure $p_2$. When the slider ram reaches its top position, valve B opens while the valves A and C close. The liquid which is now entrapped between the fixed cylinder and sliding ram enters inside the hollow sliding ram which then comes back to its starting position. This completes one cycle of intensifier.

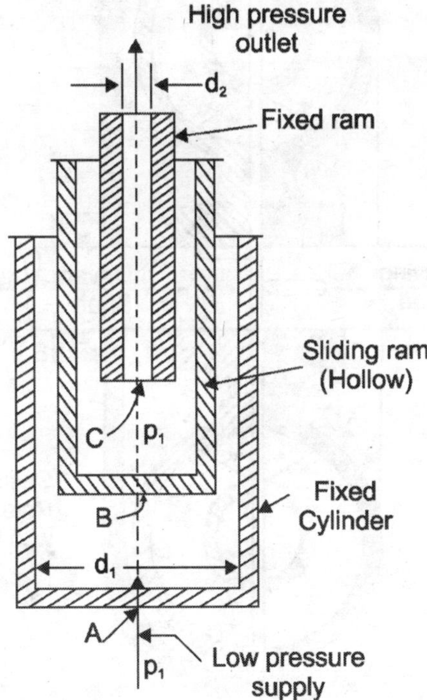


**Fig. 12.13 Modified pressure intensifier.**

## 12.12 Fluid Coupling

Fluid coupling consists of a radial pump impeller keyed to a driving shaft *A*, as shown in Fig.12.14, and a radial reaction turbine runner keyed to driven shaft *B*. There is no mechanical connection between the driving and driven shafts. The pump impeller and turbine runner together form a casing completely filled with oil as the fluid with which the shafts *A* and *B* are supposed to be coupled. The fluid is mostly ordinary mineral lubricating oil. When the shaft A is made to revolve slowly, slowly, the oil due to a forced vortex will flow out from the pump impeller and will strike the turbine runner blades. After sufficient head has been built up by increasing the speed of $A$, the fluid will drive the turbine runner and thus set the shaft $B$ in motion. When at full speed, both the shafts are supposed to be rotating at the same speed. In actual practice, owing to slip, the speed of driven shaft is about 2 % less. Thus, the efficiency of fluid coupling would be 98 %. If both the driver and follower rotate at the same speed, the circulation of oil cannot take place. The necessary reduction in speed of the driven shaft thus maintains the continuous flow of oil from pump impeller to turbine runner.

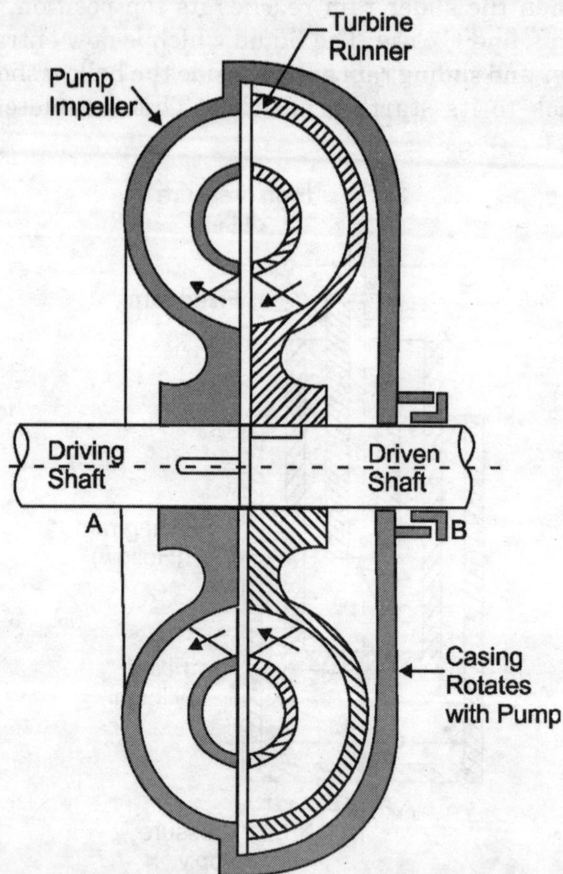


**Fig. 12.14 Fluid or hydraulic coupling.**

Let $T_1$, $T_2$ = torques applied to driving and driven shafts respectively

$N_1$, $N_2$ = rotational speeds of driving and driven shafts respectively.

Neglecting friction losses, $T_1 = T_2$

Efficiency of coupling, $\eta_c = \dfrac{\textit{power output}}{\textit{power input}}$

$$= \frac{T_2\omega_2}{T_1\omega_1} = \frac{T_2 N_2}{T_1 N_1} = \frac{N_2}{N_1} \qquad \text{... (12.13)}$$

Thus the efficiency of an ideal coupling is the ratio of the speeds of driven shaft to that of driving shaft. This is represented by Fig.12.15

The slip of fluid coupling is defined as the ratio of speed difference between the two shafts to the speed of the driven shaft.

$$\text{Slip, } s = \frac{N_1 - N_2}{N_1}$$

$$= 1 - \frac{N_2}{N_1} = 1 - \eta_c \qquad \text{...(12.14)}$$

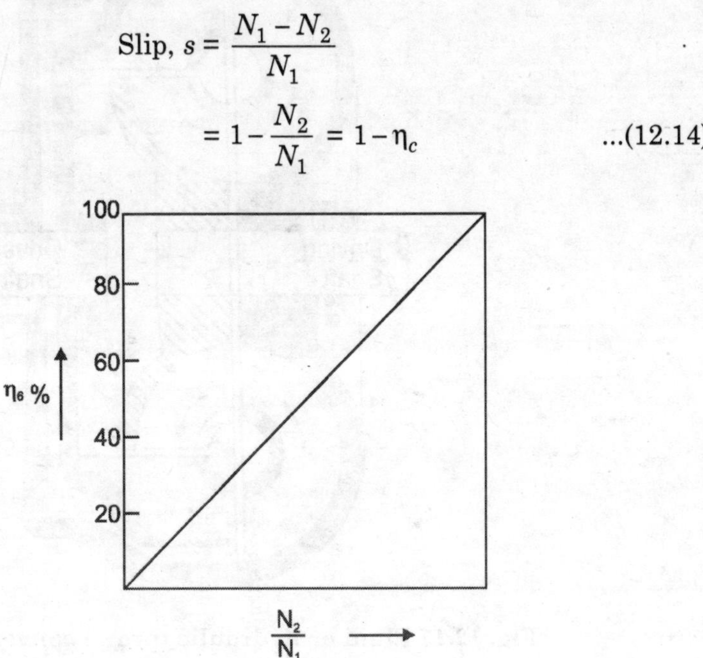


**Fig. 12.15 Efficiency of ideal coupling versus ratio of speeds of driven to driving shafts.**

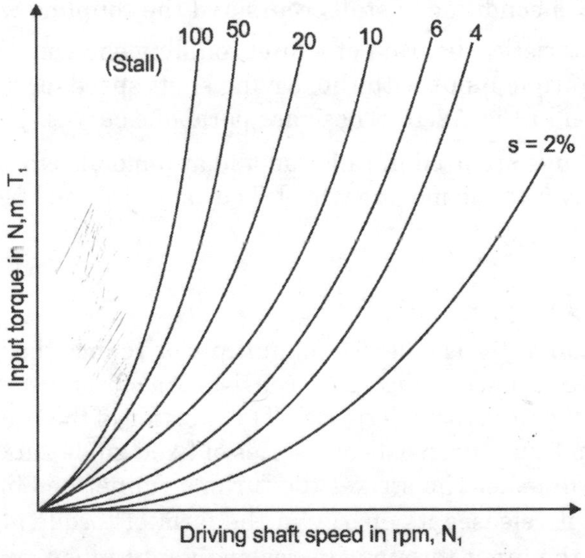


**Fig. 12.16 Characteristics diagram of fluid coupling.**

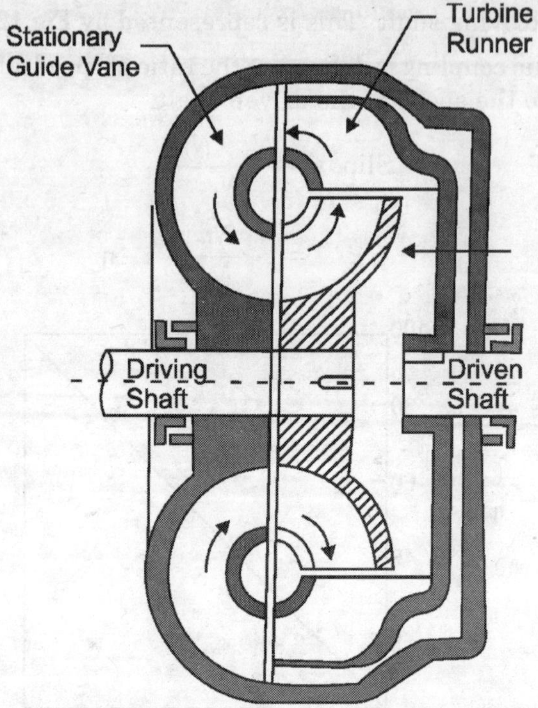


**Fig. 12.17 Fluid or hydraulic torque converter.**

$$\therefore \quad \eta_c = 1 - s \qquad \text{...(12.15)}$$

The slip in fluid coupling varies from 2 to 4 %.

When $N_2 = N_1$, a condition of stall occurs and the coupling will not work.

The characteristics diagram of a fluid coupling showing the variation of driving shaft torque input with the driving shaft speed for different values of slip is shown in Fig. 12.16. These are parabolic curves.

Fluid couplings are used in rail road and automobiles to transmit power from prime mover to the moving wheel. They are also used for power driven excavators.

## 12.13 Fluid Torque Converter

Fluid torque converter is a device to multiply or reduce the torque. A fluid torque converter, shown in Fig.12.17, consists of a stator inserted in between the pump impeller and turbine runner. The function of the stator is to change the direction of fluid. It consists of a series of fixed guide vanes. As the fluid flowing out of impeller rim strikes the turbine runner and blades mounted on the stator, it releases its energy in the form of torque and speed. As a result of this, the input torque is increasingly multiplied, and the speed of output shaft decreases. Thus the stator blades by redirecting the fluid flow multiply the input torque and correspondingly the output speed decreases. This forms the fluid torque converter. The efficiency of torque converter may be as high as 85 % to 87 %. With several stages of turbine runners and station any vanes. a variable speed ratio can be obtained.

Fig.12.18 shows the variation of fluid torque converter efficiency with the ratio of speeds of driven to driving shafts. It may be observed from Fig.12.18 that the maximum efficiency occurs at about

$$\frac{1}{2}\frac{N_2}{N_1}\cdot T_2 = T_1 + T_v$$

$T_v$ = variation of torque caused by fixed guide vanes.

$$\text{Power input, } P_1 = T_1\ \omega_1$$

$$\text{Power output, } P_2 = T_2\ \omega_2 = (T_1 + T_v)\omega_2$$

$$\text{efficiency, } \eta = \frac{P_2}{P_1} = \frac{\omega_2}{\omega_1}\left[1+\frac{T_v}{T_1}\right] \quad ...(12.16)$$

The fluid torque converter is used as a variable gear box for the automobiles and locomotive drives. It is a also used in bulldozers and scrapers.

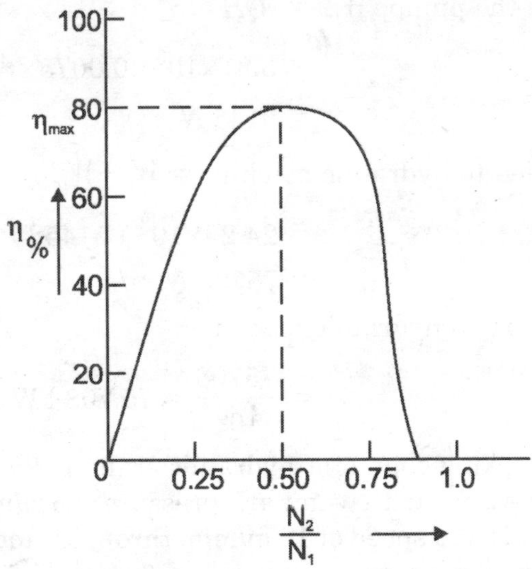


**Fig. 12.18 Efficiency of fluid torque converter versus ratio of speeds of driven to driving shaft**

**Example 12.1** An accumulator has a ram of 300 mm diameter and effective stroke of 6m. It has a total weight of 500 kN. The friction loss in ram amounts to 3 % of the total load. Calculate the total power delivered to the hydraulic machine if the ram falls steadily through its full stroke in 2 minutes. The pump delivers 0.0075 m$^3$/s.

**Solution.**

Given: $D = 0.3$ m, $L = 6$ m, $W = 500$ kN, friction loss = 3 %, $t = 120$ s, $Q_p =$ 0.0075 m³/s

Work supplied by accumulator ram per second,

$$W_r = W \times (1 - .0.03) \times \frac{L}{t}$$

$$= 500 \times 0.97 \times \frac{6}{120} = 24.25 \text{ k N. m/s}$$

Intensity if pressure in accumulator,

$$P = \frac{0.97\,W}{\frac{\pi}{4}D^2} = \frac{4 \times 0.97 \times 500}{\pi(0.3)^2}$$

$$= 6861.34 \text{ kN/m}^2$$

$$\text{Pressure head, } H = \frac{p}{\gamma}$$

$$= \frac{6861.34 \times 10^3}{9.81 \times 10^3} = 699.4 \text{ m of water}$$

Work supplied by the pump, $W_p = \gamma QH$

$$= 9.81 \times 10^3 \times 0.0075 \times 699.4$$

$$= 5145\ N.m/s$$

Total work supplied to hydraulic machine $= W_r + W_p$

$$= 24.25 \times 10^3 + 51458$$

$$= 75708\ N.m/s$$

Total power delivery to hydraulic machine

$$= \frac{75708}{10^3} = 75.708 \text{ kW}$$

**Example 12.2** Working time of a hydraulic crane is 90 s in hoisting and lowering. The crane is fed with water at a pressure of 5 N/mm². It is required to lift a load of 50 kN at a speed of 15 m/min, through a total height of 12 m. The jigger system gives a velocity ratio of 6. Calculate the stroke and diameter of ram assuming an efficiency if 60 %.

Also calculate the minimum capacity of feeding pump and the accumulator.

**Solution.**

Given: $t = 90$ s, $p = 5$ N/mm², $W = 50$ kN, $v = 15$ m/min, $h = 12$ m, $i = 6$, $\eta = 0.6$

$$\text{Load on ram, } W_r = \frac{W \times i}{\eta}$$

$$= \frac{50 \times 6}{0.6} = 500 \text{ kN}$$

Cross – sectional area of ram, $A_r = \frac{W_r}{p} = \frac{500 \times 10^3}{5} = 10^5 \; mm^2$

Ram diameter, $D = \sqrt{\frac{4 \times 10^5}{\pi}} = 356.8$ mm

Stroke length of ram, $L = \frac{h}{i} = \frac{18}{6} = 2\text{m}$

Volume of water swept by ram, $V = A_r \times L$

$$= 10^5 \times 10^{-6} \times 2 = 0.2 \; m^2$$

= Volume of water supplied by the pump

Minimum capacity of feeding pump,

$$V_p = \frac{V}{t} = \frac{0.2 \times 3600}{90} = 8 \text{ m}^3\text{/hour}$$

Operating time of crane, $t_o = \frac{h}{v} = \frac{12}{15} = 0.8$ min or 48 s

Idle time of crane, $t_i = t - t_o = 90 - 48 = 42$ s

Accumulator volume, $V_a = V_p \times t_i = \frac{8 \times 42}{3600} = 0.0933 \; m^3$

$$\text{Accumulator capacity} = p \times V_a = \frac{5 \times 0.0933 \times 10^9}{10^3}$$

$$= 466.67 \text{ k N. m}$$

**Example 12.3** An accumulator maintains a pressure of 5 N/m² in a 580 mm diameter hydraulic main. A hydraulic crane situated at a distance of 20 m from the accumulator is supplied with pressure water from this main. The ram of the hydraulic crane is 200 mm diameter. Velocity ratio of crane hook to ram is 4:1. A pressure of 0.3 N/mm² may be assumed on the ram to account for mechanical friction of ram, pulleys, etc .

Assume a coefficient of friction for the hydraulic main as 0.01. Calculate the load lifted when it is raised with a speed of 0.5 m/s.

**Solution:**

Given: $p_a = 5$ N/mm², $D_m = 50$ mm, $L_m = 200$ m, $D_r = 200$ mm, $i = 4$, $p_r = 0.3$ N/mm², $f = 0.01$, $v = 0.5 \; m/s$

Velocity of ram, $v_r = \frac{v}{i} = \frac{0.5}{4} = 0.125 \; m/s$

Let $v_m$ = velocity of water in hydraulic main

Then $$\frac{\pi}{4} D_m^2 \times v_m = \frac{\pi}{4} D_r^2 \times v_r$$

$$(50)^2 \times v_m = (200)^2 \times 0.125$$

$$v_m = 2 \text{ m/s}$$

Head of water in accumulator, $H = \frac{p_a}{\gamma} = \frac{5\times10^6}{9.81\times10^3} = 509.68$ m of water

Frictional head loss in main,

$$H_{fm} = \frac{4fLm}{D_m} \cdot \frac{v_m^2}{2g}$$

$$= \frac{4\times0.01\times200}{0.05} \times \frac{(2)^2}{2\times9.81}$$

$$= 32.6 \text{ m of water}$$

Head loss due to mechanical friction of ram, pulleys, etc,

$$H_{fs} = \frac{p_r}{\gamma} = \frac{0.3\times10^6}{9.81\times10^3} = 30.58 \text{ m of water}$$

Net head available on the ram,

$$H_{net} = H - H_{fm} - H_{fr}$$

$$= 509.68\text{-}32.6\text{-}30.58$$

$$= 446.5 \text{ m of water}$$

Net pressure on the ram, $W_r = H_{net} \times \gamma$

$$= 446.5 \times \frac{9.81\times10^3}{10^6} = 4.38 \text{ N/mm}^2$$

Load on the ram, $W_r = \frac{\pi}{4} D_r^2 \; p_n$

$$= \frac{\pi}{4} \times (200)^2 \times 4.38 = 137602 \text{ N}$$

Load lifted by crane $= \frac{W_r}{i} = \frac{137602}{4}$

$$= 34400 \text{ N or } 34.4 \text{ kN}$$

**Example 12.4** A jet pump fitted 2.5 m about the suction reservoirs and 20 m below the supply reservoir lifts water through a total height of 3 m. Calculate the efficiency of jet pump when it delivers 0.006 m³/s of water while using 0.002 m³/s from the supply reservoir.

**Solution.**

Given: $H_1 = 20$ m, $H_s = 2.5$ m, $H_s + H_d = 3$ m, $Q = 0.006$ m³/s, $Q_1 = 0.002$ m³/s

$$Q_s = Q - Q_1 = 0.006 - 0.002 = 0.004 \text{ m}^3/\text{s}$$

$$H_d = 3 - 2.5 = 0.5 \text{ m}$$

Jet pump efficiency, $$\eta = \frac{Q_s(H_s + H_d)}{Q_1(H_1 - H_d)}$$

$$= \frac{0.004\times3}{0.002(20-0.5)} = 0.3243 \text{ or } 32.43\ \%$$

**Example 12.5** The following particulars were recorded for a hydraulic ram; Supply head = 3 m, length of supply pipe = 6 m, supply pipe diameter = 25 mm, delivery head = 8 m, delivery pipe length = 8 m, delivery pipe diameter = 15 mm, water pumped = 1.6 kg, time taken = 30 s, water wasted during 30 s = 5 kg,

Assuming a frictional factor = 0.01 calculate the efficiency of ram.

**Solution.**

Given: $H = 3$ m, $L_s = 6$ m, $d_s = 25$ mm,

$h = 8$ m, $L_d = 8$ m, $d_d = 15$ mm, $q = 1.6$ kg,

$t = 30$ s, $Q = 5$ kg, $f = 0.01$

Total quantity of water supplied by the reservoir

$$= Q + q = 5 + 1.6 = 6.6 \text{ kg in } 30 \text{ s}$$

$$v_s = 4\left(\frac{Q+q}{\pi d_s^2}\right) = \frac{4\times0.22\times10^{-3}}{\pi\times(0.025)^2} = 0.448 \text{ m/s}$$

$$q = \frac{1.6}{30}\times\frac{1}{10^3} = 5.33\times10^{-5} \text{ m}^3/\text{s}$$

$$v_d = \frac{4q}{\pi d_d^2} = \frac{4\times5.33\times10^{-5}}{\pi(0.015)^2} = 0.3 \text{ m/s}$$

$$H_{ls} = \frac{4fL_s}{d_s}\times\frac{v_s^2}{2g}$$

$$= \frac{4\times0.01\times6}{0.025}\times\frac{(0.448)^2}{2\times9.81} = 0.098 \text{ m}$$

$$H_{ld} = \frac{4fL_d}{d_d}\times\frac{v_d^2}{2g}$$

$$= \frac{4\times0.01\times8}{0.015}\times\frac{(0.3)^2}{2\times9.81} = 0.098 \text{ m}$$

Efficiency of hydraulic ram, $$\eta = \frac{q(h+H_{ld})}{(Q+q)(H-H_{ls})}$$

$$= \frac{5.33\times10^{-5}(8+0.098)}{0.22\times10^{-3}(3-0.098)}$$

$$= 0.676 \text{ or } 67.6\ \%$$

**Example 12.6** A hydraulic ram with a supply pipe of 75 mm diameter has a waste valve of 100 mm diameter and 15 N weight. The length of supply pipe is 10 m. The travel of valve is 6 mm and number of beats per minute is 110. Determine the discharge through the delivery pipe against a head of 6m. Assume that there is no slip.

**Solution:**

Given: $d_s$ = 75 mm, $d_w$ = 100 mm, $L_s$ = 10 m, $s$ = 6 mm, $n$ = 110, $h$ = 6 m, $W$ = 15 N.

Pressure required to close the waste valve,

$$p_w = \frac{4W}{\pi d_w^2} = \frac{4\times15}{\pi\times100^2}$$

$$= 19.1\times10^{-4}\ N/mm^2$$

Dynamic pressure on waste valve closing

$$= \frac{\rho v^2}{2} = \frac{10^3\times v^2}{2} = 19.1\times10^{-4}\times10^6$$

$$v^2 = 3.82$$

$$v = 1.95 \text{ m/s}$$

Area of waste valve, $$A_w = \pi d_w \times s = \pi\times0.1\times0.006$$

$$= 1.885\times10^{-3}\ \text{m}^2$$

Area of supply, $$A_s = \frac{\pi d_s^2}{4} = \frac{\pi\times(0.075)^2}{4} = 4.418\times10^{-3}\ \text{m}^2$$

Maximum velocity of water in supply pipe,

$$(v_s)_{\max} = \frac{v\times A_w}{A_s} = \frac{1.95\times1.885\times10^{-3}}{4.418\times10^{-3}}$$

$$= 0.832 \text{ m/s}$$

Kinetic energy of water column in the supply pipe at the time of closing the waste valve

$$= \frac{1}{2}mv^2 = \frac{1}{2}[p\ A_sL_s]\times(v_s)_{\max}^2$$

$$= \frac{1}{2}\times10^3\times4.418\times10\times(0.832)^2$$

$$= 15.29 \text{ N. m}$$

Neglecting slip, weight of water that could be lifted against a delivery head of 6 m,

$$W_w = \frac{15.29}{6} = 2.548 \text{ N}$$

$$\text{Discharge} = n \times W_w = 110 \times 2.548$$

$$= 280 \text{ N/min or } 28.54 \text{ liters/min}$$

**Example 12.7** An accumulator is loaded with 400 kN weight. The diameter of ram is 300 mm and stroke of 6 m. Its friction may be takes as 5 %. It takes 2 minutes to fall through its full stroke. Find the total work supplied and power delivered to the hydraulic appliance by the accumulator when 0.0075 m³/s of liquid is being delivered by a pump while the accumulator descends with the stated velocity.

**Solution:**

Given: $W$ = 400 kN, $D$ = 300 mm, $l$ = 6 m, friction = 5 %, $t$ = 2 min, $Q$ = 0.0075 m³/s

Distance moved by the ram per second,

$$v = \frac{L}{t} = \frac{6}{2\times 60} = 0.05 \text{ m/s}$$

Work done by the accumulator per second = $0.95 \times W \times v$

Pressure of water, $$p = \frac{0.95 \times W \times 4}{\pi D^2} = \frac{0.95 \times 400 \times 10^3 \times 4}{\pi (0.3)^2}$$

$$= 5.3759 \text{ MPa}$$

Pressure head, $$H = \frac{p}{\gamma} = \frac{5.3759 \times 10^6}{9.81 \times 10^3}$$

$$= 548 \text{ of water}$$

Work supplied by pump per second = $\gamma QH = 9.81 \times 10^3 \times 0.0075 \times 548$

$$= 40319 \text{ N. m/s}$$

Total work supplied to hydraulic appliance

$$= 19000 + 40319$$

$$= 59319 \text{ N. m/s or } 59.319 \text{ kW}$$

**Example 12.8** A pressure intensifier receives water from a over head tank through a pipe line 60 mm in diameter and 120 m long and conveys high pressure water to a hydraulic press which has 240 m diameter ram and exerts a force of 360 kN. The diameters of ram and sliding piston of intensifier are 0.1 m and 1 m respectively. If the level of water in the overhead tank is 15 m above the inlet to the low pressure side of the intensifier, calculate the speed with which the ram of the hydraulic press moves to exert the force.

Take friction coefficient, $f = 0.0075$ for the pipe line between the supply reservoir and the intensifier.

**Solution.**

Given: $d_p = 60$ mm, $L_p = 120$ m, $D_r = 240$ mm, $F = 360$ kN, $D_1 = 1$ m, $D_2 = 0.1$ m, $H = 15$ m, $f = 0.0075$.

Let $v_p$ = velocity of water in pipe line

$$\text{Frictional head loss, } H_{fp} = \frac{4fL_pv_p^2}{2gd_p} = \frac{4\times0.0075\times120\times v_p^2}{2\times9.81\times0.06}$$

$$= 3.058\, v_p^2\ m$$

Pressure head on the low pressure side of intensifier,

$$H_1 = H - H_{fp} = 15 - 3.058\, v_p^2\ m$$

Corresponding intensity of pressure,

$$p_1 = \gamma H_1 = 9.81\times10^3\left(15-3.058\, v_p^2\right)\ \text{N/m}^2$$

Pressure in the high pressure side of intensifier,

$$p_2 = p_1 \times \frac{A_1}{A_2} = p_1\left(\frac{D_1}{D_2}\right)^2$$

$$= 9810\left(15-3.058v_p^2\right) = \left(\frac{1}{0.1}\right)^2$$

$$= 981\times10^3\left(15-3.058v_p^2\right) N/m^2$$

Force on the ram of hydraulic press,

$$F = A_r p_2 = \frac{\pi}{4}D_r^2 p_2$$

$$= \frac{\pi}{4}(0.24)^2\times981\times10^3\left(15-3.058v_p^2\right)$$

$$= 360\times10^3$$

$$44.3794\left(15-3.058v_p^2\right) = 360$$

$$15-3.058v_p^2 = 8.112$$

$$v_p = 1.5 \text{ m/s}$$

Discharge through the pipe line,

$$Q_p = \frac{\pi}{4}d_p^2\, v_p = \frac{\pi}{4}(0.06)^2\times1.5$$

$$= 4.24\times10^{-3}\ \text{m}^3/\text{s}$$

Now rate of water flow to the low pressure side of intensifier,

$$Q_1 = Q_p = 4.24 \times 10^{-3} \text{ m}^3/\text{s}$$

For continuity of flow, $$Q_r = Q_2$$

$$\frac{\pi}{4} D_r^2 \, v_r = Q_2$$

$$\frac{\pi}{4}(0.24)^2 \times v_r = 4.24 \times 10^{-5}$$

Velocity of ram, $$v_r = 9.37 \times 10^{-4} \text{ m/s}$$

**Example 12.9** The diameter of ram and plunger of a hydraulic press are 250 mm and 30 mm respectively. The leverage of handle is 10:1. With a plunger stroke of 300 mm, the press is able to lift 200 kN through 1.5 m in 2.5 minutes. Calculate:

(a) Force applied at the end of lever,

(b) Number of strokes to be performed by the plunger in one second, and

(c) Power requires to drive the plunger.

Take packing friction of plunger as well as ram as 5 % of load.

**Solution.**

Given: $D_r = 250$ mm, $d_p = 30$ mm, $\frac{l_1}{l_2} = 10$, $L_p = 300$ mm, $W_r = 200$ kN, $L_r = 1.5$ m, $T = 2.5$ min, $k = 5$ % of load

(a) For same pressure throughout, we have

$$P = \frac{F_p(1-0.01k)}{A_p} = \frac{W_r}{A_p(1-0.01k)}$$

$$\therefore \quad F_p = \left(\frac{A_p}{A_r}\right) W_r \times \frac{1}{(1-0.01k)^2}$$

$$= \left(\frac{30}{250}\right)^2 \times 200 \times \frac{1}{(1-0.01\times5)^2}$$

$$= 3.19 \text{ k N}$$

Force applied at the end on lever,

$$P = \frac{l_2}{l_1} \times F_p = \frac{3.19}{10}$$

$$= 0.319 \text{ kN or } 319 \text{ N}$$

(b) $$\text{Number of strokes} = \frac{A_r}{A_p} \times \frac{L_r}{L_p}$$

$$= \left(\frac{250}{30}\right)^2 \times \frac{1.5}{0.3} = 347$$

Strokes per second, $n = \dfrac{347}{2.5 \times 60} = 2.3 \simeq 3$

(c) Work done at plunger/s $= F_p \times L_p \times n$

$$= 3.19 \times 0.3 \times \frac{347}{2.5 \times 60} = 2.214 \text{ kW}$$

**Example 12.10** A load of 100 kN is required to be lifted by a hydraulic lift through a height of 15 m once in every 2 minutes. The lift travels up at the rate of 1.25 m/s. During the working stroke of lift, water is supplied to it from the accumulator and the pump at a pressure of 3000 kN/m². If the efficiency of the pump is 80 % and that of the lift is 75 %, determine:

(a) Power required to drive the pump, and

(b) Minimum capacity of accumulator.

**Solution:**

Given: $W = 100$ kN, $h = 15$ m, $t = 2$ min, $v = 1.25$ m/s, $p = 3000$ kN/m², $\eta_p = 0.80$, $\eta_l = 0.75$

(a) Work done by water in raising the lift,

$$P = Wv = 100 \times 1.25 = 125 \text{ kNm/s or } 125 \text{ kW}$$

$$\text{Power supplied to lift, } P_l = \frac{P}{\eta_l} = \frac{125}{0.75} = 166.67 \text{ kW}$$

$$\text{Working period of lift} = \frac{h}{v} = \frac{15}{1.25} = 12 \text{ s}$$

Idle period of lift = total time – working time

$= 2 \times 60 - 12 = 108$ s

Let $P_p$ = output of pump, kW

Energy stored in accumulator during idle time $= P_p \times 108$ kJ

$$\text{Energy supplied by the accumulator per second} = \frac{P_p \times 108}{12} = 9\,P_p \text{ kW}$$

Total power supplied by the pump $= P_p + 9\,P_p = 10\,P_p = P_l = 166.67$

$$\therefore \quad P_p = \frac{166.67}{9} = 18.52 \text{ kW}$$

$$\text{Power required to drive the pump} = \frac{P_p}{\eta_p} = \frac{18.52}{0.80} = 23.15 \text{ kW}$$

(b) Energy stored in accumulator during idle time

$= 108 \times 18.52 = 2000.16$ kJ

$= p \times$ cylinder volume of accumulator

$$\therefore \quad \text{Capacity (volume) of accumulator} = \frac{2000.16}{3000} = 0.6667 \text{ m}^3$$

## Summary

1. A gear pump is used for pumping viscous liquids by intermeshing identical spur wheels.
2. Actual discharge of a gear pump = $\frac{2AbzN}{60} \times \eta_{vol}$.
3. Screw pumps are especially suitable for operation at high speed and are free from turbulence and pulsation.
4. Vane pumps are used especially for machine tools and industrial equipment.
5. A jet pump is mainly used in mines for lifting oil and petroleum.
6. Jet pump efficiency = $\frac{Q_s(H_s + H_d)}{Q_1(H_1 - H_d)}$
7. An air lift pump is particularly suitable for draining out water in mines and for pumping water containing solids.
8. Hydraulic ram is a device to raise a part of large amount of water available at some height, to a greater height.
9. Efficiency of a hydraulic ram considering friction losses in pipes,

$$\eta = \frac{q(h + H_{ld})}{(Q+q)(H - H_{ls})}$$

D' Aubisson efficiency, $\eta_A = \frac{qh}{(Q+q)H}$

Rankine efficiency, $\eta_R = \frac{q(h-H)}{QH}$

10. Hydraulic accumulator is a device to accumulate liquid under pressure by a pump when it is not required by the hydraulic appliance. Capacity of accumulator = $\frac{\pi}{4} D^2 Lp$
11. Hydraulic press works on the Pascal's law of trans mission of equal fluid pressure in all directions.

$$\frac{F_2}{F_1} = \frac{A_2}{A_1}$$

12. Hydraulic lift is a device used for carrying persons from one floor to another floor of a building
13. Hydraulic crane is a device used for lifting heavy loads.

14. Pressure intensifier is a device to multiply the pressure supplied by the pump to suit the requirements of a high pressure machine.

$$p_2 = \left(\frac{A_1}{A_2}\right)p_1$$

15. Fluid coupling is a device which is used for power transmission from one shaft to another through a liquid medium.

$$T_1 = T_2$$

$$\text{Coupling efficiency, } \eta_c = \frac{N_2}{N_1} = \frac{\omega_2}{\omega_1}$$

$$\text{Slip, } s = 1 - \frac{\omega_2}{\omega_1} = 1 - \eta_c$$

Or $$\eta_c = 1 - s$$

16. Fluid torque converter is a device to multiply or reduce the torque from one shaft to another.

$$T_1 = T_1 + T_v$$

Torque converter efficiency,

$$\eta = \frac{\omega_2}{\omega_1}\left(1 + \frac{T_v}{T_1}\right)$$

## Multi-Choice Questions

1. The device used to store energy of liquid under pressure and subsequently make it available to a hydraulic appliance is
   (a) Hydraulic ram (b) hydraulic press
   (c) hydraulic accumulator (d) hydraulic coupling.
2. The device used to increase the pressure of liquid with the help of low pressure liquid is
   (a) hydraulic intensifier (b) hydraulic press
   (c) hydraulic crane (d) hydraulic accumulator
3. A device used for power transmission from one shaft to another through a liquid medium is
   (a) hydraulic press
   (b) hydraulic intensifier
   (c) hydraulic torque contester
   (d) fluid coupling

4. Which of the following is employed in automobile gear boxes?
   (a) Hydraulic ram (b) fluid coupling
   (c) fluid torque converter (d) hydraulic intensifier.
5. The efficiency of a fluid coupling, $\eta_c$ and slip, $s$ are related by
   (a) $\eta_c = 1 - s$ (b) $\eta_c = s$
   (c) $\eta_c = 1 - s^2$ (d) $\eta_c = s^2$
6. The fluid coupling efficiency is
   (a) $\eta_c = \frac{w_2}{w_1}$ (b) $\eta_c = 1 - \frac{w_2}{w_1}$
   (c) $\eta_c = \frac{w_1}{w_2}$ (d) $\eta_c = 1 + \frac{w_2}{w_1}$
7. The device used in hilly areas for pumping small quantities of water to higher levels from large quantities of water available at low level is called
   (a) Hydraulic accumulator (b) hydraulic ram
   (c) jet pump (d) air lift pump.
8. The working of which of the devices is based on Pascal's law?
   (a) Jet pump (b) hydraulic press
   (c) airlift pump (d) fluid coupling

## Answers

1. (c) 2. (a) 3. (d) 4. (c) 5. (a) 6. (a) 7. (b) 8. (b)

## Review Questions

1. What is a gear pump?
2. Describe the working of a hydraulic intensifier.
3. What is a hydraulic accumulator?
4. What is the principle of working of a hydraulic press?
5. Define capacity of a hydraulic accumulator.
6. Where do we mainly use air lift pump?
7. What is the principle of working of a jet pump?
8. What are the uses of fluid coupling?
9. What is a fluid torque converter?
10. Where do we use hydraulic ram?

11. What is the principle of working of a hydraulic ram?
12. What is the difference between a fluid coupling and fluid torque converter?

## Numerical Problems

1. The ram diameter for a hydraulic crane is 150 mm. the ratio between the movement of load and ram is 6:1. Water is supplied through a 40 mm diameter pipe having a length of 400 m. The pressure at the inlet end of pipe is 8 N/mm$^2$. The friction coefficient for the pipe is 0.07. A pressure of 0.6 N/mm$^2$ is required on the ram to overcome mechanical losses. Determine:
   (a) The maximum speed with which a load of 10 kN can be lifted,
   (b) The load and speed of lifting which correspond to the maximum power obtained from the crane.
2. An intensifier has a ram diameter of 150 mm and sliding cylinder piston diameter of 750 mm. Calculate the pressure of water on the low pressure side of the intensifier if the pressure of water on the high pressure side is to be 20 N/mm$^2$. The loss due to friction at each of the packings of the intensifier is 5 % of the total pressure on each piston.

   [**Ans.** 0.886 N/mm$^2$ ]
3. The ram of an accumulator is loaded with 1500 kN. It supplies power through a 100 mm diameter pipe, 1500 m long. The loss in the pipe is estimated to be 2 % of the power supplied by the accumulator. The friction coefficient in the pipe is 0.01. The diameter of the ram, working friction less is 0.75. What is the maximum power it can supply?
4. Calculate the displacement of the ram of a hydraulic crane, whose efficiency is 50 % in order that with a water pressure of 5 N/mm$^2$, it may raise a load of 250 kN. Find the diameter of ram, if the stroke is 5 times the diameter.
5. An accumulator has a ram 300 mm diameter and a stroke of 5 m. It is loaded with 800 kN. The friction of glands consumes 1.5 % of this load. If the ram falls steadily in 2 minutes while the pumps are delivering 0.032 m$^3$/s, what power is being delivered to the fluid machine by the hydraulic system?
6. As hydraulic lift raises a load of 90 kN at a speed of 0.5 m/s through a height of 15 m once in 2 minutes. The lift is being worked from an accumulator which is fed by a pump having an efficiency of 80 %. If

the pressure of water is 7.5 N/mm$^2$ and the efficiency of the lift 70 %, calculate the power required to drive the pump and the minimum capacity of the accumulator.

7. A hydraulic crane has a ram 200 mm diameter and water is supplied to the crane at 5 N/mm$^2$ at the rate of $0.9 \times 10^{-3}$ m$^3$ /s. The pulley system of the crane has a velocity ratio of 6. If the height to which the load is to be lifted is 15 m, find the load that can be lifted and the time taken to lift the same.

8. The ram of a hydraulic accumulator is 400 mm in diameter. The stroke is 6 m and the water pressure is 8 MPa. If the useful work given by the accumulator during one full downward stroke is utilized in raising a load to a height of 15 m by means of a hydraulic crane, whose efficiency is 60 %, find the load lifted. If this work is dome in 3 minutes, what is the gross power of the crane?

# CHAPTER 13 Fluid Machinery Laboratory Practice

**Experiment 13.1** *To determine the "coefficient of impact" of water jet on different types of vanes.*

***Apparatus.*** Impact of jet test set up, stop watch, weights and Vernier call iper.

***Theory.*** When a jet issues form a nozzle of area $A$, then the velocity $v$ of the impinging jet can be calculated from

$$Q = C_c vA \text{ m}^3/\text{s} \quad \text{....(a)}$$

where $C_c$ = coefficient of contraction of the nozzle = 0.97

$Q$ = quantity of fluid falling on the plate.

If the jet impinges on a flat plate normal to jet, then the force $F_x$ exerted by the fluid on the plate is [see Fig. 13.1(a)].

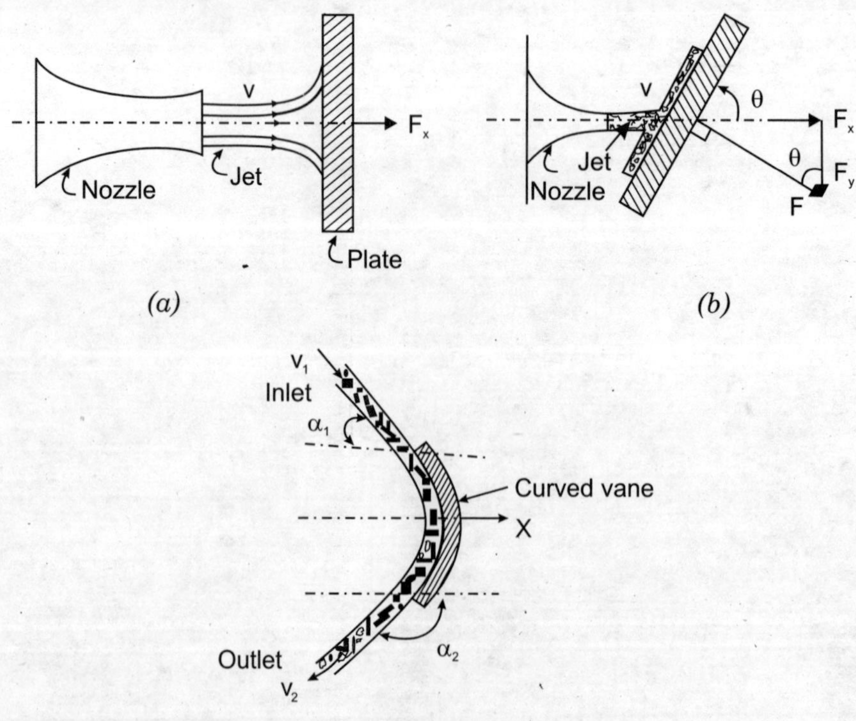


**Fig. 13.1**

$$F_x = \rho Q v \quad \text{...(b)}$$

where $\rho$ = density of fluid, kg/m$^3$

When the jet impinges on an inclined plate as shown in Fig.13.1 (b), the dynamic force acting normal to the plate is:

$$F = \rho Q v \sin\theta$$

The component of force $F$ in the direction of jet is,

$$F_x = F \sin\theta = \rho Q v \sin^2 \theta \quad \text{...(c)}$$

When the jet impinged on a curved stationary vane as shown in Fig. 13.1(c), the force on the vane in the x-direction is,

$$F_x = \rho Q v (\cos \alpha_1 - \cos \alpha_2) \quad \text{...(d)}$$

If $\alpha_1 = 0$, then

$$F_x = \rho Q v (1 - \cos \alpha_2) \quad \text{...(e)}$$

The actual force $F_a$ on the vane is determined by balancing the steel yard. Then, we have

Coefficient of impact $K$

$$= \frac{F_a}{F_x} \quad \text{...(f)}$$

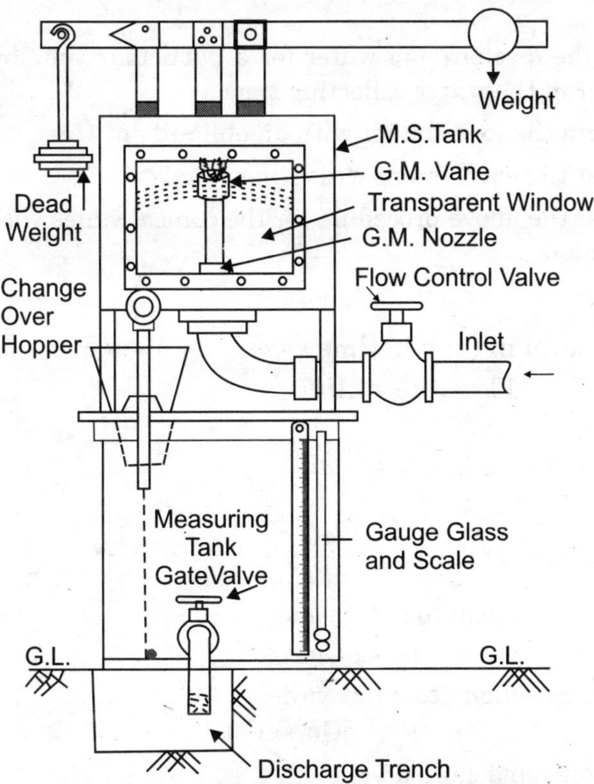


**Fig. 13.2 Impact of jet apparatus.**

***Experimental set-up.*** The apparatus (see Fig. 13.2) consists of nozzle housing, nozzle, vane, weighing mechanism, transparent tank and water collecting tank. Different types of vanes with normal and oblique impact facility are provided. The weighing mechanism consists of a steel yard of 1 m length, an anti-friction bearing fulcrum, a dead weight measuring device, a trace weight, a stopper and a horizontal position indicator. To observe the flow and the jet deflection, the tank is fitted with transparent perspex sheet on the two opposite sides. The other sides are provided with sliding door facilitating easy change of the vanes and nozzles. The water collecting tank is of $50 \times 60 \times 80 \text{ cm}^3$ size with a bye-pass arrangement having partition wall and change over hopper and is fitted with a glass scale for quick and easy measurement of water level. A gate valve is provided to drain out water when the unit is not in use.

**Procedure**

1. Measure the diameter of nozzle and fix it in the nozzle housing.
2. Fix a flat vane at normal jet impact position and balance the steel yard.
3. Open the valve in the delivery pipe of the overhead constant head tank.
4. Note the discharge of water for a particular time by shifting the hopper in the water collecting tank.
5. Perform the experiment with an obliquity of 15°.
6. Repeat the experiment with other nozzles.
7. Repeat the above procedure for the conical vanes with different jet deflections.

**Calculations**

Rise of water level in tank in time $t = h$

$$\text{Flow rate} = Ah/t$$

$$\text{Area of nozzle} = \frac{\pi}{4} d^2$$

$$\text{Velocity of flow, } v = \frac{4Q}{\pi d^2 C_c}$$

For jet impinging normal to a flat vane

$$F_x = \rho Q v$$

For jet impinging oblique to a flat vane

$$F_x = \rho Q v \sin^2 \theta$$

For jet impinging on a curved vane

$$F_x = \rho Q v \left(\cos\alpha_1 - \cos\alpha_2\right).$$

**Observations**

Density of fluid, =

Area of tank, $A$ =

| Sl. No. | Type of vane. | Nozzle dia. d | Nozzle area, m$^2$ | $\theta$, Deg. | Time, $t$ s | Water Level rise, $h$ m | Flow rate, m$^3$/s | Flow Velocity, v m/s | $F_x$ | $F_a$ | $K = \frac{F_a}{F_x}$ |
|---|---|---|---|---|---|---|---|---|---|---|---|
| | | | | | | | | | | | |

**Precautions**

1. The head in the overhead water tank should be kept constant.
2. The opening of the valve should not be changed during and after taking the water level reading in the water collecting tank.
3. The steel yard should be allowed to stabilize before taking the reading.

**Experiment 13.2** *To determine the efficiency of a Pelton turbine and to plot its characteristic curves.*

***Apparatus.*** Pelton wheel, stop watch, tachometer, pressure gauge, weight, brake dynamometer, venturimeter and centrifugal pump.

***Theory.*** Pelton turbine is an impulse turbine which is used to utilize high head for generation of power by converting it into kinetic energy by means of a spear and nozzle arrangement. The water leaves the nozzle in the form of a jet and then strikes the buckets of the turbine. After striking the buckets the jet of water is deflected through an angle of 160 to 170 deg. While passing over the buckets the velocity of water is reduced and hence an impulsive force is supplied to them which in turn start rotating.

The water impining on the bucket of a Pelton turbine is shown in Fig. 13.3.

Let $u_1$ = peripheral speed of wheel.

$v_1$ = jet velocity.

The relative velocity of water at inlet, $v_{r1}$ is given by

$$v_{r1} = v_1 - u_1$$

(since $v_1$ and $u_1$ are taken as collinear).

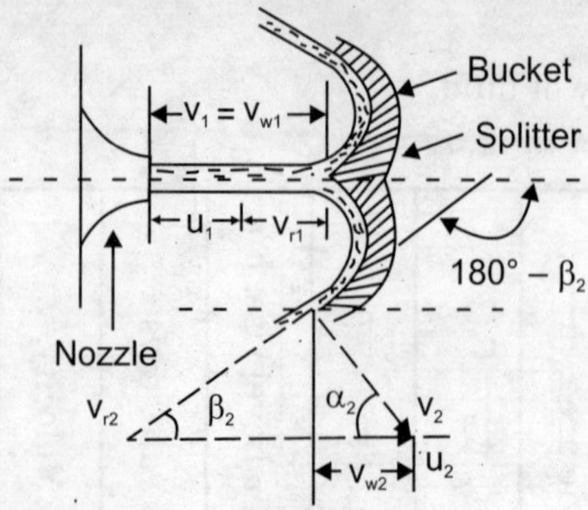


**Fig. 13.3**

Peripheral component of absolute velocity of water is given by whirl velocity,

$$v_{w1} = v_1 \cos\alpha_1 = v_1 \qquad (\because \beta_1 = 180°, \alpha_1 = 0°)$$
$$= v_{r1} + u_1$$

The relative velocity at outlet, $v_{r2}$ is slightly less than $v_{r1}$ due to frictional loss over the inner surface of the bucket and the loss due to the jet striking over the splitter of bucket. Hence $v_{r2} = kv_{r1}$, where $k$ is a coefficient which takes account of friction loss over the bucket.

Form outlet velocity triangle, we have

$$v_{\omega 2} = u_2 - v_{r2} \cos\beta_2$$
$$= u_2 - kv_{r1} \cos\beta_2$$

$$\text{Force of jet, } F = \rho Q\left(v_{w1} - v_{w2}\right)$$
$$= \rho Q\left[\left(v_{r1} + \mathrm{u}_1\right) - \left(\mathrm{u}_2 - kv_{r1} \cos\beta_2\right)\right]$$

Now $\qquad u_1 = v_2 = u = \dfrac{\pi \mathrm{DN}}{60}$

where $D$ = mean diameter of wheel

$N$ = rpm of wheel.

Hence $\qquad F = \rho Q v_{r1}\left(1 + k\cos\beta_2\right)$ ...(a)

Power output $\qquad \rho_t = F \times u$ ...(b)

Power input $\qquad \rho_t = \dfrac{1}{2}\rho Q v_1^2$ ...(c)

Wheel efficiency $\qquad \eta_t = \dfrac{P_t}{P_i}$ ...(d)

The unit quantities are defined as follows

Unit rate of flow $\qquad Q_1 = \dfrac{\mathrm{Q}}{\sqrt{H}}$ ...(e)

Unit speed $\qquad N_1 = \mathrm{N}/\sqrt{H}$ ...(f)

Unit power $$P_1 = \frac{P_t}{H^{3/2}} \quad ...(g)$$

Unit force $$F_1 = F/H \quad ...(h)$$

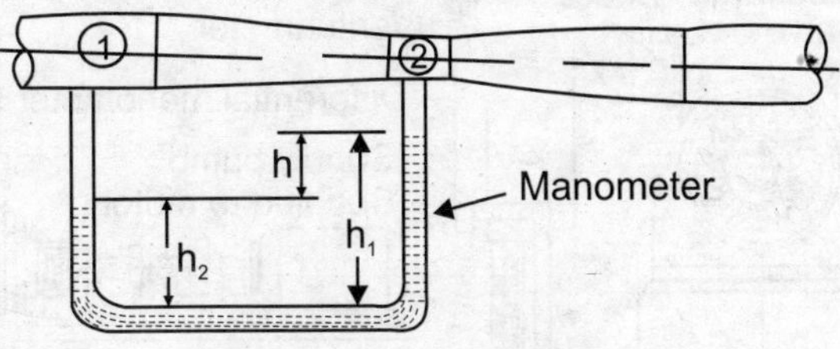


**Fig. 13.4 Venturimeter**

In the experimental set up, the discharge is measured by means of a venturimeter (see Fig. 13.4). If $a_1$ and $a_2$ are the area of cross-section at locations 1and 2 respectively and $C$ is the cocfficient of venturmeter, then it can be shown that

$$Q = \frac{C\, a_1 a_2}{\sqrt{a_1^2 - a_2^2}} \sqrt{2gH_1} \quad ...(i)$$

Where $H_1$ = 12.6 h is the equivalent water column.

For $$\frac{a_2}{a_1} = 0.36,\ C = 0.962 \text{ and } d_1 = 65 \text{ mm}$$

$$Q = 0\,0055\sqrt{H_1}\ m^3/s \quad ...(j)$$

If $H$ is the effective turbine head in m, then

$$P_i = \rho Q\, gH \text{. Watts.} \quad ...(k)$$

The power output is measured by a brake dynamometer. If $W$ is the dead weight on the brake and $S$ be the reading of the spring balance in kg, then

$$P_t = \frac{\pi DN}{60}(W - S)\, g \text{ watts} \quad ...(l)$$

where $D$ = dia. of brake drum + dia. of rope

$N$ = rpm

$g$ = 9.80665 m/s$^2$

Speed factor (ratio) $K_{u1}$

$$= \frac{u}{\sqrt{2gH}} \quad ...(m)$$

***Experimental set-up.*** The experimental set-up as shown in Fig. 13.5 consists of a Pelton turbine supplied with water under high

pressure by a centrifugal pump. The water is conveyed through a venturimeter having manometer connection to determine the flow rate of water. The nozzle opening can be increased or decreased by operating the spear wheel. The spear can be positioned at 8 places viz., $\frac{1}{8}, \frac{2}{8}, \ldots\ldots, \frac{8}{8}$

of nozzle opening. The turbine shaft is coupled to a brake dynamometer to measure its output.

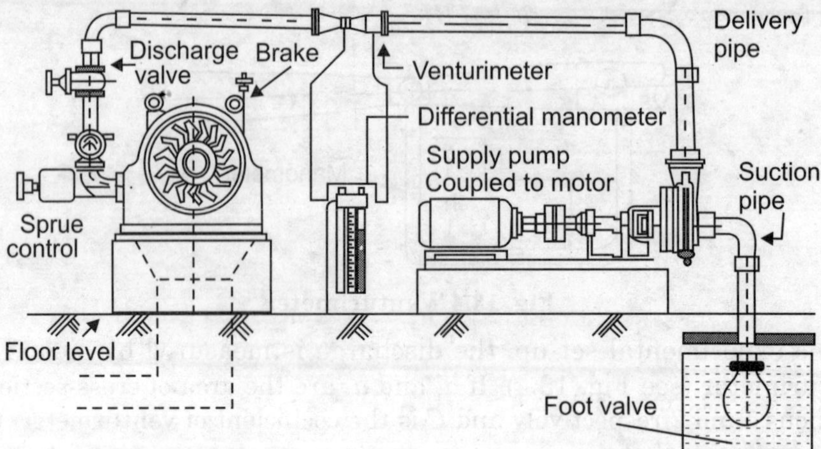


**Fig. 13.5 Experimental set-up for Pelton turbine.**

## Procedure

1. Keep the nozzle opening at 75 percent of full opening.
2. Start the centrifugal pump.
3. Allow the water to flow into the turbine.
4. Note the speed of the turbine by a tachometer.
5. Take the readings of the manometer.
6. Note the pressure of water in the pressure gauge to determine the head generated by the pump.
7. In order to plot the main characteristics, keep the head constant and to plot the operating characteristics, keep the speed constant.
8. Load the turbine by weight on the brake and note the spring reading.
9. Repeat the experiment for different loads, speeds and gate openings.
10. Stop the pump motor.
11. Calculate the power input, power output and efficiency.
12. plot the main and operating characteristics.

## Calculations

$$H = H_g + z + \frac{\sqrt{v^2}}{2g}$$

$$h_1 = ...., h_2 = ....$$

$$h = h_1 - h_2$$

$$H_1 = 12.6\text{ h}$$

$$Q = 0.0055\sqrt{H_1}$$

$$v = \frac{4Q}{\pi d^2}$$

$$P_i = \rho Q\, gH$$

$$W = \ldots.,\ S = \ldots.,\ N = \ldots.,$$

$$u_0 = \frac{\pi D_0 N}{60}$$

$$P_t = u_0 (W - S) g$$

$$\eta = \frac{P_t}{P_i}$$

## Observations

Brake drum dia =

Rope dia =

Effective brake drum dia $D_0$ =

Dia of inlet pipe = $d$

Distance from centre line of turbine to pressure gauge = $z$

| Sl No. | Nozzle opening | Head, $H$ $m$ | Manometer readings | | | | $Q$, m³/s | $P_i$ watts | $N$, rpm | $u_0$, m/s | $W$, kg | $S$, kg | $P_t$ = watts | Efficiency % | $P_1 = \frac{P_t}{H^{3/3}}$ | $Q_1 = \frac{Q}{\sqrt{H}}$ | $N_1 = \frac{N}{\sqrt{H}}$ |
|---|---|---|---|---|---|---|---|---|---|---|---|---|---|---|---|---|---|
| | | | $h_1$, m | $h_2$, m | $h$, m | $H_1$ m | | | | | | | | | | | |
| | | | | | | | | | | | | | | | | | |
| | | | | | | | | | | | | | | | | | |

***Results.*** The main characteristics are obtained by keeping the head constant and varying the speed by allowing a variable quantity of water to flow through the inlet opening. Thus a series of $Q$ and $N$ values are obtained. For each value, $P_t$ is calculated. The values of $Q$, $N$ and $P_t$ are reduced to unit head and plotted. A typical plot is shown in Fig. 13.6(a). For the operating characteristics speed is kept constant and $Q$ and $H$ is varied. A typical operating characteristics are shown in Fig. 13.6(b).

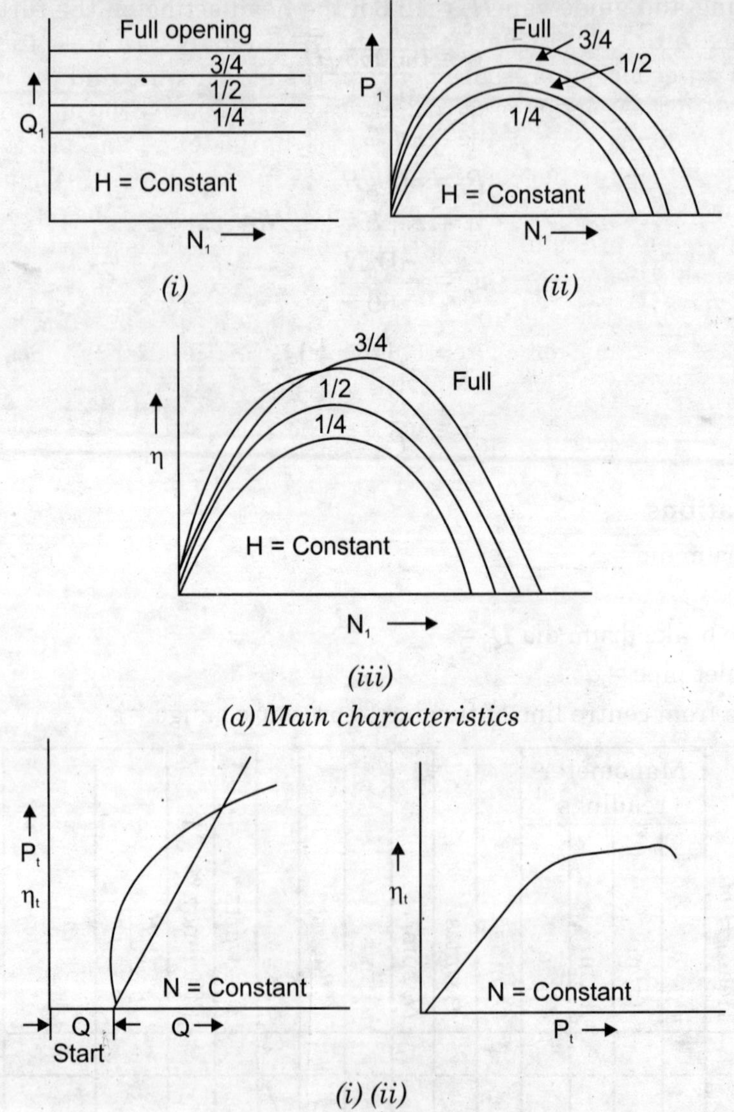


(a) Main characteristics



*(b) Operating characteristics*

**Fig. 13.6 Characteristics of Pelton turbine.**

**Experiment 13.3** *To draw the characteristics curve of a Francis turbine.*

***Apparatus.*** Francis turbine, tachometer, brake dynamometer, venturimeter, centrifugal pump, pressure gauge, vacuum gauge, weights and manometer.

***Theory.*** A Francis turbine is an inward radial flow reaction turbine. It operates under medium heads and requires medium quantity of water. The water is led to the turbine though scroll casing and directed towards a number of stationary guide (wicket) vanes. While passing though the spiral casing and guide vanes, a part of the head acting on the turbine is transformed into kinetic energy and the rest remains as pressure head. There is a difference of pressure between the guide vanes and the runner which is called the reaction pressure, and is responsible for the motion of the runner. After doing its work the water is discharge to the tail race through a draft tube, which further converts the kinetic head to pressure head. Thus the pressure at the runner at the exit is reduced below atmosphere. This makes it possible to install the turbine above tail race level without any loss in available head.

A typical velocity triangles for a Francis turbine are shown in Fig. 13.7. The resultant dynamic force $F_u$ exerted by water on the runner vanes in the direction of rotation is given by

$$F_u = \rho Q(v_{w1} - v_{w2}) \qquad \text{...(a)}$$

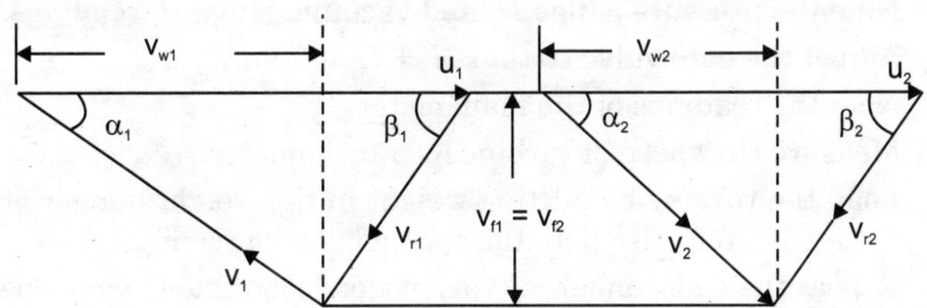


**Fig. 13.7 Velocity triangles for Francis turbine.**

If $R_1$ and $R_2$ are the radii of runner at inlet and outlet respectively, then torque $T$ becomes:

$$T = \rho Q(v_{w1}.R_1 - v_{w2}.R_2) \qquad \text{...(b)}$$

$$\text{Power developed } \rho H = \rho Q(v_{w1}.u_1 - v_{w2}.u_2) \qquad \text{...(c)}$$

$$\text{Head efficiency } \eta H = \frac{v_{w1}.u_1 - v_{w2}.u_2}{gH} \qquad \text{...(d)}$$

Now $$Q = (\pi D_1 - z_2 t)B_0.v_{f1} \qquad \text{...(e)}$$

Where $D_1$ = inlet dia of runner

$z_2$ = number of runner blades

$t$ = thickness of blades

$B_0$ = height of guide (runner) blades.

The behaviour of the turbine may be exhibited by main characteristics (constant head curves), operating characteristics (constant speed curves) and Muschel curves (constant efficiency curves).

***Experimental set-up.*** The turbine consists essentially of a runner, a ring of adjustable guide vanes, a volute casing and a draft tube. The runner consists of two shrouds with a number of fixed curved vanes in between.

The guide vanes can be rotated about their axis by means of a head wheel, and their position is indicated by a pair of dummy guide vanes fixed outside the turbine casing. A centrifugal pump supplies water under pressure to the turbine, and the flow can be controlled by means of the gate valve fitted in the pipe line. The discharge in the pipe line and hence through the turbine can be measured with a venturimenter.

The head of water is measured with a pressure gauge and vacuum gauge. The output of the turbine is determined by means of a rope brake dynamometer. The experimental set-up is shown in Fig. 13.8.

**Procedure**

1. Keep the guide vanes fully opened.
2. Prime the pump and start it.
3. Vent the manometer.
4. Note the pressure gauge $H_g$ and vacuum gauge $H_v$ readings.
5. Adjust the gate valve so that $H_g + H_v = 15$m.
6. Note the readings of the manometer.
7. Measure the speed of turbine by a tachometer.
8. Load the turbine by putting weight in the weight hanger of brake dynamometer. Also note the spring balance reading.
9. Repeat the experiment for various loads and guide vane openings.
10. Stop the pump motor.
11. Calculate the power input, power output, efficiency, specific and unit quanities.
12. Plot the characteristic curves.

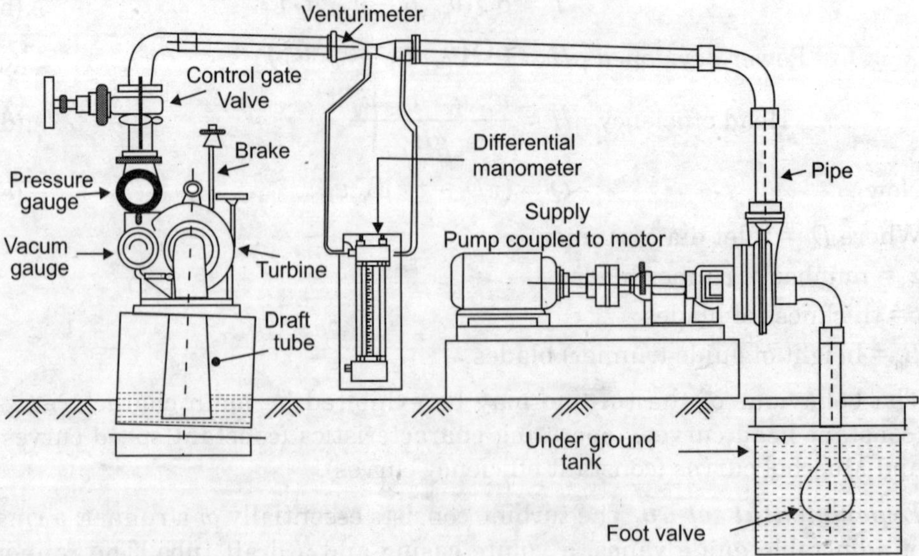


**Fig. 13.8 Experimental set-up for Francis turbine.**

**Observations**

Brake drum dia =

Rope thickness =

Dia of inlet pipe = $d_i$

Effective brake drum dia,

$D_0$ = Dia of exit pipe = $d_e$

$$u_0 = \pi D_0 N/60$$

| SI. No. | Gate opening | $H_g$, m | $H_v$, m | $H$, m | Manometer readings | | | | $Q$, m$^3$/s | $Pi$, watts | $N$, rpm | $u_0$, m/s | $W$, kg | $S$, kg | $Pt$, watts | η % | Unit speed, $N/\sqrt{H}$ | Unit power, $P_t/H^{3/2}$ | Unit discharge $Q/\sqrt{H}$ |
|---|---|---|---|---|---|---|---|---|---|---|---|---|---|---|---|---|---|---|---|
| | | | | | $h_1$, m | $h_2$, m | $h$, m | $H_1$, m | | | | | | | | | | | |
| | | | | | | | | | | | | | | | | | | | |

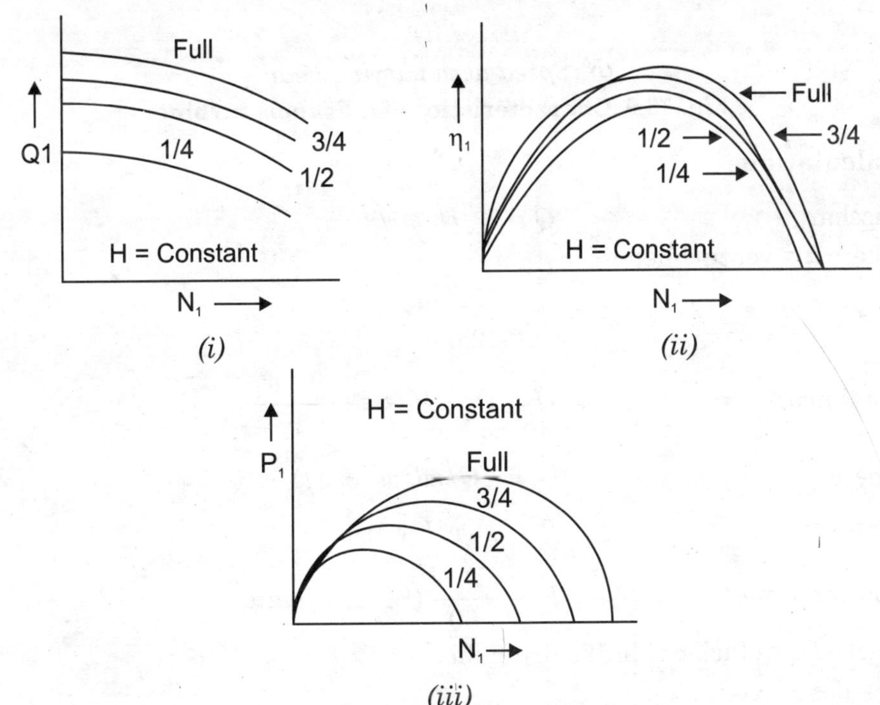


*(a) Main characteristics*

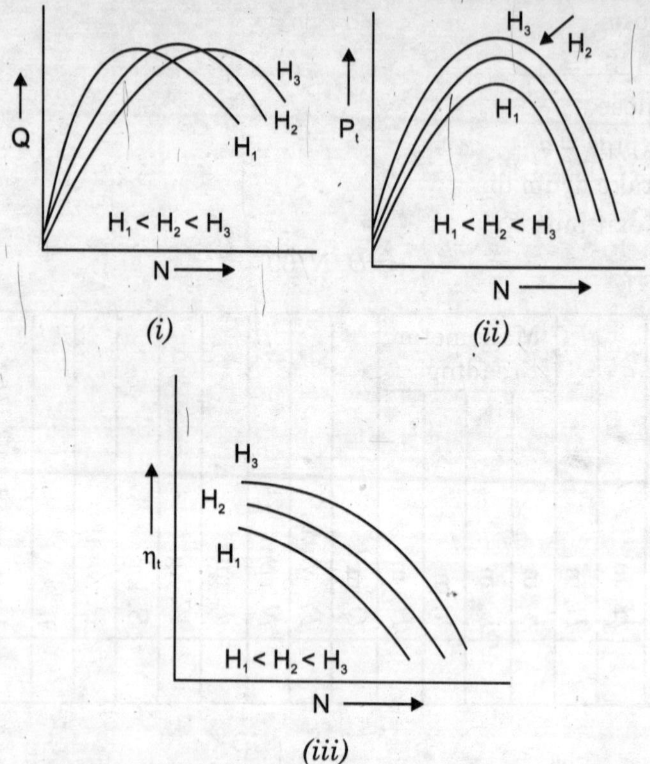


*(b) Speed with varying head*

**Fig. 13.9 Characteristics of a Francis turbine.**

## Calculations

Discharge $Q = k\sqrt{H_1}\ m^3/s$

where $k$ = venturimeter constant

$$H_1 = 12.6\ (h_1 - h_2)$$
$$= 12.6\ \text{h}.$$

Total head $H = H_g + H_v + z + \frac{v_i^2 - v_e^2}{2g}$

where $v_i = 4Q/\pi d_i^2, v_e = 4Q/\pi d_e^2$

Power input $P_i = \rho QgH$ watts.

Power output $P_t = \frac{\pi D_0 N}{60}(W - S)g$ watts

where $D_0$ = efficiency brake drum dia.

$W$ = dead weight, kg.

$S$ = spring balance reading, kg.

Efficiency, $\eta = \frac{P_t}{P_i}$

Specific flow $$Q_{11} = \frac{Q}{D_1^2\sqrt{H}}$$

Specific power, $$\rho_{11} = \frac{P_t}{D_1^2 H^{3/2}}$$

Specific speed $$N_s = \frac{N\sqrt{P_t}}{H^{5/4}}$$

***Results.*** The main characteristics for a Francis turbine are shown in Fig.13.9 (a) and speed with varying head characteristics in Fig. 13.9 (b).

**Experiment 13.4** To draw the characteristic curves of a Kaplan turbine.

***Apparatus.*** A Kaplan turbine, brake dynamometer, tachometer, centrifugal pump, venturimeter, manometer, pressure gauge, vacuum gauge and weights.

***Theory.*** The Kaplan turbine operates under low heads and large quantities of flow. The water from the pump enters through the spiral casing into the guide vanes and then to the runner. While passing through the spiral casing and guide vanes a portion of the pressure head is converted into kinetic energy. The water enters the runner at a higher velocity than in the pipe line. The velocity triangles for a Kaplan turbine are shown in Fig.13.10. It can be shown that:

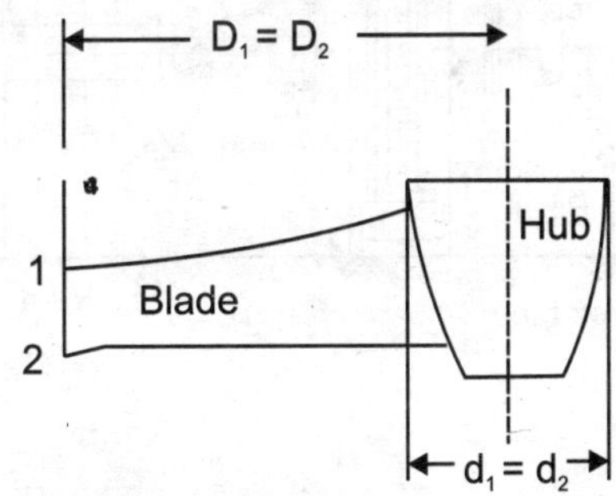


*(a) Runner of Kaplan turbine.*

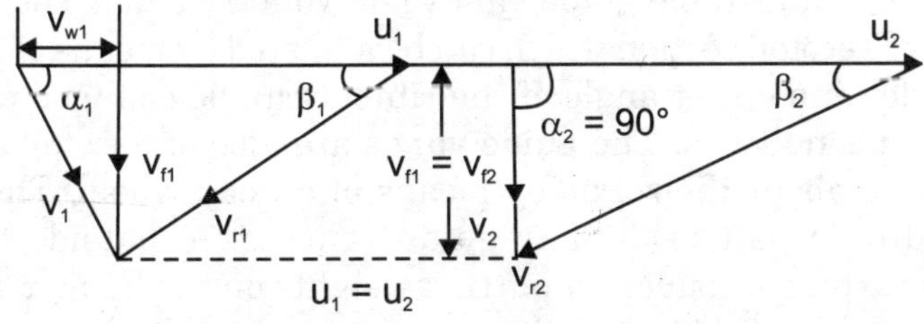


*(b) Velocity triangles.*

**Fig. 13.10**

Specific flow $$Q_{11} = \frac{Q}{(D_1^2 - d_1^2)\sqrt{H}} \quad \text{...(a)}$$

Where $Q = k_t \frac{\pi}{4}\left(D_1^2 - d_1^2\right)v_{f1}$ and $k_t$ accounts for the area accupied by the blades.

Power developed $$P_H = \rho Q\left(v_{w1}.u_1 - v_{w2}.u_2\right) \qquad \text{...(b)}$$

Here $$u_1 = u_2 = \frac{\pi D_1 N}{60} \text{ and } v_{u2} = 0$$

Head efficiency $$\eta H = \frac{u_1.v_{w1}}{gH} \qquad \text{...(c)}$$

Experimentally, the efficiency is determined in the same way as for Pelton or Francis turbines. Characteristic curves may also be drawn in the same fashion.

***Experimental Set-up.*** A Kaplan turbine is an axial flow reaction turbine. It consists essentially of a runner with adjustable

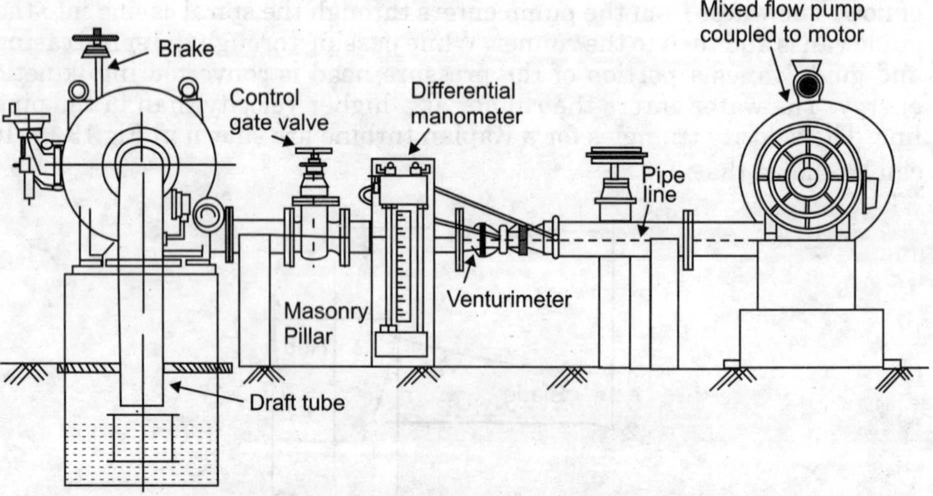


**Fig. 13.11 Experiment setup for Kaplan turbine.**

Blade, a ring of adjustable guide vanes and volute casing. The runner has four blades of aerofoil section which can be adjusted by means of a regulator. Thus the inlet and outlet angle of the runner can be changed according to the operating conditions. The guide vanes are also of aerofoil section and can be rotated about their axis by means of a head wheel. Their position is indicated by a pair of dummy guide vanes fixed outside the turbine casing. The turbine is loaded by putting weights on the hanger of the brake dynamometer. Fig.13.11 shows the experimental set-up.

Water under pressure is supplied to the turbine by a mixed flow centrifugal pump. The discharge from the pump is controlled by means of a gate valve fitted in the pipe line. The discharge in the pipe line and hence through the turbine can be measured with the help of a venturimeter. The head of water is measured with the help of a pressure gauge and vacuum gauge.

## Procedure

1. Keep the guide vane and runner at 4/8 opening.
2. Prime and start the pump.
3. Open the gate valve slowly.
4. Note the reading of pressure and vacuum gauge.
5. Note the reading of monometer.
6. Measure the speed of turbine by a tachometer.
7. Load the turbine by putting weights on the brake dynamometer and note the reading of spring balance.
8. Repeat the experiment for various brake loads and different opening of guide vanes.
9. Stop the pump motor.
10. Calculate the power input, power output, efficiency and unit quantities.
11. Plot the characteristic curves.

$$\text{Total head, } H = (H_g + H_v) + z + \frac{v_i^2 - v_e^2}{2g}$$

Where $\upsilon_i = 4Q/\pi d_i^2, v_e = 4Q/\pi d_e^2$

and $z$ = vertical distance between the gauges.

Equivalent head of water $H_1 = 12.6\,(h_1 - h_2)$

$= 12.6$ m

Discharge $Q = k\sqrt{H_1}$ m³/s

where $k$ = venturimeter constant.

## Observations

Brake drum dia =

Rope dia =

Equivalent drum dia,

$D_0$ = Dia of inlet pipe = $d_i$

Dia of exit pipe = $d_e$

| Sl. No. | Gate opening | $H_g$, m | $H_v$, m | $H$, m | Manometer readings | | | | $Q$, m³/s | $Pi$, watts | $N$, rpm | $u_0$, m/s | $W$, kg | $S$, kg | $Pt$, watts | $\eta$ % | $N_1 = \frac{N}{\sqrt{H}}$ | $P_1 = \frac{P_t}{H^{3/2}}$ | $Q_1 = \frac{Q}{\sqrt{H}}$ |
|---|---|---|---|---|---|---|---|---|---|---|---|---|---|---|---|---|---|---|---|
| | | | | | $h_1$, m | $h_2$, m | $h$, m | $H_1$, m | | | | | | | | | | | |
| | | | | | | | | | | | | | | | | | | | |
| | | | | | | | | | | | | | | | | | | | |

Power input $\qquad P_i = \rho QgH$ watts

Power output $\qquad P_t = \dfrac{\pi D_0 N}{60}(W - S)g$ watts

Efficiency $\qquad \eta = \dfrac{P_t}{P_i}$

***Result.*** The main characteristics for a Kaplan turbine are shown in Fig.13.12 (a) and operating characteristics in Fig.13.12(b).

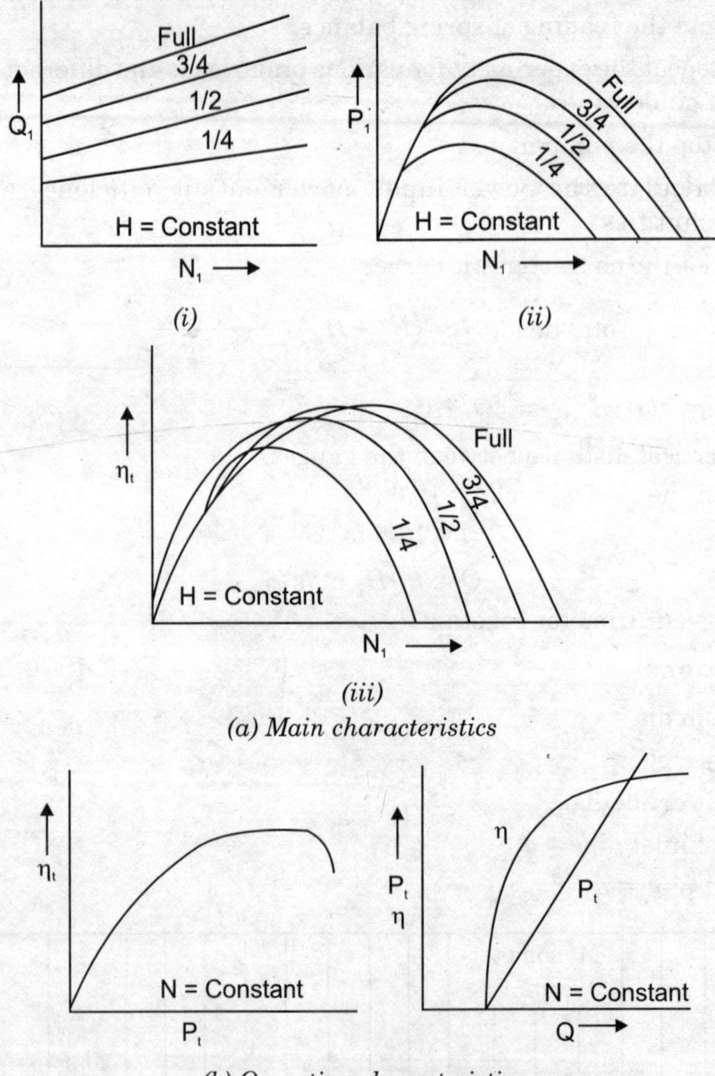


*(a) Main characteristics*

*(b) Operating characteristics*

**Fig. 13.12 Characteristics for Kaplan turbine.**

**Experiment 13.5** *To draw the characteristics curves of a centrifugal pump.*

***Apparatus.*** Centrifugal pump, tachometer, stop watch, pressure gauge, vacuum gauge, water collecting tank and energy meter.

***Theory.*** A centrifugal pump is a contrivance to raise liquids form a lower to a higher level by creating the required pressure with the help of centrifugal action. It imparts whirling motion to the liquid to increase its pressure energy by means of curved blades mounted on the impeller. The liquid enters the impeller at its centre (eye) and discharge in a gradually increasing passage of the casing known as the volute chamber, which serves to convert a part of the velocity head into pressure head.

Before starting the pump, its casing together with impeller, and suction pipe must be filled with liquid in order to remove air, gas or vapour in that region. The removal of air by filling the pump with water is known as priming.

The total head against which the pump has to work is given by

$$H = H_s + H_d + z + \frac{v_d^2 - v_s^2}{2g} \quad \text{...(a)}$$

where $H_s$ = suction head measured by vacuum gauge.

$H_d$ = delivery head measured by pressure gauge.

$z$ = vertical distance between the pressure tappigns for pressure and vacuum gauge in the delivery and suction pipes respectively.

$v_d$, $v_s$ = velocity of liquid in delivery and suction pipes respectively.

The power required to drive the pump is given by

$$P = \frac{\rho Q g H_{mano}}{\eta_{overall}} \text{ watts} \quad \text{...(b)}$$

where $H_{mano} = H_s + H_d + H_L$ ...(c)

$H_L$ = total loss of head.

Fig.13.13 show the velocity triangles for a centrifugal pump. It can be shown that

$$\frac{H_{mano}}{\eta_{mano}} = \frac{u_2 v_{w2} - u_1 v_{w1}}{g} \quad \text{...(d)}$$

Generally $\alpha_1 = 90°$, than

$$\frac{H_{mano}}{\eta_{mano}} = \frac{u_2 . v_{w2}}{g} \quad \text{...(e)}$$

The characteristic curves for the pump show manometric head, brake power and efficiency of the pump as ordinates with its discharge as the abscissa. Pump characteristic can be divided into three categories: (i) Main characteristic including operating characteristic (ii) Constant efficiency curves and (iii) Constant head and constant discharge curves. Main characteristic are obtained by fixing the pump speed at some arbitrary value and plotting $H$, $P$ and $\eta$ against $Q$. during operation the pum must run constantly with the seed of the driving unit. Normally, this is the designed speed. The particular set of main characteristics which corresponds to the designed speed are called the operating characteristics.

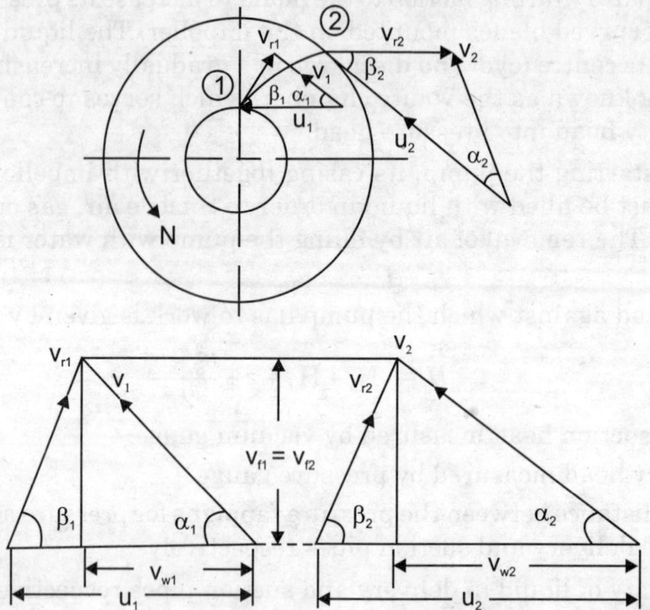


**Fig. 13.13 Velocity triangles for a centrifugal pump.**

***Experimental set-up.*** The experimental set-up for the centrifugal pump is shown in Fig.10.14. It consists of a single stage centrifugal pump drive by squirrel-cage induction motor with the help of a stepless V-belt drive. Different speed of the pump can be obtained by changing the centre distance between the pump and the motor. The motor is fitted on a base which can be moved forward and backward by means of a handwheel and screw. The motor V-pulley is spring loaded to provide variable diameter. The discharge is measured by collecting water in the measuring tank for a specified time. The pressure in the delivery and suction pipes is measured by the pressure and vacuum gauge respectively. The input to the pump is measured by the energymeter by counting its revolutions for a specified time.

**Procedure**

1. Prime the pump and start the motor.
2. Rotate the handwheel unit the required speed is obtained.
3. Note the reading of the pressure and vacuum gauge.
4. Note the time '$t$' taken by the energy meter to complete 10 revolution (say) by means of a stop watch.
5. Note the rise of water level '$h$' in the water collecting tank for the same time '$t$' or for some other time.

6. Repeat the experiment, varying the head from maximum at shut-off to the minimum where gate valve is fully opened.
7. Repeat the experiment for four different rpm.
8. Stop the motor.
9. Make necessary calculations and plot the characteristic curves.

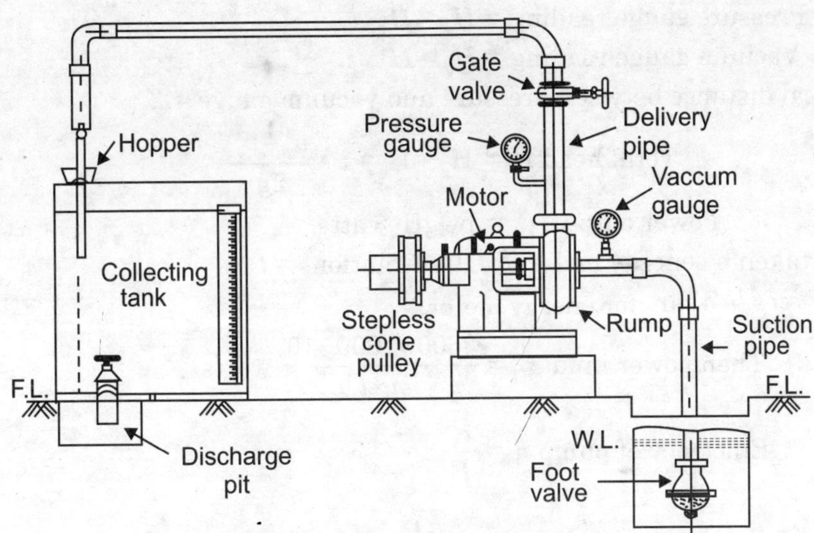


**Fig. 13.14 Experimental set-up for centrifugal pump.**

## Observations

Sunction pipe dia. $d_s$ =

Delivery pipe dia. $d_d$ =

Energy meter constant = $n$ revs/kg hr.

Area of tank = $A$

$z$ =

| Sl. No. | $N$, rpm | $H_g$, m | $H_v$, m | $h$, m | $t$, s | $Q$, m³/s | $\frac{v_d^2}{2g}$, m | $\frac{v_s^2}{2g}$, m | $H$, m | $P_o$ watts | $P_i$ watts | η % |
|---|---|---|---|---|---|---|---|---|---|---|---|---|
| | | | | | | | | | | | | |

## Calculations

Area of tank = $A$ m²

Rise of water level in time $t$ = $h$ m

$$\text{Discharge, } Q = \frac{Ah}{t} \; m^3/s$$

Area of delivery pipe = $a_d$

Area of suction pipe = $a_s$

Velocity in delivery pipe, $v_d = \frac{Q}{a_d} m/s$

Velocity in suction pipe, $v_s = \frac{Q}{a_s} m/s$

Pressure gauge reading = $H_g = H_d$

Vacuum gauge reading = $H_\upsilon = H_s$

Vertical distance between pressure and vacuum gauge = $Z$

$$\text{Total head } H = \mathrm{H}_g + \mathrm{H}_\upsilon + z + \frac{v_d^2 - v_s^2}{2g}$$

$$\text{Power output, } \rho_o = \rho \mathrm{QgH} \text{ watts.}$$

Time taken by energy meter for 10 revolutions = $t$

If $n$ = *revs./kW* hr. for energy meter,

$$\text{Then power input } \rho_i = \frac{3600 \times 1000 \times 10}{n \times t} \text{ watts.}$$

$$\text{Efficiency of pump } \eta = \frac{P_o}{P_i}$$

$$\text{Specific speed of pump } N_s = \frac{\mathrm{N}\sqrt{Q}}{H^{3/4}}$$

***Results.*** The typical characteristics curves for a centrifugal pump are shown in Fig.13.15.

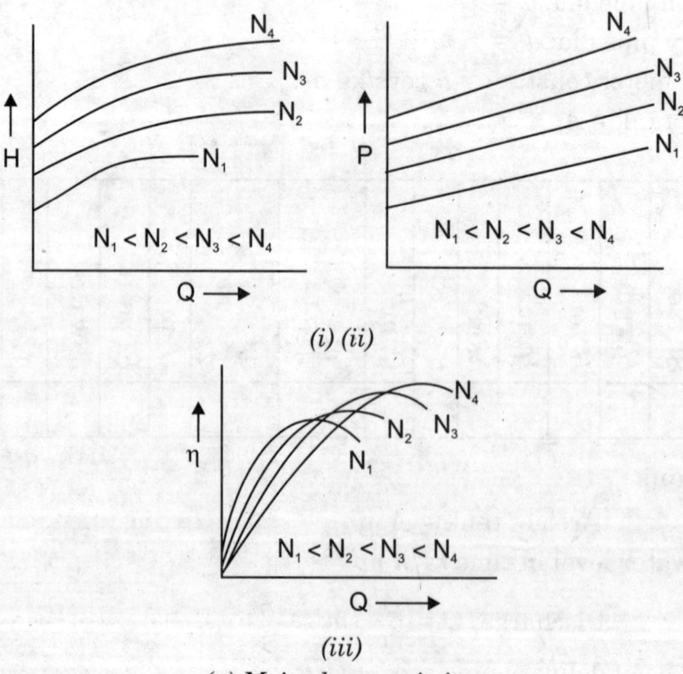


*(i) (ii)*

*(iii)*

*(a) Main characteristics.*

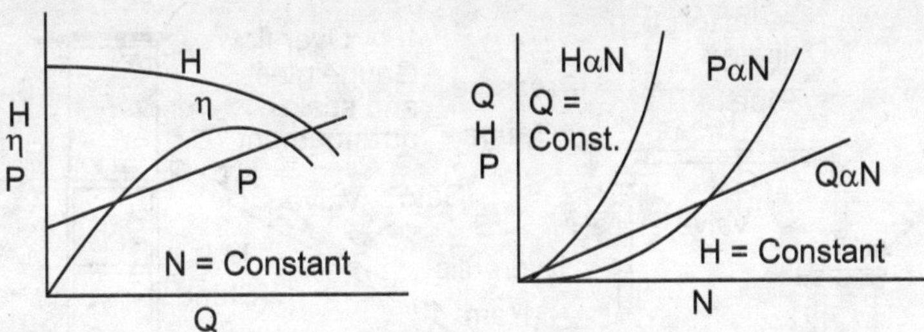


*(b) Operating characteristics.* *(c) Q, H, P v's N curves.*

**Fig. 13.15 Characteristics curves for a centrifugal pump.**

**Experiment 13.6** *To draw the characteristics curves of a Hydraulic ram at constant valve lift and constant supply head.*

***Apparatus.*** Hydraulic ram, supply tank, supply pipe, gate valve, shifting valve, delivery pipe, pressure gauge, V-notch wire, collecting tank and stopwatch.

***Theory.*** Hydraulic ram is a contrivance to rise a part of large amount of water available at some height, to a greater height. This is employed when some natural source of water like a spring or a stream is available at some altitude. This is also known as impulse pump. The impulse is developed at the expense of dynamic inertia possessed by a moving column of water, i.e., the momentum of a long column of water flowing through the supply pipe is mode to force part of water to a level higher than that of the supply source itself.

Let $H_s$ = supply head above waste valve, m

$H_d$ = delivery head above waste valve, m

$Q$ =the quantity of waste water, m³/s

$q$ = the quantity of water pumped by ram, m³/s.

$$\text{Rankine efficiency } \eta_R = \frac{q(H_d - H_s)}{Q \times H_s} \quad \text{...(a)}$$

$$D' \text{ Aubisson efficiency } \eta_D = \frac{q.H_d}{(Q+q)H_s} \quad \text{...(b)}$$

***Experimenter set-up.*** The experimenter set-up for the hydraulic ram is shown in Fig.13.16. It consists of a water supply tank *A* connected to a supple pipe *S* having a gate valve *G*. Water flows through the supply pipe to a waste valve *W*. When the waste valve closes, the water hammer force open the delivery valve. The delivery head is measured by a pressure gauge. The quantity of useful water is measured in a collecting tank and the waste water is measured by a 60° V-notch weir.

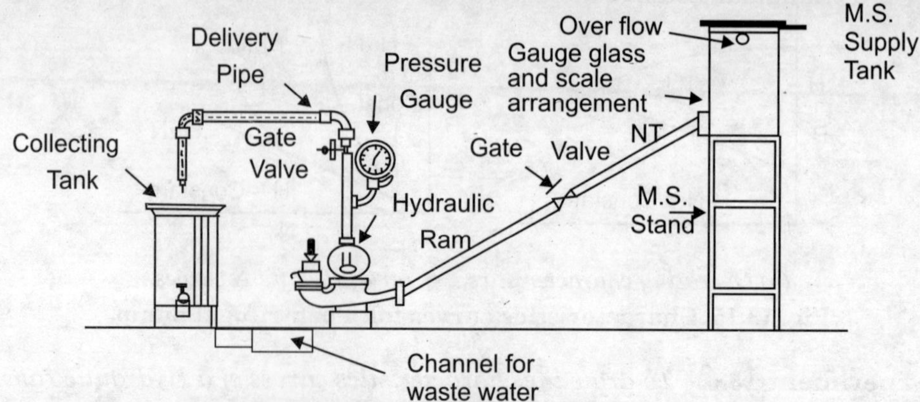


**Fig. 13.16 Experimenter set-up for the hydraulic ram.**

## Procedure

1. Admit water into the supply tank and keep the supply head constant by using an overflow pipe.
2. Keep the delivery valve closed.
3. Start the hydraulic ram by reciprocating the waste valve twice or thrice.
4. Adjust the delivery valve so that pressure gauge reads about 20 m of water head (say).
5. Calculate the waste water rate $Q$ by measuring the head over the V-notch and taking $C_d = 0.62$.
6. Determine the useful water rate q being pumped by the ram from the collecting tank.
7. Note the time for 50 beats and calculate the number of beats per minute.
8. Repeat the experiment for six different values of delivery head at an interval of 3 to 4 m of water.
9. Calculate the Rankine and $D'$ Aubisson efficiency.
10. Plot beats per minute υ's $\eta_R$, $\eta_D$, $q$ and $Q$.

## Observations

Area of collection tank $A$ =

| Sl. No. | $H_s$ m | $H_d$ m | Time for 50 beats mts | Beats fir minute | Head over notch $h_1$ m | $Q^*$ m³/s | Water level in Collecting tank $h_2$ m | Time of collecting water $t$ s | $Q = \frac{Ah_2}{t}$ m³/s | $\eta_R$ % | $\eta D$ % |
|---|---|---|---|---|---|---|---|---|---|---|---|
| | | | | | | | | | | | |

*For a triangular notch of included angle θ

$$Q = \frac{8}{15} C_d \sqrt{2g}.\ \tan\frac{\theta}{2}.h_1^{5/2}$$

For $\qquad C_d = 0.62\ g = 9.81\ \text{m/s}^2,\ \theta = 60°$

$$Q = 0\cdot 84563 h_1^{5/2}$$

Calculations

$$h_1 = Q = 0\cdot 84563 h_1^{5/2}\ \text{m}^3/\text{s}$$

Area of collection tank = A m²

Height of water collected in time *t*.

$$= h_2\ \text{m}$$

$$q = \frac{Ah_2}{t}\ \text{m}^3/\text{s}$$

$$H_d = ,\ H_s =$$

Rankine efficiency, $\qquad \eta_R = \dfrac{q(H_d - H_s)}{Q \times H_s}$

*D'* Aubisson efficiency, $\qquad \gamma_D = \dfrac{q.H_d}{(Q+q)H_s}$

***Results.*** The delivery head will be found to increase with the increase in number of beats. The efficiency of the ram is maximum when $H_d/H_s = 4$. A typical plot of number of beats per minute υ's $Q$, $q$ and $\eta_D$ is shown in fig. 13.17.

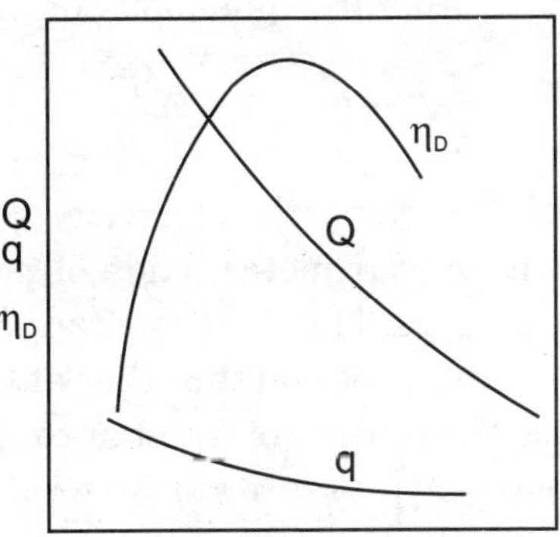


**Fig. 13.17**

**Experiment 13.7** *To study the cavitation phenomenon and plot cavitation curves using open circuit cavitation test rig.*

***Apparatus.*** Venturimeter, collecting tank, pump-motor-set, manometer, stop watch, vacuum gauge and pressure gauge.

***Theory.*** Whenever the pressure in any turbine or pump part falls below the evaporation pressure , the liquid boils and a large number of small bubbles of vapour are formed. It may happen that a stream of water cuts short of its path giving rise to eddies and vortices which may contain voids or bubbles. These bubbles mainly formed on account of low pressure, are carried by the stream to high pressure zones where the vapours condense and the bubbles suddenly, collapse, as the vapours are condensed to liquid again. This results in the formation of a cavity and the surrounding liquid rushes to fill it. The stream of liquid coming from all directions collide at the centre of cavity giving rise to a very high pressure. Formation of cavity and high pressure are repeated many thousand times a second causing pitting on the metallic surface of runner blades or impeller. This phenomenon is called cavitation.

According to Prof. Dietrich Thoma the critical cavitation factor is given by:

$$\sigma_{crit} = \frac{H_b - H_s}{H} = \frac{(H_a - H_v) - H_s}{H} \quad \text{...(a)}$$

Where $H_b$ = barometric pressure

$= H_a - H_v$

$H_a$ = average atmospheric pressure

$H_v$ = vapour pressure

$H_s$ = suction pressure

$H$ = working head of turbine.

According to Prof. F.H. Roger the maximum permissible specific speed of a Francis turbine for a cavitation free phenomenon is given by:

$$N_s = 562\sqrt{\frac{H_b - H_s}{H}} \quad \text{...(b)}$$

***Experimental set-up.*** The cavitation test rig is shown in Fig.13.18. It consists of a transparent venturimeter made of perspex sheet fitted in a piping system of 60 cm dia pipe. There are two tappings in the venturimeter for the manometer one at the inlet and the other at the throat. As the throat section has reduced area, the velocity of water at the throat section is greater then in the pipe and hence the pressure at the throat section is smaller than in the pipe. The pressure at the throat section can be decreased by regulating the discharge by means of a gate valve. When the pressure attains the value of vapour pressure of water, the inception of cavitation begins at the throat. As we further decrease the pressure the cavitation grows to full developed cavitation. The phenomenon can be observed clearly in the set-up and at the same time the rattling noise of cavitation can be noticed. With the help of a special manometer the actual to pressure at the throat and the inlet of venturimeter with respect to atmospheric pressure can be measured.

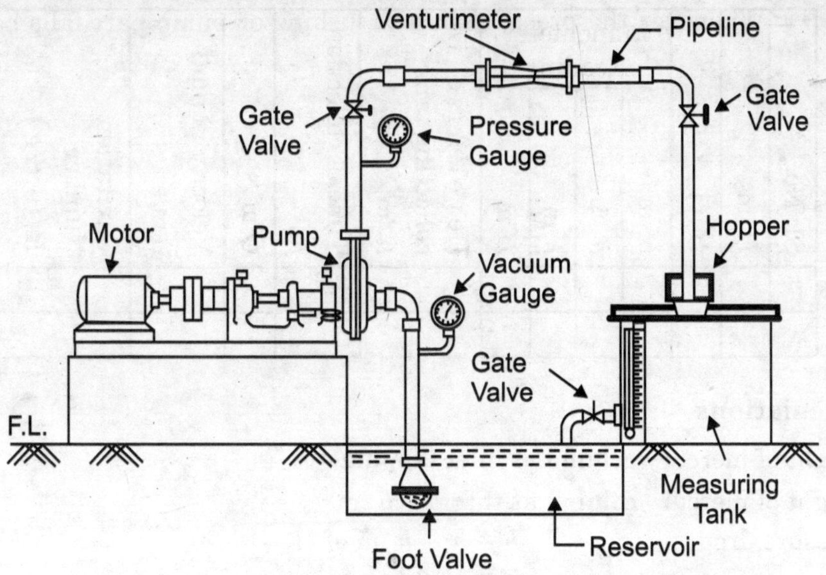


**Fig. 13.18 Cavitation test rig.**

The water flow rate is measured in a graduated collecting tank. A centrifugal pump-motor set supplies the water. A special three limb manometer whose two limbs are connected to pressure tappings in the venturimeter and the third limb exposed to atmospheric pressure, is used to measure pressure.

## Procedure

1. Keep the delivery valve closed.
2. Close the pump outlet valve.
3. Prime the centrifugal pump and start it.
4. Open the delivery valve slightly.
5. Open the pump outlet valve gradually to full open position.
6. Vent the air off from the manometer.
7. Note the reading of vacuum gauge, pressure gauge and manometer.
8. Note the time for 20 cm level rise of water in the collection tank by means of a stop watch.
9. Increase the opening of delivery valve slowly and take ten reading for different valve openings. Observe the cavitation phenomenon in the venturimeter.

## Observations

Area of collection tank $A$ =

Inlet pipe dia. $d$ =

| Sl. No. | $H_s$ m | $H_s$ m | Manometer Reading $h_1$ m | $h_2$ m | $h$ m | $H$ m | Level rise in collecting tank $h_3$ m | Time of collection $t$ s | $Q$ m³/s | Velocity in pipe $\upsilon$ m/s | Cavitation Number $H/(\upsilon^2/2g)$ |
|---|---|---|---|---|---|---|---|---|---|---|---|
| | | | | | | | | | | | |
| | | | | | | | | | | | |

**Calculations**

Height of mercury column at inlet tapping = $h_1$ m

Height of mercury column at throat = $h_2$ m

Pressure drop $h = h_1 - h_2$ m of $H_g$.

or $H = 12.6\ h$ m of water.

Let $h_3$ = level of water rise in collecting tank in time $t$.

$$Q = \frac{Ah_3}{t} m^3/s$$

Area of inlet pipe of dia $d$.

$$= \frac{\pi d^2}{4}$$

Velocity of water at inlet $\upsilon = \frac{4Q}{\pi d^2} m/s$.

$$\text{Cavitation number} = \frac{\text{H}}{(\upsilon^2/2g)}$$

The incipient cavitation index $\sigma_i = \frac{p_o - p_\upsilon}{\frac{1}{2}\rho\upsilon^2}$

where $p_o$ = pressure at throat

$p_\upsilon$ = vapour pressure of water at that temperature

$p_o$ = velocity at throat.

***Results.*** Plot cavitation number $\upsilon$'s pressure drop and pressure gauge readings.

**Experiment 13.8** *To study the characteristics of a gear pump at constant speed.*

***Apparatus.*** Gear pump coupled to an electric motor, pressure gauge, vacuum gauge, water collecting tank, stop watch, energy-meter and metre rod.

***Theory.*** A single stage external type gear pump consists of two identical intermeshing spur gear with a fine clearance inside a suitably shaped casing as shown in Fig.13.19. Liquid is entrained in the spaces between the teeth and the casing and carried round between the gears from the suction port to the discharge port. The volumetric efficiency of the pump is given by

$$\eta_Q = \frac{\text{Actual discharge}}{\textit{Ideal}\ \text{discharge}}$$

$$= \frac{Q}{2\,a\,l\,z\,N} \qquad \text{...(a)}$$

where $a$ = area enclosed between two adjacent teeth and casing.

$l$ = axial length of teeth.

$z$ = number of teeth in each gear

$N$ = rpm.

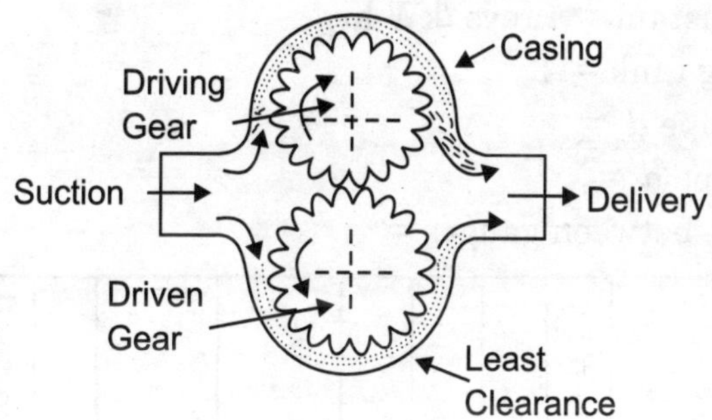


**Fig. 13.19 Gear pump.**

$$\text{Ideal discharge} \simeq 0{\cdot}95\,C(D-C)l \qquad \text{...(b)}$$

Where $C$ = centre distance between axes of gears

$D$ = outside dia of gears.

***Experimental set-up.*** The experimental set-up shown in Fig. 13.20 consists of a gear pump coupled to an electric motor. The supply pipe to the pump is dipped in the supply tank. The delivery pipe is fitted with a gate valve to control the delivery head. A pressure gauge and a vacuum gauge are fitted in the delivery and suction pipes respectively to measure the pressure. Discharge is measured by means of a collecting tank. The power input to the pump is measured by an energymeter.

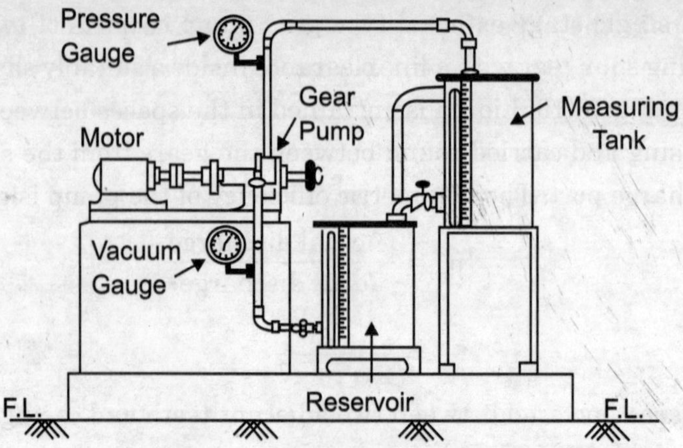


**Fig. 13.20 Experimental set-up for gear pump.**

## Observations

Energymeter constant = $n$ revs./kW hr.

Area of collecting tank = $A$

Dia of delivery pipe $d_d$ =

Dia of suction pipe $d_s$ =

Vertical distance between gauges =

| Sl. No. | $H_g$ m | $H_v$ m | Level rise in tank $h$ m | Time taken $t$ s | $Q$ m³/s | $v_d$ m/s | $v_s$ m/s | $\frac{v_d^2}{2g}$ m | $\frac{v_s^2}{2g}$ m | $H$ m | Time for 10 revs. $T$ s | $Pi$ watts | $Po$ watts | η % |
|---|---|---|---|---|---|---|---|---|---|---|---|---|---|---|
| | | | | | | | | | | | | | | |

## Procedure

1. Fill the supply tank to the required height.
2. Open the gate valve in the delivery pipe fully.
3. Start the motor.
4. Throttle the gate valve to get required head.
5. Note the pressure and vacuum gauge readings.
6. Note the time for 10 revolutions (say) of the energymeter.
7. Note the time for a specified water level rise in the collecting tank.
8. Measure the vertical height of pressure gauge above vacuum gauge.

9. Repeat the experiment at various heads by throttling the gate valve.

**Calculations**

Pressure gauge reading = $H_g$

Vacuum gauge reading = $H_v$

Vertical distance between = $Z$

$$\text{Total head } H = H_g + H_v + Z + \frac{v_d^2 - v_s^2}{2g}$$

Area of collecting tank = A m²

Level rise in tank in t seconds = $h$ m

$$\text{Discharge } Q = \frac{A\,h}{t} m^3/s$$

Energy meter constant = $n$ revs./kW hr.

Time for 10 revolutions = $T$ seconds.

Power input $P_i = \dfrac{3600 \times 1000 \times 10}{n \times T}$ watts.

Power input $P_o = \rho g\, QH$ watts.

Pump efficiency $\eta = \dfrac{P_o}{P_i}$

---

## Quiz Questions

1. Discuss the dependence of the vane coefficient on the vane geometry.
2. Indicate, with reasons the type of vane you would recommended for use in a water turbine for power generation.
3. State the impulse momentum equation.
4. Define the term runaway speed as applied to a turbine. What is its significance?
5. What measures are needed to improve the efficiency of the turbine?
6. Would the insulation of a draft tube be helpful in improving the head or efficiency of a Pelton turbine?
7. Would characteristic plots on the basis of specific quantities be as good as those on the basis of unit quantities?
8. How many buckets are there in the Pelton turbine? Are these the minimum buckets?
9. What is the function of the casing of the Pelton turbine?
10. When do we use multijets in a Pelton turbine?

11. State the relationship between the jet speed and bucket speed for maximum efficiency of a Pelton wheel.
12. Define jet ratio and speed ratio.
13. If air leaks in at the draft tube inlet, how will it affect the head on the turbine and its efficiency?
14. Will the points of placement of the pressure gauge and the vacuum gauge make any difference in the estimation of the total head across the turbine?
15. Will the efficiency of operation of a reaction turbine decrease if its draft tube is remove and it is installed just above the tail race?
16. Can the turbine be operated as a pump if power is supplied to it as the water supply and discharge lines are interchanged functionally?
17. Why are the characteristic curves plotted between the unit quantities and not between the absolute quantities?
18. Compare the performance curves of the Kaplan turbine with those of the other turbine, e.g., Pelton and Francis turbines and comment.
19. What is the difference between Kaplan and propeller turbine
20. Can cavitaion take place in the pump?
21. What is function of the volute casing in a centrifugal pump?
22. What is the meaning of priming of pump?
23. Define specific speed of a centrifugal pump and write its expression.
24. Why in some cases the pump has to be installed in a well?
25. Why centrifugal pump is recommended to the farmers for irrigation?
26. When and where do we use hydraulic ram?
27. Why the length of the suction pipe has to be very long?
28. Define various efficiencies of a hydraulic ram.
29. What is the principle of working a of a hydraulic ram?
30. Define cavitation.
31. Define cavitation index.
32. How cavitation can be avoided?
33. How cavitation is different from separation?
34. On what factors does the cavitation in water turbines depend?
35. What is a gear pump? How does this operate?

# CHAPTER 14 Conventional Questions from IES

**Example 14.1** A 1/5 scale model of a centrifugal pump absorbs 20 kW when pumping against a test head of 8 m at its best speed of 400 rpm. If the actual pump works against 32 m head, find the speed and power required for the actual pump. Determine also the quantities of water discharged by the two pumps. [IES, 1993]

**Solution.**

Given: $P_m = 20$ kW, $\dfrac{D_m}{D_\rho} = \dfrac{1}{5}$, $H_m = 8$ m, $N_m = 400$ rpm, $H_p = 32$ m

For a centrifugal pump,

1. $\dfrac{\sqrt{H}}{DN}$ = Const.

$$\therefore \quad \left(\frac{\sqrt{H}}{DN}\right)_m = \left(\frac{\sqrt{H}}{DN}\right)_\rho$$

$$N_\rho = N_m \left(\frac{D_m}{D_\rho}\right)\sqrt{\frac{H_\rho}{H_m}} = 400 \times \frac{1}{5} \times \sqrt{\frac{32}{8}} = 160 \text{ rpm}$$

2. $\dfrac{P}{D^s N^3}$ = Const.

$$\therefore \quad P_p = P_m \left(\frac{D_\rho}{D_m}\right)^5 \left(\frac{N_\rho}{N_m}\right)^3$$

$$= 20\ (5)^5 \left(\frac{160}{400}\right)^3 = 4000 \text{ kW}$$

3. $\dfrac{Q}{ND^3}$ = Const.

$$\frac{Q_\rho}{Q_m} = \left(\frac{N_\rho}{N_m}\right)\left(\frac{D_\rho}{D_m}\right)^3 = \left(\frac{160}{400}\right)(5)^3 = 50$$

Now $P = \dfrac{\gamma Q H}{10^3}$ kW

$$Q_m = \frac{20 \times 10^3}{9.81 \times 10^3 \times 8} = 0.2548 \text{ m}^3/\text{s}$$

$$Q_p = 50 \times 0.2548 = 12.742 \text{ m}^3/\text{s}$$

**Example 14.2** A single acting reciprocating pump operating at 120 rpm has a piston diameter of 200 mm and stroke of 300 mm. The suction and delivery heads are 4m and 20 m, respectively. If the efficiency of both suction and delivery strokes is 75 percent, determine the power required by the pump. [IES, 1993]

**Solution.**

Given: $N = 120$ rpm, $D = 20$ mm, $L = 300$ mm, $H_s = 4$m, $H_d = 20$m, $\eta = 0.75$

$$\text{Discharge,} \quad Q = \frac{\text{ALN}}{60} = \frac{\frac{\pi}{4}(0.2)^2 \times 0.3 \times 120}{60}$$

$$= 18.85 \times 10^{-3} \text{ m}^3/\text{s}$$

$$\text{Power, } P = \frac{\gamma QH}{\eta \times 10^3}$$

$$= \frac{9.81 \times 10^3 \times 18.85 \times 10^{-3} \times (4+20)}{0.75 \times 10^3}$$

$$= 5.917 \text{ kW}$$

**Example 14.3** An inward flow reaction turbine, operating under 30 m head, develops 4000 kW while running at 300 rpm. The overall efficiency of the water turbine is 0.85; the hydraulic efficiency is 0.9; and the radial velocity of flow at inlet is 7 m/s. The inlet guide vane angle at full gate opening is 30°. Calculate the diameter and width of the runner at inlet. Blade thickness coefficient is 5%. [IES,1994]

**Solution.**

Given : $H = 30$ m, $P = 400$ kW, $N = 300$ rpm, $\eta_o = 0.85$, $\eta_{hyd} = 0.90$.

$v_{f1} = 7$ m/s, $\alpha = 30°$, $k_t = 0.95$

$$\text{Overall efficiency, } \eta_o = \frac{P}{\gamma QH}$$

$$0.85 = \frac{400 \times 10^3}{9.81 \times 10^3 \times Q \times 30}$$

$$Q = 15.99 \text{ m}^3/\text{s}$$

$$\text{Hydraulic efficiency, } \eta_{hyd} = \frac{u_1 v_{w1}}{gH}$$

$$v_{f1}/v_{w1} = \tan\alpha_1$$

$$v_{w1} = \frac{7}{\tan 30^\circ} = 12.12 \text{ m/s}$$

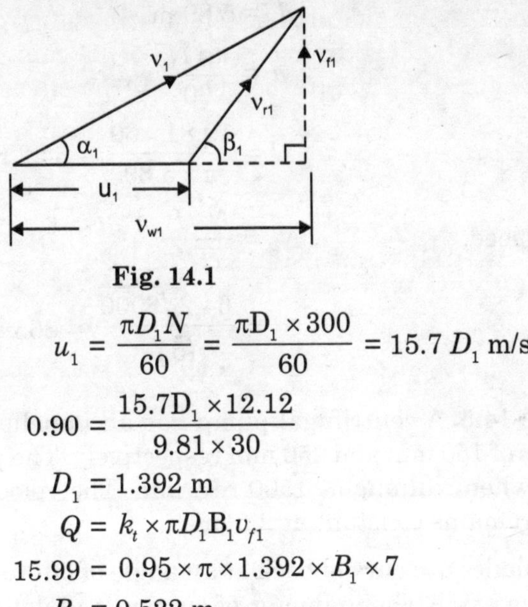


**Fig. 14.1**

$$u_1 = \frac{\pi D_1 N}{60} = \frac{\pi D_1 \times 300}{60} = 15.7\, D_1 \text{ m/s}$$

$$0.90 = \frac{15.7 D_1 \times 12.12}{9.81 \times 30}$$

$$D_1 = 1.392 \text{ m}$$

Now
$$Q = k_t \times \pi D_1 B_1 v_{f1}$$

$$15.99 = 0.95 \times \pi \times 1.392 \times B_1 \times 7$$

$$B_1 = 0.522 \text{ m}$$

**Example 14.4** Calculate the diameter and speed of the runner of a Kaplan turbine developing 600 kW under an effective head of 5m. Overall efficiency of the turbine is 90%. The diameter of the boss is 0.4 time the external diameter of the runner. The turbine speed ratio is 2.0 and flow ratio 0.6. What is the specific speed of the turbine? [IES, 1995]

**Solution.**

Given: $P = 600$ kW, $H = 5$m, $\eta_o = 0.9$, $d_6 = 0.4\, d_o$, speed ratio = 2,

Flow ratio = 0.6

$$\text{Speed ratio} = \frac{u}{\sqrt{2gH}} = 2$$

$$u = 2 \times \sqrt{2 \times 9.81 \times 5} = 19.81 \text{ m/s}$$

$$\text{Flow ratio} = \frac{v_f}{\sqrt{2gH}} = 0.6$$

∴
$$v_f = 0.6 \times \sqrt{2 \times 9.81 \times 5} = 5.94 \text{ m/s}$$

Power developed,
$$P = \frac{\gamma Q H}{10^3} \times \eta_o$$

$$Q = \frac{6000 \times 10^3}{9.81 \times 10^3 \times 5 \times 0.9} = 135.9 \text{ m}^3\text{/s}$$

Also
$$Q = \frac{\pi}{4}\left(d_o^2 - d_b^2\right)v_f$$

$$135.9 = \frac{\pi}{4}\left[1 - (0.4)^2\right] d_o^2 \times 5.94$$

$$d_o = 5.89 \text{ m}$$

$$u = \frac{\pi d_o N}{60}$$

$$N = \frac{19.81 \times 60}{\pi \times 5.89} = 64.2 \text{ rpm}$$

Specific speed, $$N_s = \frac{N\sqrt{P}}{H^{5/4}}$$

$$= \frac{64.2\sqrt{6000}}{(5)^{5/4}} = 665$$

**Example 14.5** A centrifugal pump has an impeller with inner and outer diameters of 150 mm and 250 mm respectively. The pump delivers 0.05 m³/s of water when running at 1500 rev./min. The velocity of flow through the impeller remains constant at 2.5 m/s.

The blades are curved back at an angle of 30° to the tangent at exit and the inlet is axial. The diameter of suction and delivery pipes are 150 mm and 100 mm respectively. The pressure head at suction is 4 m below and that at delivery is 18 m above atmosphere. The power required to drive the pump is 18 kW. Find

(a) the vane angle at inlet
(b) the overall efficiency, and
(c) the manometer efficiency. [IES,1996]

**Solution.**

Given: $D_1 = 150$ mm, $D_2 = 250$ mm, $Q = 0.05$ m³/s, $N = 1500$ rpm, $v_f =$ 2.5 m/s, $d_s = 150$ mm, $d_d = 100$ mm, $H_s = 4$ m, $H_d = 18$ m, $P = 18$ kW, $\beta_2 = 30°$, $\alpha_1 = 90°$

$$u_1 = \frac{\pi D_1 N}{10^3 \times 60} = \frac{\pi \times 150 \times 1500}{10^3 \times 60} = 11.78 \text{ m/s}$$

$$u_2 = \frac{\pi D_2 N}{10^3 \times 60} = \frac{11.78 \times 250}{150} = 19.63 \text{ m/s}$$

$$\tan \beta_1 = \frac{v_{f1}}{u_1} = \frac{2.5}{11.78} = 0.21222$$

$$\beta_1 = 11.98°$$

The velocity triangles are shown in fig.14.2

Now $$v_{f1} = v_{f2} = v_f = 2.5 \text{ m/s}$$

Velocity of water in suction pipe,

$$v_S = \frac{4Q}{\pi d_s^2} = \frac{4 \times 0.05}{\pi (0.150)^2} = 2.83 \text{ m/s}$$

Velocity of water in delivery pipe,

$$\upsilon_d = \frac{4Q}{\pi d_d^2} = \frac{4 \times 0.05}{\pi (0.1)^2} = 6.37 \text{ m/s}$$

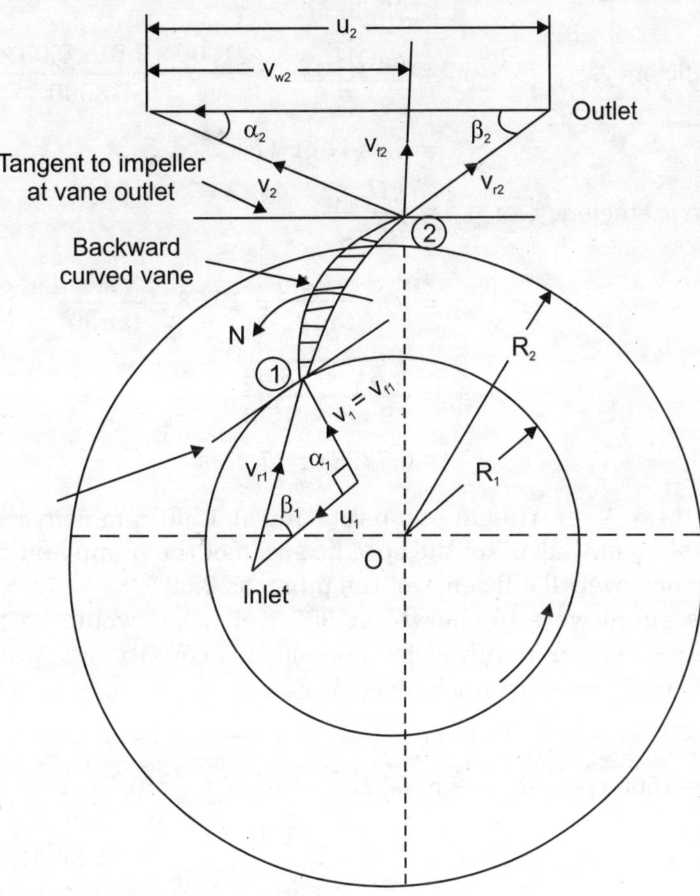


**Fig. 14.2**

Applying Bernoulli's theorem at inlet (point 1) and outlet (point 2) of impeller, we have

$$\frac{p_2}{\rho} + \frac{\upsilon_2^2}{2} = \frac{p_1}{\rho} + \frac{\upsilon_1^2}{2} + g.H_{mano}$$

$$g.H_{mano} = \frac{p_2 - p_1}{\rho} + \frac{1}{2}\left(\upsilon_2^2 - \upsilon_1^2\right)$$

Now $\frac{p_1}{\rho g} = \text{H}_{atm} - 4 = 10.3 - 4 = 6.3 \text{ m}$

$$\frac{p_2}{\rho g} = \text{H}_{atm} + 18 = 10.3 + 18 = 28.3 \text{ m}$$

Also $\upsilon_2^2/2 = \upsilon_d^2/2$

And $\quad v_1^2/2 = v_s^2/2$

$$\therefore \quad H_{mano} = (28.3 - 6.3) + \frac{1}{2 \times 9.81}\left[(6.37)^2 - (2.83)^2\right]$$

$$= 23.66 \text{ m}$$

Overall efficiency,
$$\eta_o = \frac{\rho g Q H_{mano}}{P} = \frac{1000 \times 9.81 \times 0.05 \times 23.66}{18000}$$

$$= 0.6447 \text{ or } 47\%$$

Mano metric efficiency,
$$\eta_{mano} = \frac{gH_{mano}}{u_2 v_{w2}}$$

$$v_{w2} = u_2 - \frac{v_f}{\tan\beta_2} = 19.63 - \frac{2.5}{\tan 30°} = 15.3 \text{ m/s}$$

$$\eta_{mano} = \frac{9.81 \times 23.66}{19.63 \times 15.3}$$

$$= 0.7728 \text{ or } 77.28\%$$

**Example 14.6** A centrifugal pump rotating at 1500 rpm delivers 0.2 m³/s at a head of 15 m. Calculate the specific speed of the pump and the power input. Assume overall efficiency of the pump as 0.68.

If this pump were to operate at 900 rpm what would be the head, discharge and power required for homologous conditions? Assume that overall efficiency remains unchanged at new rpm. [IES, 1998]

**Solution.**

Given: $N = 1500$ rpm, $Q = 0.2$ m³/s, $H = 15$ m, $\eta_{overall} = 0.68$

Specific speed,
$$N_s = \frac{NQ^{1/2}}{H^{3/4}} = \frac{1500 \times (0.2)^{1/2}}{15^{3/4}} = 88.01 \text{ rpm}$$

Power input,
$$P = \frac{\rho g Q H}{\eta_{overall} \times 10^3} kW$$

$$= \frac{1000 \times 9.81 \times 0.2 \times 15}{0.68 \times 10^3} = 43.28 \text{ kW}$$

$$H_2 = H_1\left(\frac{N_2}{N_1}\right)^2 = 15\left(\frac{900}{1500}\right)^2 = 5.4 \text{ m}$$

$$Q_2 = Q_1\left(\frac{N_2}{N_1}\right) = 0.2\left(\frac{900}{1500}\right) = 0.12 \text{ m}^3/\text{s}$$

$$P_2 = P_1\left(\frac{N_2}{N_1}\right)^3 = 43.28\left(\frac{900}{1500}\right)^3 = 9.3485 \text{kW}$$

**Example 14.7** A centrifugal pump has an impeller of 80 cm in diameter and it delivers 1 m³/sec against a head of 80 m. The impeller runs at 1000 rpm and the width at outlet is 8 cm. If the leakage loss is 3 percent of the discharge, external mechanical loss is 10 kW and the hydraulic efficiency is 80 percent, calculate the blade angle at outlet, the power required and overall efficiency of the pump. [IES, 2000]

**Solution.**

Give: $d_2 = 80$ cm, $b_2 = 8$ cm, $Q_a = 1$ m³/s, $H_{mano} = 80$ m, $N = 1000$ rpm

Leakage loss = 3% of discharge,

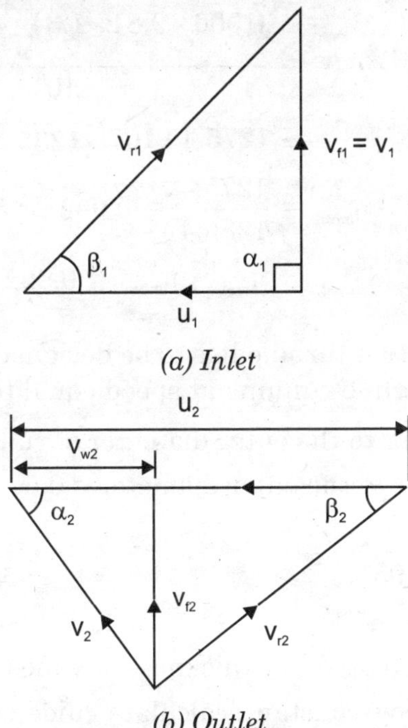


**Fig. 14.3 Velocity triangles.**

$$\text{Mechanical loss} = 10 \text{ kW}, \; h_{hyd} = 0.8$$

$$Q_{th} = 1.3\, Q_a = 1.3 \text{ m}^3\text{/s}$$

$$u_2 = \frac{\pi d_2 N}{60} = \frac{\pi \times 0.8 \times 1000}{60} = 41.8 \text{ m/s}$$

$$Q_{th} = \pi d_2 b_2 v_{f2}$$

The velocity triangles are shown in Fig. 14.3.

$$v_{f2} = \frac{1.3}{\pi \times 0.8 \times 0.08} = 64.7 \text{ m/s}$$

$$\eta_{mano} = \frac{gH_{mano}}{u_2 v_{w2}}$$

$$v_{w2} = \frac{9.81 \times 80}{41.8 \times 0.8} = 23.47 \text{ m/s}$$

$$\tan\beta_2 = \frac{v_{f2}}{u_2 - v_{w2}} = \frac{64.7}{41.8 - 23.47} = 3.535$$

$$\beta_2 = 74.2^\circ$$

$$\text{Power required} = \frac{(\rho g Q)\left(\frac{u_2 v_{w2}}{g}\right)}{10^3} + 10$$

$$= \frac{(1000 \times 9.81 \times 1.3)\left(\frac{41.8 \times 23.47}{9.81}\right)}{10^3} + 10$$

$$= 1275.4 + 10 = 1285.4\ kW$$

$$\eta_{mech} = \frac{1275.4}{1285.4} = 0.992 \text{ or } 99.2\%$$

$$\eta_{overall} = \eta_{mech} \times \eta_{mano} = 0.992 \times 0.8 = 0.79 \text{ or } 79\%$$

**Example 14.8** A reaction turbine has to be designed to developed 370 kW under a head of 70 m while running at speed equal to 750 rpm.

Ratio of width of runner to the outer diameter of runner = 0.1.

Ratio of inner diameter to the outer diameter = 0.5

Flow ratio = 0.15

Hydraulic efficiency = 95%

Mechanical efficiency = 84%

Circumferential area occupied by thickness of vanes = 5%

Assuming velocity of flow constant, calculate guide vane angle at inlet and runner vane angles at inlet and outlet.

The radial velocity is equal to 3.5 m/sec. [ISE, 2003]

**Solution.**

Given: $P = 370$ kW, $H = 70$ m, $N = 750$ rpm, $\frac{B_2}{D_2} = 0.1$, $\frac{D_1}{D_2} = 0.5$, $K_f = 0.15$,

$\eta_{hyd} = 0.95$, $\eta_{mech} = 0.84$, $k_t = 1 - 0.05 = 0.95$, $v_{f1} = v_{f2} = 3.5$ m/s

Overall efficiency, $\eta_\circ = \eta_{mech}\ \eta_{hyd} = 0.84 \times 0.95 = 0.798$

$$Q = \frac{p}{\gamma H \eta_\circ} = \frac{370 \times 10^3}{9810 \times 70 \times 0.798} = 0.675 \text{ m}^3\text{/s}$$

$$v_{f1} = v_{f2} = K_f\,[2gH]^{0.5}$$

$$= 0.15\left[2 \times 9.81 \times 70\right]^{0.5} = 5.56 \text{ m/s}$$

$$Q = \eta_{hyd}\ \pi\ D_2 B_2 v_{f2}$$

$$0.675 = 0.95 \times \pi \times D_2 B_2 \times 5.56$$

$$D_2 B_2 = 0.040$$

$$0.1\ D_2^2 = 0.040$$

$$D_2 = 0.631 \text{ m}$$

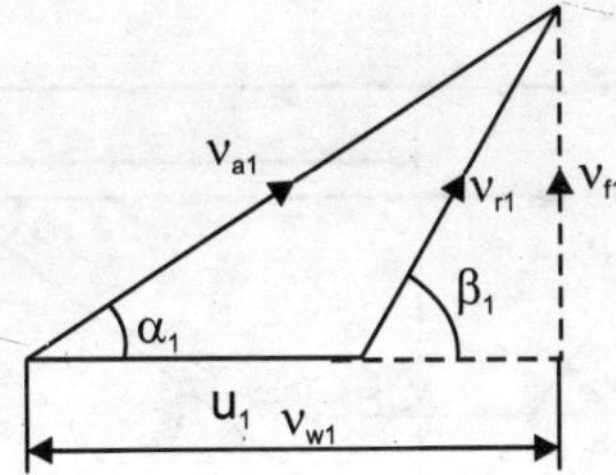

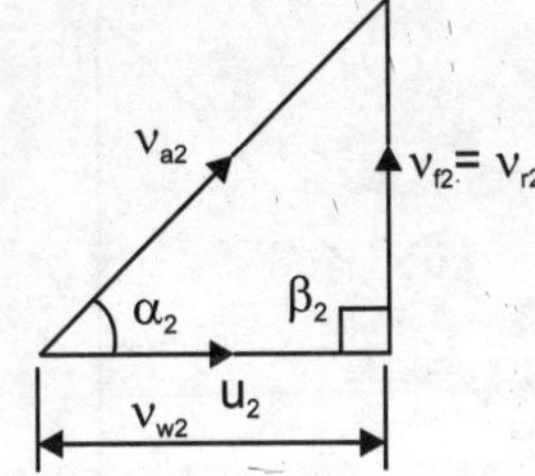


**Fig. 14.4 Velocity triangles.**

$$D_1 = \frac{D_2}{2} = 0.316\text{m}$$

$$u_1 = \pi \times 0.316 \times \frac{750}{60} = 12.41 \text{ m/s}$$

$$u_2 = 2u_1 = 24.82 \text{ m/s}$$

$$\eta_{hyd} = \frac{u_1 v_{w1}}{gh}$$

$$0.95 = \frac{12.41 v_{w1}}{9.81 \times 70}$$

$$v_{w1} = 52.56 \text{ m/s}$$

$$\tan\alpha_1 = \frac{v_{f1}}{v_{w1}} = \frac{5.56}{52.56}, \ \alpha_1 = 6.03°$$

$$\tan\beta_1 = \frac{v_{f1}}{v_{w1} - u_1} = \frac{5.56}{52.56 - 12.41}, \ \beta_1 = 7.88°$$

$$\tan\alpha_2 = \frac{v_{f2}}{v_{w2}} = \frac{v_{f1}}{u_2} = \frac{5.56}{24.82} \ \alpha_2 = 12.63°$$

**Example 14.9** A centrifugal pump is to discharge 0.2 m³/s at a speed of 1400 rpm against a head of 25 m. The impeller diameter is 0.3 m, its width at outlet is 0.05 m and the manometric efficiency is 80%.

Determine the vane angle at the outer periphery of the impellor and lable the angle by sketching the impeller. [IES, 2004]

**Solution.**

Given: $Q = 0.2\ \text{m}^3/\text{s}$, $N = 1400$ rpm, $H = 25$ m, $D_2 = 25$ m, $D_2 = 0.3$, $B_2 = 0.05$.

$$\eta_{mano} = 0.80$$

The velocity triangles are shown in Fig.14.5.

$$\upsilon_{f2} = \frac{Q}{\pi D_2 B_2} = \frac{0.2}{\pi \times 0.3 \times 0.05} = 4.25\ \text{m/s}$$

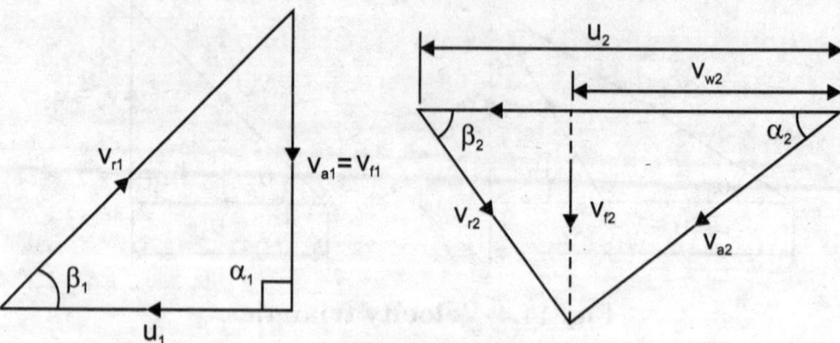


**Fig. 14.5**

$$u_2 = \frac{\pi D_2 N}{60} = \frac{\pi \times 0.3 \times 1400}{60} = 22\ \text{m/s}$$

$$\upsilon_{w2} = \frac{gH}{u_2 \eta_{mano}} = \frac{9.81 \times 25}{9.8 \times 22} = 14\ \text{m/s}$$

$$u_2 - \upsilon_{f2} \cot \alpha_2 = \upsilon_{\omega 2}$$

$$22 - 4.25 \cot \alpha_2 = 14$$

$$\alpha_2 = 28°$$

$$\tan \beta_2 = \frac{\upsilon_{f2}}{u_2 - \upsilon_{w2}}$$

$$= \frac{4.25}{22 - 14} = \frac{4.25}{8} = 0.53125$$

$$\beta_2 = 27.98°$$

**Example 14.10** A centrifugal pump is required to discharge 600 liters of water pr second and develop a head of 15 m when the impeller rotates at 750 rpm. The manometric efficiency is 80%. The loss of head in the pump due to fluid resistance being assumed to be 0.027 $V^2$ m of water, where $V$ is the velocity with which water leaves the impeller. Water enters the impeller without shock or whirl and the velocity of flow is 3.2 m/sec.

Determine:

(a) the impeller diameter and outlet area,

(b) blade angle at the outlet edge. [IES, 2006]

**Solution.**

Given: $Q = 0.6$ m³/s, $H_{mano} = 15$ m, $N = 750$ rpm, $\eta_{mano} = 0.8$, $v_{f2} = 3.2$ m/s, $h_f = 0.027$ V²

(a) Impeller diameter $D_2$:

$$\frac{u_2 v_{w2}}{g} = \frac{H_{mano}}{\eta_{mano}} = \frac{15}{0.8} = 18.75 \text{ m}$$

Also, $$H_{mano} = \frac{u_2 v_{w2}}{g} - \text{loss of head in the pump, } h_f$$

Loss of head in the pump, $h_f = 18.75 - 15 = 3.75$m

$$0.027\, V^2 = 3.75$$

$$V^2 = \frac{3.75}{0.27} = 138.38$$

$$V = 11.78 \text{ m/s}$$

Velocity of flow at outlet, $$v_{f2} = \frac{Q}{\text{area of flow}}$$

$$3.2 = \frac{0.6}{\text{area of area}}$$

Area of flow, $$A_f = \frac{0.6}{3.2} = 0.1875 \text{ m}^2$$

Tangential velocity of impeller at outlet,

$$u_2 = \frac{\pi D_2 N}{60} = \pi \times D_2 \times \frac{750}{60} = 39.27\, D_2$$

Also, $$\frac{u_2 v_{w2}}{g} = 1875$$

$$v_{w2} = \frac{9.81 \times 18.75}{39.27 D_2} = \frac{4.684}{D_2}$$

From Fig.14.6, we have $$v_{f2}^2 = v_{a2}^2 - v_{w2}^2$$

$$(3.2)^2 = (11.785)^2 - \left(\frac{4.684}{D_2}\right)^2$$

$$10.24 = \frac{138.89 - 21.94}{D_2^2}$$

$$D_2^2 = 0.17054$$

$$D_2 = 0.413 \text{ m or } 413 \text{ mm}$$

$$v_{\omega 2} = \frac{4.684}{0.413} = 11.342 \text{ m/s}$$

$$u_2 = 39.27 \times 0.413 = 16.218 \text{ m/s}$$

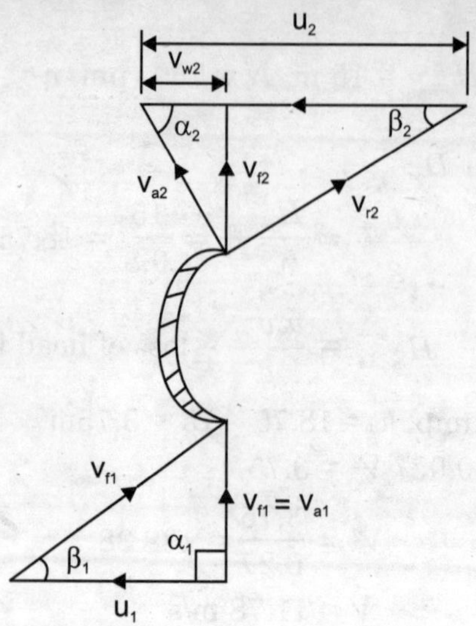


**Fig. 14.6 Velocity triangles.**

(b)
$$\tan\beta_2 = \frac{v_{f2}}{(u_2 - v_{w2})}$$

$$= \frac{3.2}{(16.218 - 11.342)} = 0.656$$

$$\beta_2 = 33.276°$$

**Example 14.11** Laboratory tests are required to be performed on a 1/10 scale model to find the best design for a hydraulic turbine. The turbine is required to develop 30,000 kW under a net head of 50 m running at 300 rpm. The head available at the laboratory is 12 m. Determine.

(a) The corresponding speed
(b) Power, and
(c) Discharge.

The overall efficiency is 88%. [IES, 2006]

**Solution.**

Given: $\frac{D_m}{D_\rho} = \frac{1}{10}$, $P_\rho = 30{,}000$ kw, $H_\rho = 50$ m, $N_p = 300$ rpm, $H_m = 12$ m

Overall efficiency,
$$\eta_\circ = \frac{\text{Shaft power}}{\text{Water power}} = \frac{P}{\gamma QH}$$

$$Q_\rho = \frac{P_p}{\eta_\circ \gamma H} = \frac{30000 \times 10^3}{0.88 \times 9.81 \times 10^3 \times 50} = 69.5\text{m}^3$$

Specific speed,
$$N_s = N\sqrt{\frac{P}{H^{5/4}}} = 300\sqrt{\frac{30000}{50^{5/4}}} = 390.8$$

(a)
$$\left[\frac{H}{N^2D^2}\right]_m = \left[\frac{H}{N^2D^2}\right]_\rho$$

$$N_m = N_\rho \frac{(D_\rho)}{D_m} \frac{(H_m)^{1/2}}{H_\rho}$$

$$= 300 \times 10 \times \frac{(12)^{1/2}}{50}$$

$$= 1469.7\text{rpm}$$

(b)
$$\left[\frac{P}{N^3D^5}\right]_m = \left[\frac{P}{N^3D^5}\right]_\rho$$

$$P_m = P_\rho \left[\frac{D_m}{D_\rho}\right]^5 \left[\frac{N_m}{N_\rho}\right]^3$$

$$= 30{,}000\left(\frac{1}{10}\right)^5 \left(\frac{1469.7}{300}\right)^3 = 35.273 \text{ kw}$$

(c)
$$\left(\frac{Q}{ND^3}\right)_m = \left(\frac{Q}{ND^3}\right)_\rho$$

$$Q_m = Q_\rho \left(\frac{N_m}{N_\rho}\right)\left(\frac{D_m}{D_\rho}\right)^3$$

$$= 69.5\left(\frac{1469.7}{300}\right)\left(\frac{1}{10}\right)^3 = 0.34 \text{ m}^3/\text{s}$$

**Example 14.12** A centrifugal pump is required to lift 9100 litres/min of water against a head of 6 m while running at 500 rpm. The velocity of flow through the wheel is 2 m/s and the manometric efficiency is 60%. The angle of vane tip with the directive of motion is 150°.

Work out the following:

(a) Sketch an impeller of centrifugal pump and draw the velocity triangle,

(b) The diameter of impeller,

(c) The width of impeller. [IES, 2007]

**Solution.**

Given: $Q = 9100$ rpm or 0.151 m³/s, $H = 6$ m, $N = 500$ rpm, $v_f = 2$ m/s, $\eta_{mano} = 0.6$, $\beta_2 = 180° - 150° = 30°$.

(a) The sketch of the impeller and velocity triangles are shown in Fig.14.7.

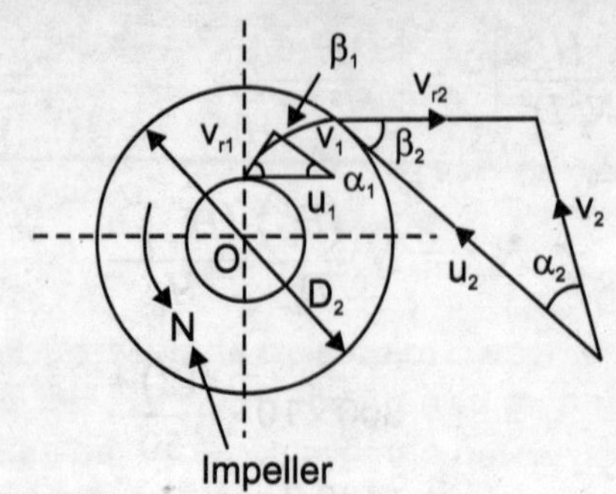


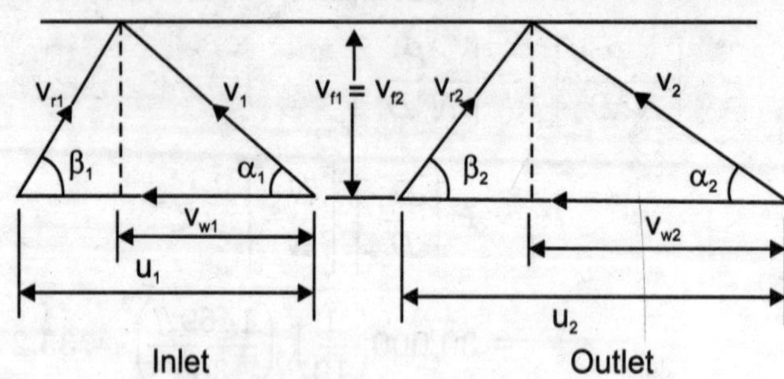


**Fig. 14.7 Velocity triangles.**

(b)
$$\frac{H_{mano}}{\eta_{mano}} = \frac{u_2 v_{\omega 2}}{g} \text{ for } \alpha_1 = 90^\circ$$

$$u_2 v_{\omega 2} = \frac{6 \times 9.81}{0.6} = 98.1 \quad \text{...(1)}$$

$$u_2 = \frac{\pi D_2 N}{60}$$

$$= \frac{\pi D_2 \times 500}{60} = 26.18\ D_2$$

$$\beta_2 = 180 - 150^\circ$$

$$= 30^\circ,\ v_{f2} = 2 \text{ m/s}$$

$$\tan \beta_2 = \frac{v_{f2}}{u_2 - v_{w2}}$$

$$\tan 30^\circ = \frac{2}{26.18 D_2 - v_{w2}}$$

$$v_{u2} = 26.18 D_2 - 3.464 \quad \text{...(2)}$$

Substituting eqn. (2) in (1), we get,

$$26.18 D_2 (26.18 D_2 - 3.464) = 98.1$$

$$D_2^2 - 0.1323 D_2 - 0.1431 = 0$$

$$D_2 = 0.1323 + \sqrt{(0.1323)^2 + 4 \times 1 \times 0.1431}$$

$$= 0.45 \text{ m or } 45 \text{ cm.}$$

(c)
$$Q = \pi D_2 B_2 v_{f2}$$
$$B_2 = \frac{9100 \times 10^{-3}}{60} \times \frac{1}{\pi \times 0.45 \times 2}$$
$$= 0.054\ m \text{ or } 5.4\ cm$$

**Example 14.13** A jet of water makes an angle with the direction of motion of a series of moving blades. If the blade angle at inlet and outlet are θ and ϕ respectively with the direction of motion, derive an expression for the work done. If there is on shock at entry, show that the maximum efficiency that can be achieved from the system is

$$\eta_{max} = \frac{\cos^2\alpha}{2}\left[1 + \frac{K\cos\phi}{\cos\theta}\right]$$

Assume $K = \left(\frac{v_{r2}}{v_{r1}}\right)$. Also state what would be the maximum efficiency for semi-circular blades, when $K = 1$. [IES, 2008]

**Solution.**

(a) Velocity triangle at inlet and outlet

(b) Combined velocity triangle

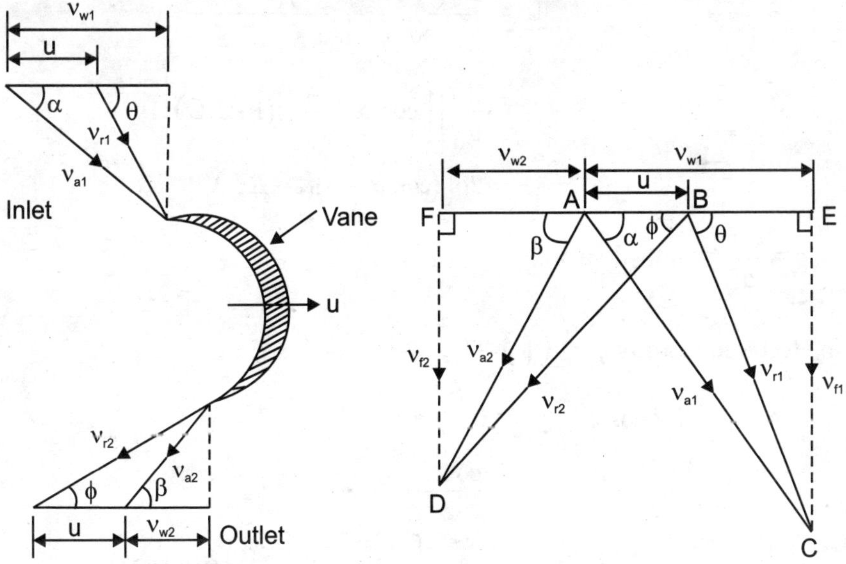


*(a) Velocity triangle of inlet and outlet* *(b) Combined velocity triangle*

**Fig. 14.8**

Work done per kg of water, $w$ = Rate of change of momentum × Distance travelled

$$= \left[v_{w1} - (-v_{w2})\right]u = (v_{w1} + v_{w2})u$$

Blade efficiency,

$$\eta_b = \frac{\textit{Work} \text{ done per kg of water}}{\textit{energy} \text{ supplied per kg of water}}$$

$$= \frac{(v_{w1} + v_{w2})u}{\frac{1}{2}v_{\alpha 1}^2} = \frac{2u(v_{w1} + v_{w2})}{v_{\alpha 1}^2}$$

From Fig.14.8 (b), we have

$$v_{\omega 1} + v_{\omega 2} = \text{AE+AF} = (AB + BE) + (BF - AB)$$

$$= BE + BF$$

$$= v_{r1} \cos\theta + v_{r2} \cos\phi$$

$$= v_{r1} \cos\theta \left[1 + \frac{v_{r2}}{v_{r1}} \cdot \frac{\cos\phi}{\cos\theta}\right]$$

$$= v_{r1} \cos\theta [1 + KC]$$

where $K = \frac{v_{r2}}{v_{r1}}$ and $C = \frac{\cos\phi}{\cos\theta}$

Now

$$\text{BE} = \text{AE-AB} = v_{\upsilon 1} \cos\alpha - u$$

$$\eta_b = \frac{2\text{u}(v_{\alpha 1} \cos\alpha - u)(1 + KC)}{v_{\alpha 1}^2}$$

$$= \frac{2\text{u}}{v_{\alpha 1}}\left[\left(\cos\alpha - \frac{4}{v_{\alpha 1}}\right)(1 + KC)\right]$$

$$= 2\rho[(\cos\alpha - \rho)(1 + KC)]$$

where $\rho = \frac{\text{u}}{v_{\alpha 1}}$ = speed ratio.

For $\eta_b$ to be maximum, $\frac{d}{d_\rho}(\eta_b) = 0$

$$\cos\alpha \times 2\rho = 0$$

$$\rho = \frac{\cos\alpha}{2}$$

$$(\eta_b)_{\max} = \cos\alpha\left[\left(\cos\alpha - \frac{\cos\alpha}{2}\right)(1 + KC)\right]$$

$$= \frac{\cos^2\alpha}{2}(1 + KC)$$

$$= \frac{\cos^2\alpha}{2}\left[1 + K\frac{\cos\phi}{\cos\theta}\right]$$

Hence proved .

For $K = 1$:

$$(\eta_b)_{max} = \frac{\cos^2 \alpha}{2}\left[1+\frac{\cos\phi}{\cos\theta}\right].$$

**Example 14.14** During a trial run on a pump, cavitation occurred when the sum of static pressure and velocity head at inlet was reduced to 3.2 m. The pump total head was 35 m at a discharge of 0.045 m³/sec. The vapour pressure of water = 1.8 kPa and the barometer pressure = 750 mm of Hg. If the pump is to operate at some other location where the atmospheric pressure was reduced that the vapour pressure of water is 830 Pa, what is the value of cavitation parameter when the pump develops the same total head and discharge? Is it necessary to reduce the height of the pump and if so, by how much? [IES, 2008]

**Solution.**

Given: $H_S + \frac{v_S^2}{2g} = 3.2$ m, $H_{mano} = 35$ m, $Q = 0.045$ m³/s, $P_{vap} = 1.8$ kPa, $P_{atm}$ = 750 mm of Hg

$$H_{vap} \frac{1.8 \times 10^3}{9.81 \times 10^3} = 0.1835\text{m of } H_2O$$

$$H_{atm} = 750 \times 10^{-3} \times 13.6 = 10.2 \text{ m of } H_2O$$

$$\text{Covitation parameter} = \frac{H_{atm} - \left[H_{vap} + H_s + \frac{v_S^2}{2g}\right]}{H_{mano}}$$

$$= \frac{10.2 - (0.1835 + 3.2)}{35} = 0.19476$$

At the other location, we have

$$P_{atm} = 620 \text{ mm of Hg}$$

$$H_{atm} = 620 \times 10^{-3} 13.6 = 8.432 \text{ m of } H_2O$$

$$P_{vap} = 830 \text{ Pa}$$

$$H_{vap} = \frac{830}{9.81 \times 10^{-3}} = 0.0846 \text{ m of } H_2O$$

$$0.19476 = \frac{8.432 - (0.0846 + H)}{35}$$

$$H = 1.53 \text{ m}$$

It is necessary to reduce to height of pump.

$$\Delta H = 3.20\text{-}1.53 = 1.67 \text{ m}$$

**Example 14.15** A three jet pelton turbine is required to generate 10000 kW under a net head of 400 m. The blade angle at the outlet is 15°and the decrease in the relative velocity while passing over the blade is 5%. Determine,

(a) The diameter of the jet

(b) The force exerted by a jet on the buckets.

Given: Overall efficiency of turbine = 80%

Co-efficient of velocity = 0.98

Speed ratio = 0.46

Further, if the jet ratio is not to be less than 10, calculate

(a) Speed of the wheel for a frequency of $50\, hertz/\sec$.

(b) Corresponding wheel diameter. [IES,2009]

**Solution.**

Given: $\eta = 3$, $P = 10{,}000$ kW, $H = 400$ m, $\beta_2 = 15°$, $v_{r2} = 0.95\, v_{r1}$, $\eta_o = 0.80$, $C_v = 0.98$, $\rho = \dfrac{u}{v_1} = 0.46$, $m = \dfrac{D}{d} = 10$, $f = 50$ Hz

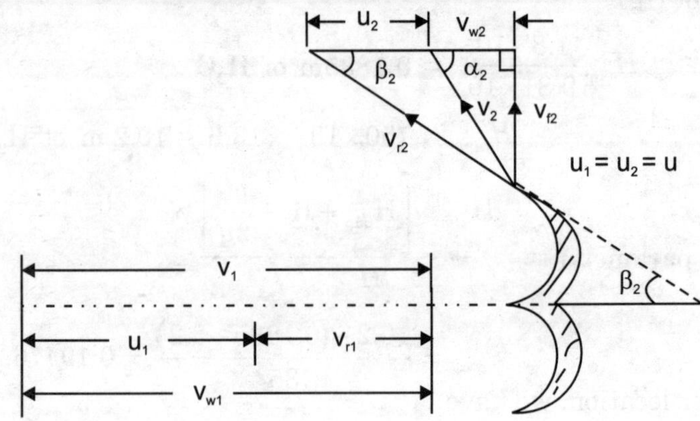


**Fig. 14.9 Velocity triangles.**

(a) Refer to Fig.14.9

Jet velocity, $v_1 = C_v\sqrt{2gH} = 0.98\sqrt{2\times 9.81\times 400}$

$= 86.82$ m/s

Wheel velocity, $u = \rho v_1 = 0.46 \times 86.82 = 39.93$ m/s

Overall efficiency, $\eta_o = \dfrac{P}{\gamma Q H}$

$$0.8 = \frac{10{,}000\times 10^3}{9.81\times 10^3 \times Q \times 400}$$

$$Q = 3.186 \text{ m}^3/\text{s}$$

Discharge per jet $= \dfrac{Q}{\eta} = \dfrac{3.186}{3} = 1.062$ m³/s

Now $\frac{Q}{n} = \frac{\pi}{4}d^2 v_1$

$$1.062 = \frac{\pi}{4}d^2 \times 86.82$$

$$d = 0.125 \text{ m or } 125 \text{ mm}$$

(b) Wheel diameter, $D = 10d = 10 \times 125 = 1250 \text{ mm or } 1.25 \text{ m}$

$$v_{r1} = v_1 - u = 86.82 - 39.93 = 46.89 \text{ m/s}$$

$$v_{r2} = 0.95 \times 46.89 = 44.54 \text{ m/s}$$

$$v_{\omega 1} = v_1 = 86.82 \text{ m/s}$$

$$v_{w2} = v_{r2} \cos \beta_2 - u = 44.54 \cos 15° - 39.93$$

$$= 3.097 \text{ m/s}$$

Force exerted by 3 jets on the buckets,

$$F_u = \rho Q(v_{w1} + v_{w2})$$

$$= 10^3 \times 3.186(86.82 + 3.097)$$

$$= 286.47 \text{ kN}$$

Force exerted by one jet $= \frac{286.47}{3} = 95.492$ kN

(c) Peripheral speed, $u = \frac{\pi DN}{60} = \frac{\pi \times 1.25 \times N}{60} = 39.93$

$$N = 610 \text{ rpm}$$

Synchronous speed, $N_{syn} = \frac{60 \text{ f}}{p} = \frac{60 \times 50}{5} = 600 \text{ rpm}$

Revised diameter of wheel, $D_1 = \frac{610}{600} \times 1.25 = 1.27$ m

**Example 14.16** A centrifugal pump is to deliver 4.5 m³/s when running at 750 rpm. The diameter of the impeller at inlet is 53 cm and at outlet is 76 cm. It may be assumed that the air enters radially with speed of 15 m/s. The vanes are set back wards at outlet at 70° to the tangent, and width at outlet is 10 cm. The volute casing gives at 30% recovery of the outlet velocity head. The losses in the impeller may be taken as equivalent to 25% of the outlet velocity head. Blade thickness effects may be neglected. Determine.

(a) The manometric efficiency and

(b) The pressure at the discharge. [IES, 2009]

**Solution.**

Given: $Q = 4.5$ m³/s, $N = 750$ rpm, $D_1 = 53$ cm, $D_2 = 76$ cm, $\alpha_{12} = 90°$, $\upsilon_1 = 15$ m/s, $\beta_2 = 70°$, $\beta_2 = 10$ cm, $h_{volute} = 0.3\ \frac{v_2^2}{2g}$, $H_\lambda = 0.25\ \frac{v_2^2}{2g}$

Refer to Fig.14.10.

$$u_1 = \frac{\pi D_1 N}{60} = \frac{\pi \times 0.53 \times 750}{60} = 20.81 \text{ m/s}$$

$$u_2 = \frac{\pi D_2 N}{60} \; \frac{\pi \times 0.76 \times 750}{60} = 29.84 \text{ m/s}$$

$$Q = \pi D_1 B_1 v_{f1} = \pi D_2 B_2 v_{f2}$$

$$4.5 = \pi \times 0.76 \times 0.1 \times v_{f2}$$

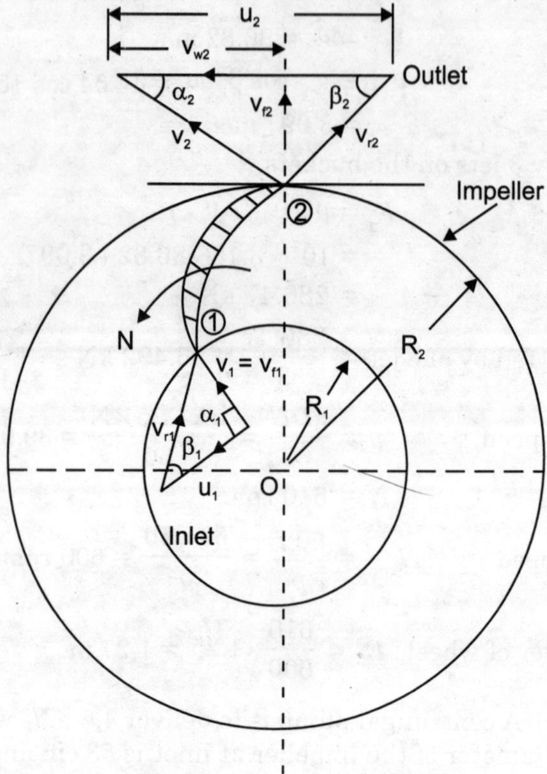


**Fig. 14.10 Velocity triangles.**

$$v_{f2} = 18.85 \text{ m/s}$$

$$\tan \beta_2 = \frac{v_{f2}}{u_2 - v_{w2}}$$

$$\tan 70° = \frac{18.85}{29.84 - v_{w2}}$$

$$2.74748 = \frac{18.85}{29.84 - v_{w2}}$$

$$v_{w2} = 22.98 \text{ m/s}$$

$$\tan \alpha_2 = \frac{v_{f2}}{v_{\omega 2}} = \frac{18.85}{22.98} = 0.8203$$

$$\alpha_2 = 39.36°$$

$$\cos \alpha_2 = \frac{v_{w2}}{v_2} = 0.77316$$

$$v_2 = 29.72 \text{ m/s}$$

Applying Bernoulli's theorem at inlet and of impeller, me have

$$\frac{p_1}{\gamma} + \frac{v_1^2}{2g} = \frac{p_2}{\gamma} + \frac{v_2^2}{2g} - \frac{u_2 v_{w2}}{g}$$

$$\frac{p_2 - p_1}{\gamma} = \frac{v_1^2 - v_2^2}{2g} + \frac{u_2 v_{w2}}{g}$$

$$= \frac{15^2 - (29.72)^2}{2 \times 9.81} + \frac{29.84 \times 22.98}{9.81}$$

$$= -33.55 + 69.90 = 36.35 \text{ m of air}$$

Velocity head at outlet $= \frac{v_2^2}{2g} = \frac{(29.72)^2}{2 \times 9.81} = 45.02$ m of air

Head recovered in velocity casing $= 0.3 \times 45.02 = 13.51$ m of air

Head loss in impeller $= 0.25 \times 45.02 = 11.26$ m of air

Net pressure rise $= 36.35 + 13.51 - 11.26 = 38.6$ m of air

$$\eta_{mano} = \frac{\textit{net pressure } \text{rise}}{\frac{u_2 v_{\omega 2}}{g}}$$

$$= \frac{38.6}{69.90} = 05522 \text{ or } 55.22\%$$

# Chapter 15 Multiple-Choice Questions from IES

## (a) Turbines

1. Two Pelton wheels $A$ and $B$ have the same specific speed and are working under the same head. Wheel $A$ produces 400 kW at 1000 rpm. If $B$ produces 100 kW, then its rpm is
   (a) 4000 (b) 2000
   (c) 1500 (d) 1250

2. Consider the following energies associated with a Pelton turbine :
   1. Mechanical energy
   2. Kinetic energy
   3. Potential energy

   The correct sequence of energy conversion starting from the entry of fluid is
   (a) 1-2-3 (b) 2-3-1
   (c) 3-2-1 (d) 1-3-2

3. Consider the following types of water turbines:
   1. Bulb 2. Francis
   3. Kaplan 4. Pelton

   The correct sequence of order in which the operating head decreases while developing the same power is
   (a) 4-2-3-1 (b) 3-4-1-2
   (c) 2-1-4-3 (d) 1-3-2-4

4. Which one of the following graphs correctly represents the relations between Head and Specific speed for Kaplan and Francis turbine?

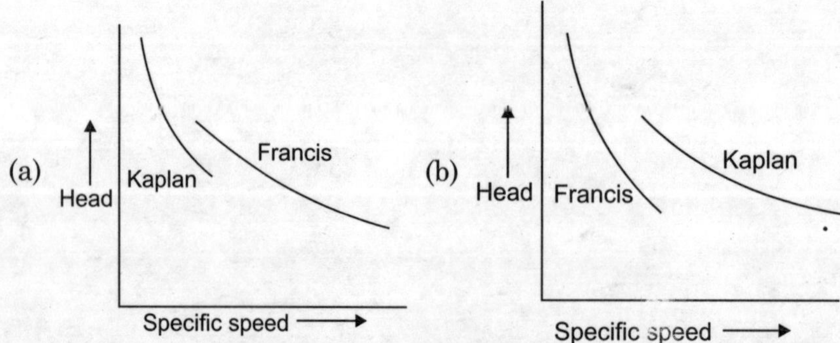

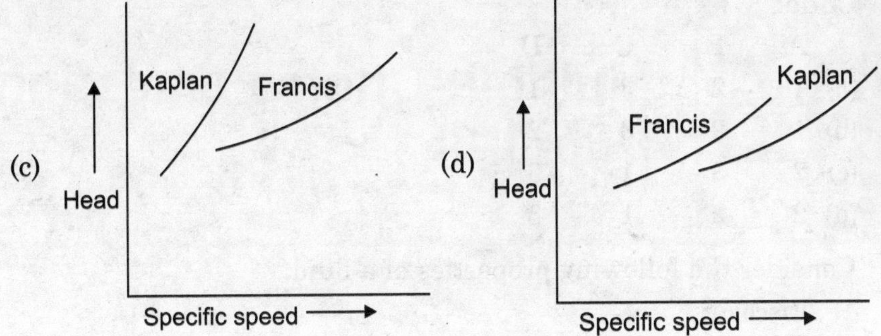


5. Euler equation of turbine giving energy transfer per unit mass $E_0$ (where $U$, $V_\omega$, $V_r$ and $V$ represents the peripheral, whirl, relative and absolute velocities respectively. Suffix 1 and 2 refer to the turbine inlet and outlet respectively) is given by

(a) $E_0 = U_1V_{\omega 1} - U_2V_{\omega 2}$ (b) $E_0 = U_1V_{r1} - U_2V_{r2}$

(c) $E_0 = U_1V_1 - U_2V_2$ (d) $E_0 = V_1V_{\omega 1} - V_2V_{\omega 2}$

6. What is the range of the speed radio $\phi$ for Kaplan turbine for its most efficient operation?

(a) 0.10 to 0.30 (b) 0.43 to 0.65

(c) 0.85 to 1.20 (d) 1.40 to 2.00

7. An impulse turbine operating with a single nozzle has a specific speed of 5. What will be the approximate specific speed of the turbine if the turbine is operated with one more additional nozzle of the same size?

(a) 4 (b) 6

(c) 7 (d) 10

8. A Francis turbine is coupled to an alternator to generate electricity with a frequency of 50 Hz. If the alternator has 12 poles, then the turbine should be regulated to run at which one of the following constant speeds?

(a) 250 rpm (b) 500 rpm

(c) 600 rpm (d) 1000 rpm

9. Match List-I (Flow Parameter) with List-II (Type of Turbine) and select the correct answer using the codes given below the Lists:

| List-I (Flow Parameter) | List-II (Type of Turbine) |
|---|---|
| A. High head | 1. Francis turbine |
| B. Axial flow | 2. Pelton wheel |
| C. Mixed flow | 3. Kaplan turbine |
| D. High specific speed | |

**Code:**

| | A | B | C | D |
|---|---|---|---|---|
| (a) | 1 | 2 | 3 | 1 |
| (b) | 1 | 2 | 3 | 2 |
| (c) | 2 | 3 | 1 | 3 |
| (d) | 2 | 2 | 1 | 3 |

10. Consider the following properties of a fluid:
    1. Viscosity
    2. Surface tension
    3. Capillarity
    4. Vapour pressure

    Which of the above properties can be attributed to the flow of jet of oil in an unbroken stream?

    (a) 1 only (b) 2 only
    (c) 1 and 3 (d) 2 and 4

11. A Pelton wheel with single jet rotates at 600 rpm. The velocity of the jet from the nozzle is 100 m/s. If the ratio of the vane velocity to jet velocity is 0.44, what is the diameter of the Pelton wheel?

    (a) 0.7 m (b) 1.4 m
    (c) 2.1 m (d) 2.8 m

12. Which of the following types of turbine is/are suitable for tidal power plants?
    1. Tubular turbine
    2. Kaplan turbine
    3. Bulb turbine
    4. Francis turbine

    Select the correct answer using the code given below:

    (a) 1 only (b) 1 and 3
    (c) 2 and 4 (d) 4 only

13. What should be the ratio of blade speed to jet speed for the maximum efficiency of a Pelton wheel?

    (a) $\frac{1}{4}$ (b) $\frac{1}{2}$
    (c) $\frac{3}{4}$ (d) 1

14. Which one of the dimensionless numbers identifies the compressibility effect of a fluid?

    (a) Euler number (b) Froude number
    (c) Mach number (d) Weber number

15. When a hydraulic turbine is operated, it is fount that it has a high design efficiency and this efficiency remains constant over a wide range of regulation from the design condition. What is the type of this turbine?

(a) Pelton (b) Francis
(c) Kaplan (d) Propeller

16. If α is the blade angle at the outlet, then the maximum hydraulic efficiency of an ideal impulse turbine is

(a) $\frac{1+\cos\alpha}{2}$ (b) $\frac{1-\cos\alpha}{2}$
(c) $\frac{1-\sin\alpha}{2}$ (d) $\frac{1+\sin\alpha}{2}$

17. Which one of the following turbines exhibits a nearly constant efficiency over a 60% to 140% of design speed?

(a) Pelton turbine
(b) Francis turbine
(c) Deriaz turbine
(d) Kaplan turbine

18. In the statement, "in a reaction turbine installation, the head of water is decreased and the rpm is also decreased at a certain condition of working. The effect of each of these changes will be to *X* power delivered due to decrease in head and to *Y* power delivered due to decrease in rpm". $\frac{N\sqrt{Q}}{H^{3/4}}$ *X* and *Y* stand respectively for

(a) decrease and increase
(b) increase and increase
(c) decrease and decrease
(d) increase and decrease

19. Which of following are the beneficial effects of air vessel fitted to delivery side of a reciprocating pump?

1. Constant rate of discharge can be ensured
2. Power consumption can be reduced
3. Discharge can be increased
4. Constant velocity of the piston can be ensured

Select the correct answer using the codes given below:

**Codes:**

(a) 1 and 4 (b) 1 and 2
(c) 2 and 4 (d) 1 and 3

20. The following lists refer to fluid machinery Match List I with List II select the correct answer using the codes given below the Lists:

| List I | List II |
|---|---|
| A. Draft tube | 1. Impulse turbine |
| B. Surging | 2. Reciprocating pump |
| C. Air vessel | 3. Reaction turbine |
| D. Nozzle | 4. Centrifugal pump |

**Codes:**

| | A | B | C | D |
|---|---|---|---|---|
| (a) | 4 | 3 | 2 | 1 |
| (b) | 3 | 4 | 2 | 1 |
| (c) | 3 | 4 | 1 | 2 |
| (d) | 4 | 3 | 1 | 2 |

21. Chances of occurrence of cavitation are high if the
   (a) Local pressure becomes very high
   (b) Local temperature becomes low
   (c) Thoma cavitation parameter exceeds a certain limit
   (d) Local pressure falls below the vapour pressure

22. The movable wicket gates of a reaction turbine are used to
   (a) control the flow of water passing through the turbine.
   (b) control the pressure under which the turbine is working.
   (c) strength the casing of the turbine.
   (d) reduce the size of the turbine.

23. Match List I with List II and select the correct answer using the codes given below the Lists:

| List I (Types of turbine) | List II (Characteristics of turbines) |
|---|---|
| A. Propeller | 1. Inward flow reaction |
| B. Francis | 2. Tangential flow impulse |
| C. Kaplan | 3. Axial flow reaction with fixed vanes |
| D. Pelton | 4. Axial flow reaction with adjustable vanes |

**Codes:**

| | A | B | C | D |
|---|---|---|---|---|
| (a) | 2 | 4 | 1 | 3 |
| (b) | 3 | 4 | 1 | 2 |
| (c) | 2 | 1 | 4 | 3 |
| (d) | 3 | 1 | 4 | 2 |

24. Consider the following statement:
Cavitation in hydraulic machines occurs at the
1. exit of a pump
2. entry of the pump
3. exit of a turbine
of these statements
(a) 1 and 2 are correct (b) 1 and 3 are correct
(c) 1, 2 and 3 are correct (d) 2 and 3 are correct

25. Euler equation for water turbine is derived on the basis of
(a) conservation of mass.
(b) rate of change of linear momentum.
(c) rate of change of angular momentum
(d) rate of change of velocity.

26. A hydraulic power station has the following major items in the hydraulic circuit:
1. Draft tube 2. Runner
3. Guide wheel 4. Penstock
The correct sequence of these items in the direction of flow is
(a) 4, 2, 3, 1, 5 (b) 4, 3, 2, 5, 1
(c) 1, 2, 3, 5, 4 (d) 1, 3, 2, 4, 5

27. The Specific speed of a turbine is defined as the speed of a member of the same homologous series of such a size that it.
(a) Delivers unit discharge at unit head.
(b) Delivers unit discharge at unit power.
(c) Delivers unit power at unit discharge.
(d) Produces unit power under the unit head.

28. Mach List I with List II and select the correct answer

| List I | List II |
|---|---|
| A. Pelton wheel (single jet) | 1. Medium discharge, low head |
| B. Francis Turbine | 2. High discharge, low head. |
| C. Kalpan Turbine | 3. Medium discharge, medium head. |
| | 4. Low discharge, high head |

**Codes:**

| | A | B | C |
|---|---|---|---|
| (a) | 1 | 2 | 3 |
| (b) | 1 | 3 | 4 |
| (c) | 4 | 1 | 3 |
| (d) | 4 | 3 | 2 |

29. A reaction turbine discharges 30 m³/s of water under a head of 10 m with an overall efficiency of 92 %. The power developed is:

(a) 295.2 kW (b) 287.0 kW

(c) 279.0 kW (d) 265.2 kW

30. Given that

$N$ = speed, $P$ = power, $H$ = head

The specific speed of a hydraulic turbine is given by

(a) $\frac{N\sqrt{P}}{H^{4/5}}$ (b) $\frac{N\sqrt{P}}{H^{5/4}}$

(c) $\frac{P\sqrt{N}}{H^{4/5}}$ (d) $\frac{P\sqrt{N}}{H^{5/4}}$

31. As water flows through the runner of a reaction turbine, pressure acting on it would vary from:

(a) more than atmospheric pressure to vacuum

(b) less than atmospheric pressure to zero gauge pressure

(c) atmospheric pressure to more than atmospheric pressure

(d) atmospheric pressure to vacuum

32. Which one of the following statements regarding an impulse turbine is correct?

(a) There is no pressure variation in flow over the buckets and the fluid fills the passageway between the buckets

(b) There is no pressure variation in flow over the buckets and the fluid does not fill the passageway between the buckets

(c) There is pressure drop in flow over the buckets and the fluid fills the passageway between the buckets

(d) There is pressure drop in flow over the buckets and the fluid does not fill the passageway between the buckets

33. In reaction turbines, the draft tube is used

(a) for the safety of the turbine.

(b) to convert the kinetic energy of flow by a gradual expansion of the flow cross-section.

(c) to destroy the undesirable eddies.

(d) for none of the above purposes.

34. Which one of the following forms of draft tube will NOT improve the hydraulic efficiency of the turbine?

(a) Straight cylindrical (b) Conical type

(c) Bell-mouthed (d) Bent tube

35. Which one of the following turbines is used in underwater power stations?
(a) Pelton turbine (b) Deriaz turbine
(c) Tubular turbine (d) Turgo-impulse turbine

36. A pelton wheel is ideally suited for
(a) High head and low discharge
(b) High head and high discharge
(c) Low head and low discharge
(d) Medium head and medium discharge

37. Consider the following turbine:
1. Kaplan.
2. Pelton wheel
3. Francis
The correct sequence in increasing order of the specific speeds of these turbines is
(a) 2, 3, 1 (b) 2, 1, 3
(c) 3, 1, 2 (d) 1, 2, 3

38. Consider the following turbine/wheels:
1. Francis turbine
2. Pelton wheel with two or more jets
3. Pelton wheel with a single jet
4. Kaplan turbine
The correct sequence of these turbine/wheels in increasing order of their specific speeds is
(a) 2, 3, 1, 4 (b) 3, 2, 1, 4
(c) 2, 3, 4, 1 (d) 3, 2, 4, 1

39. The gross head available to a hydraulic power plant is 100 m. The utilized head in the runner of the hydraulic turbine is 72 m. If the hydraulic efficiency of the turbine is 90%, the pipe friction head is estimated to be
(a) 20 m (b) 18 m
(c) 16.2 m (d) 1.8 m

40. If the full-scale turbine is required to work under a head of 30 m and to run at 428 r.p.m., then a quarter-scale turbine model tested under a head of 10 m must run at
(a) 143 r.p.m. (b) 341 r.p.m.
(c) 428 r.p.m. (d) 988 r.p.m

41. The correct sequence of the given hydraulic turbines in decreasing order of their specific speeds is:
    (a) Pelton wheel, Francis turbine, Kaplan turbine
    (b) Propeller turbine, Francis turbine, Pelton wheel
    (c) Kaplan turbine, Pelton wheel, Francis turbine
    (d) Francis turbine, Kaplan turbine, Pelton wheel

42. The cavitation number of any fluid machinery is defined as $\sigma = \dfrac{p - p'}{(\rho V^2 / 2)}$
    (p is absolute pressure, ρ is density and $V$ is free stream velocity)
    The symbol $p'$ denotes
    (a) static pressure of fluid (b) dynamic pressure of fluid
    (c) vapour pressure of fluid (d) shear stress of fluid

43. Consider the following statements:
    A water turbine governor
    1. helps in staring and shutting down the turbo unit.
    2. controls the speed of turbine set to match it with the hydroelectric system.
    3. Sets the amount of load which a turbine unit has to carry.
    Which of these statements are correct?
    (a) 1, 2 and 3 (b) 1 and 2
    (c) 2 and 3 (d) 1 and 3

44. Consider the specific speed ranges of the following types of turbine:
    1. Francis
    2. Kaplan
    3. Pelton
    The sequence of their specific speed in increasing order is
    (a) 1, 2, 3 (b) 3, 1, 2
    (c) 3, 2, 1 (d) 2, 3,1

45. A symmetrical stationary vane experiences a force '*F*' of 100 N as shown in the given figure, when the mass flow rate of water over the vane is 5 kg/s with a velocity '*V*' 20 m/s without friction. The angle '*a*' of the vane is
    (a) zero
    (b) 30°
    (c) 45°
    (d) 60°

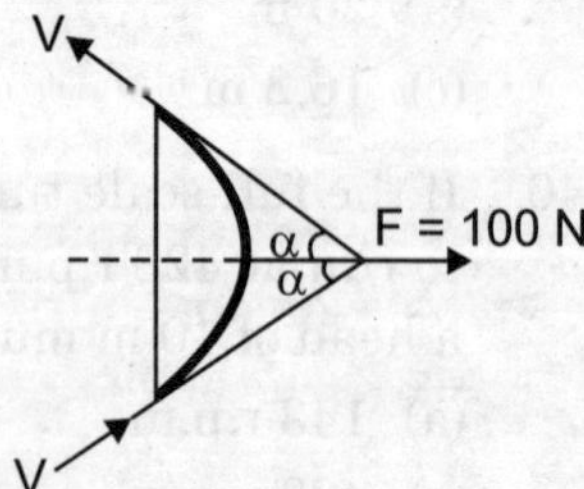

46. The gross head on a turbine is 300 m. The length of penstock supplying water from reservoir to the turbine is 400 m. The diameter of the penstock is 1 m and velocity of water through penstock is 5 m/s. If coefficient of friction is 0.0098, the net head on the turbine would be, nearly
(a) 310 m (b) 295 m
(c) 200 m (d) 150 m

47. The level of runner exit is 5 m above the tail race, and atmospheric pressure is 10.3 m. The pressure at the exit of the runner for a divergent draft tube can be
(a) 5 m (b) 5.3 m
(c) 10 m (d) 10.3 m

48. Consider the following statements:
A surge tank provided on the penstock connected to a water turbine
1. helps in reducing the water hammer.
2. stores extra water when not needed.
3. provides increased demand of water.
Which of these statements are correct?
(a) 1 and 3 (b) 2 and 3
(c) 1 and 2 (d) 1, 2 and 3

49. For a water turbine, running at constant head and speed, the operating characteristic curves in the given figure show, that upto a certain discharge '*q*' both output power and efficiency remain zero. The discharge '*q*' is required to

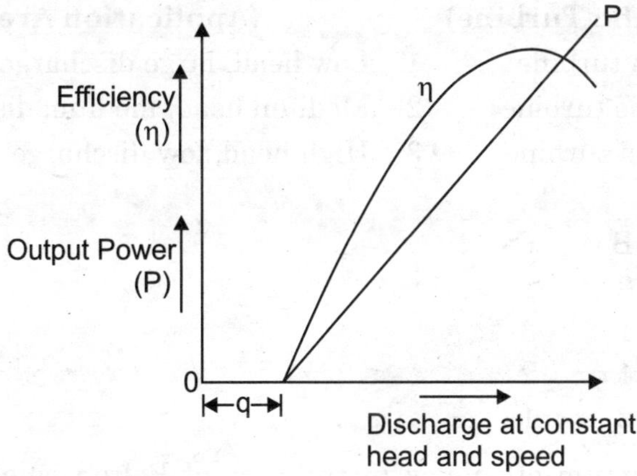


(a) overcome initial inertia
(b) overcome initial friction
(c) keep the hydraulic circuit full
(d) keep the turbine running at on load

50. The maximum number of jets generally employed in an impulse turbine without jet interference is

(a) 4 (b) 6

(c) 8 (d) 12

51. Consider the following statement:

1. Pelton wheel is a tangential flow impulse turbine.
2. Francis turbine is an axial flow reaction turbine.
3. Kaplan turbine is a radial flow reaction turbine.

Which of the above statement is/are correct?

**Codes:**

(a) 1 and 3 (b) 1 alone

(c) 2 alone (d) 3 alone

52. In fluid machinery, the relationship between saturation temperature and pressure decides the process of

(a) flow separation (b) turbulent mixing

(c) cavitation (d) water hammer

53. Efficiency of Pelton wheel shall be maximum if the ratio of jet velocity to tangential velocity of the wheel is

(a) 1/2 (b) 1

(c) 2 (d) 4

54. Match List-I (Hydraulic turbine) with List-II (Application Area) and select the correct answer using the codes given below the lists:

| List-I (Hydraulic Turbine) | List-II (Application Area) |
|---|---|
| A. Pelton turbine | 1. Low head, large discharge |
| B. Francis turbine | 2. Medium head, medium discharge |
| C. Kaplan turbine | 3. High head, low discharge |

**Codes:**

| | A | B | C |
|---|---|---|---|
| (a) | 2 | 3 | 1 |
| (b) | 2 | 1 | 3 |
| (c) | 3 | 1 | 2 |
| (d) | 3 | 2 | 1 |

55. The maximum efficiency in the case of Pelton wheel is (angle of deflection of the jet = 180° – β)

(a) $\frac{1-\cos\beta}{2}$ (b) $\frac{1+\cos\beta}{2}$

(c) $\frac{\cos\beta}{2}$ (d) $\frac{1+\cos\beta}{2}$

56. In the phenomenon of cavitation, the characteristic fluid property involved is
(a) surface tension (b) viscosity
(c) bulk modulus of elasticity (d) vapour pressure

57. If $H$ is the head available for a hydraulic turbine, the power, speed and discharge, respectively are proportional to
(a) $H^{1/2}, H^{1/2}, H^{3/2}$ (b) $H^{3/2}, H^{1/2}, H^{1/2}$
(c) $H^{1/2}, H^{3/2}, H^{1/2}$ (d) $H^{3/2}, H^{1/2}, H$

58. A draft tube is used in a reaction turbine
(a) to guide water downstream without splashing
(b) to convert residual pressure energy into kinetic energy
(c) to convert residual kinetic energy into pressure energy
(d) to streamline the flow in the tailrace

59. Which of the following water turbines maintain a high efficiency over a long range of the part load?
1. Francis turbine 2. Kaplan turbine
3. Pelton turbine 4. Propeller turbine
Select the correct answer using the code given below:
(a) 1 and 4 (b) Only 2 and 3
(c) 1, 2 and 3 (d) 2, 3 and 4

60. Which of the following advantages is/are possessed by a Kaplan turbine over a Francis turbine?
1. Low frictional losses.
2. Part load efficiency is considerably high.
3. More compact and smaller in size.
Select the correct answer using the code given below:
(a) Only 1 (b) Only 1 and 2
(c) Only 2 and 3 (d) 1, 2 and 3

61. Which of the following water turbines does not require a draft tube?
(a) Propeller turbine (b) Pelton turbine
(c) Kaplan turbine (d) Francis turbine

62. Which of the following statements are false regarding the relation specific speed, $N_s = \dfrac{N\sqrt{Q}}{H^{5/4}}$
(a) The relation is valid for both impulse as well as reaction turbine
(b) The value of specific speed depends on the system of units
(c) The value of specific speed for Pelton wheel will be lower than that for Kaplan turbine for any system of units
(d) The dimensions of specific speed are: $(FLT^{-1})$

63. A circular jet of water impinges on a vertical flat plate and bifurcates into two circular jets of half the diameter of the original. After hitting the plate

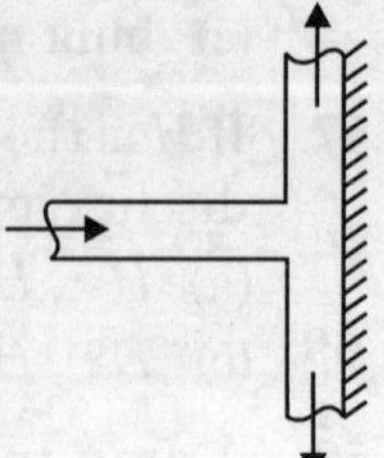

(a) The jets move at equal velocity which is twice of the original velocity
(b) The jets move at equal velocity which is 3 times of the original velocity
(c) Data given is insufficient to calculate velocities of the two outgoing jets
(d) The jets move at equal velocity which is equal to the original velocity

### *(b) Reciprocating Pumps*

64. Which one of the following pairs is not correctly matched?
(a) *Centrifugal pump:* Rotating blades in the motor create centrifugal head
(b) *Reciprocating pump:* Positive displacement pump
(c) *Turbine pump:* Centrifugal pump with guide vanes
(d) *Gear pump:* Gear teeth work like rotating blades to create centrifugal head

65. Consider the following statements regarding air vessels provided in reciprocating pump installations:
1. The air vessels are fitted both on suction and delivery sides.
2. The air vessels are fitted far from the pump cylinder.
3. The air vessels save energy by reducing the friction loss.
Which of these statements are correct?
(a) 1, 2 and 3 (b) 1 and 2
(c) 2 and 3 (d) 1 and

66. Which one of the following is correct?
In positive displacement pumps, the slip can sometimes be negative when the actual discharge is greater than the theoretical discharge. This happens in
(a) small suction pipes coupled with a low delivery head
(b) small suction pipes coupled with a medium delivery head
(c) long suction pipes coupled with a low delivery head
(d) long suction pipes coupled with a medium delivery head

67. If a reciprocating pump having a mechanical efficiency of 80% delivers water at the rate of 80 kg/s with a head of 30 m, the brake power of the pump is
(a) 29.4 kW (b) 20.8 kW
(c) 15.4 kW (d) 10.8 kW

68. Air vessel is used in a reciprocating pump to obtain
   (a) reduction of suction head
   (b) rise in delivery head
   (c) continuous supply of water at uniform rate
   (d) increase in supply of water

### *(c) Centrifugal pumps*

69. Match List-I (Type of pumps) with List-II (Associated features) and select the correct answer using the codes given below the Lists:

| List-I (Type of pumps) | List-II (Associated features) |
|---|---|
| A. Centrifugal pump | 1. Air vessel |
| B. Gear pump | 2. Draft tube |
| C. Reciprocating pump | 3. Guide vanes |
| D. Turbine pump | 4. Rotary pump |
| | 5. Rotor having blades |

**Codes:**

| | A | B | C | D |
|---|---|---|---|---|
| (a) | 4 | 2 | 5 | 3 |
| (b) | 5 | 4 | 1 | 2 |
| (c) | 4 | 2 | 3 | 1 |
| (d) | 5 | 4 | 1 | 3 |

70. On the assumption that a double suction impeller is the equivalent of two single suction impellers placed 'back to back', it is customary to base the specific speed of the double suction pump on
   (a) One half of the total capacity
   (b) Three fourth of the total capacity
   (c) Full total capacity
   (d) Double the total capacity

71. Match List-I (Industrial Needs) with List-II (Type of Pump) and select the correct answer using the codes given below the Lists:

| List-I (Industrial Needs) | List-II (Type of Pump) |
|---|---|
| A. Combustible fluid to be pumped | 1. Single stage centrifugal |
| B. High head but small discharge needed | 2. Multi- stage centrifugal |
| C. Low head but large discharge needed | 3. Positive displacement |
| D. High head and high discharge needed | 4. Jet pump |

**Codes:**

| | A | B | C | D |
|---|---|---|---|---|
| (a) | 3 | 2 | 1 | 4 |
| (b) | 4 | 3 | 1 | 2 |
| (c) | 3 | 1 | 4 | 2 |
| (d) | 4 | 3 | 2 | 1 |

72. The power ratio of a pump and its $\frac{1}{4}th$ scale model, if the ratio of the heads is 5:1, will be
(a) 100 (b) 3.2
(c) 179 (d) 12.8

73. Priming is necessary in
(a) Centrifugal pumps to lift water from a greater depth
(b) Centrifugal pumps to remove air in the suction pipe and casing
(c) Hydraulic turbine to remove air in the turbine casing
(d) Hydraulic turbine to-increase the speed of turbine and to generate more power

74. Two centrifugal pumps have impeller with outer dimensions of each equal to twice the inner dimensions. The inner diameter of the second impeller is three times the inner diameter of the first one. What is the speed ratio $N_2/N_1$ of pumps, if the pumps are required to develop the same manometric head to start delivery of water?
(a) 9 (b) 4
(c) $\frac{1}{2}$ (d) $\frac{1}{3}$

75. Water is required to be lifted by a 10kW pump from a depth of 100 m. If the pump is unable to lift the water, then which one of the following is correct?
(a) A greater capacity pump has to be used
(b) A larger diameter delivery pipe has to be used
(c) A larger diameter suction pipe has to be used
(d) A multistage pump has to be used

76. Which one of the following statements is relevant to the specific speed of a centrifugal pump?
(a) Head developed is units and discharge is unity
(b) Head developed is units and power absorbed is unity
(c) Discharge is unity and power absorbed is unity
(d) Each of head developed , power absorbed and discharge is equal to unity

77. Match List-I (Type of Pump) with List-II (Liquid handled) and select the correct answer using the codes given below the Lists:

| List-I (Type of Pump) | List-II (Liquid handled) |
|---|---|
| A. Closed impeller pump | 1. Sandy water |
| B. Semi-open impeller pump | 2. Acids |
| C. Open impeller pump | 3. Sewage water |

**Codes:**

| | A | B | C |
|---|---|---|---|
| (a) | 1 | 3 | 2 |
| (b) | 3 | 1 | 2 |
| (c) | 2 | 3 | 1 |
| (d) | 1 | 2 | 3 |

78. Two centrifugal pumps '*A*' and '*B*' operate at their maximum efficiencies at 1000 rpm and 500 rpm respectively. Against the same delivery head, pump '*A*' discharge 1 m³/s and pump *B* discharges 4 m³/s respectively. What is the ratio of specific speeds $(N_S)_A : (N_S)_B$?

(a) 1 : 2 (b) 1 : 1
(c) 1 : 4 (d) 4 : 1

79. A centrifugal pump needs 1000 W of power when operating at 1500 rpm. What is the power requirement if the speed of the pump is increased to 3000 rpm?

(a) 2000 W (b) 4000 W
(c) 6500W (d) 8000 W

80. Match List-I (Specific Speed) with List-II (Expression/Magnitude) and select the correct answer using the codes given below the lists:

| List-I (Specific Speed) | List-II (Expression/Magnitude) |
|---|---|
| A. Specific speed of turbine | 1. $N\sqrt{Q}/H^{3/4}$ |
| B. Specific speed of pump | 2. $N\sqrt{P}/H^{5/4}$ |
| C. Specific speed of Pelton wheel | 3. 50 – 250 |
| D. Specific speed of Francis turbine | 4. 10 – 5 |

**Codes:**

| | A | B | C | D |
|---|---|---|---|---|
| (a) | 3 | 4 | 1 | 2 |
| (b) | 3 | 2 | 1 | 4 |
| (c) | 2 | 1 | 4 | 3 |
| (d) | 1 | 2 | 4 | 3 |

81. Why is a minimum of Net Positive Suction Head required for a hydraulic pump?
(a) To prevent cavitation (b) To increase discharge
(c) To increase suction head (d) To increase efficiency

82. A centrifugal pump operating at 1000 rpm develops a head of 30 m. If the speed is increased to 2000 rpm and the pump operates with the same efficiency, what is the head developed by the pump?
(a) 60 m (b) 90 m
(c) 120 m (d) 150 m

83. Which one of the following statements is correct?
When a fluid passes from the inlet to exit of the rotor in a centrifugal pump, tangential momentum
(a) Increases and energy increases
(b) Decreases and energy increases
(c) Remains unchanged and energy decreases
(d) Increases and energy remains unchanged

84. A pump is installed at a height of 5 m above the water level in the sump. Frictional loss on the suction side is 0.6 m. If the atmospheric pressure is 10.3 m of water and vapour pressure head is 0.4 m (abs), the NPSH (Net positive suction head) will be
(a) 3.7 m (b) 4 m
(c) 4.3 m (d) 4.6 m

85. Consider the following pumps:
1. Centrifugal pump, single-stage
2. Centrifugal pump, multi-stage
3. Reciprocating pump.
4. Jet pump.
The pump (s) which can be used to lift water through a suction head of 12 m from a well would include
(a) 2 alone (b) 1, 3 and 4
(c) 4 alone (d) 1 and 3

86. Which one of the following pairs of formulae represents the specific speeds of turbine and pump respectively? (Notations have their usual meanings).
(a) $\frac{NQ^{1/2}}{H^{3/4}}$ and $\frac{NP^{1/2}}{H^{5/4}}$ (b) $\frac{NQ^{1/2}}{H^{3/4}}$ and $\frac{NP^{1/2}}{H^{3/4}}$
(c) $\frac{NP^{1/2}}{H^{3/4}}$ and $\frac{NQ^{1/2}}{H^{5/4}}$ (d) $\frac{NP^{1/2}}{H^{3/4}}$ and $\frac{NQ^{1/2}}{H^{3/4}}$

87. Consider the following data for the performance of a centrifugal pump: Speed: 1200 rpm, flow rate: 30 l/s, head : 20 m, power: 5kW. If the speed is increased to 1500 rpm, the power will be nearly equal to
(a) 6.5 kW (b) 8.7 kW
(c) 9.8 kW (d) 10.9 kW

88. Consider the following statements regarding the specific speed of a centrifugal pump:
1. Specific speed is defined as the speed of a geometrically similar pump developing unit power under unit head.
2. At the same specific speed, the efficiency is greater with larger capacity.
3. The specific speed increases with increase in outer blade angle.
4. The specific speed varies directly as the square root of the pump discharge.

Of these statements
(a) 1 and 2 are correct
(b) 2 and 4 are correct
(c) 3 and 4 are correct
(d) 2 and 3 are correct

89. A centrifugal pump is started with its delivery valve kept
(a) fully open (b) fully closed
(c) partially open (d) 50% open

90. Consider the following statement:
If pump NPSH requirements are not satisfied, then
1. It will not develop sufficient head to raise water.
2. Its efficiency will be low.
3. It will deliver very low discharge.
4. It will be cavitated.

Of these statements
(a) 1, 2 and 3 are correct
(b) 1 and 4 are correct
(c) 2, 3 and are 4 correct
(d) 1, 2, 3 and 4 are correct

91. Which of the following are the functions of a volute casing in a centrifugal pump?
1. To collect water from the periphery of the impeller and to transmist it to the delivery pipe at a constant velocity.
2. To increase the discharge of the pump.
3. To increase the efficiency of the pump.
4. To reduce the loss of head in discharge.

Select the correct answer from the codes given below:
(a) 1, 2 and 3
(b) 2, 3 and 4
(c) 1, 3 and 4
(d) 1 and 2

92. For attaining a non-overloading characteristic in centrifugal pumps.
(a) backward bent vanes are preferred over forward bent vanes.
(b) forward bent vanes are preferred over backward bent vanes.
(c) forward bent vanes are preferred over vanes radial at outlet.
(d) vanes radial at outlet are preferred over backward vanes.

93. Specific speed of a pump and specific speed of a turbine are (symbols have the usual meaning)
(a) $\frac{N\sqrt{Q}}{H^{3/4}}$ and $\frac{N\sqrt{P}}{H^{5/4}}$ respectively
(b) $\frac{N\sqrt{Q}}{H^{3/4}}$ and $\frac{N\sqrt{P}}{H^{3/4}}$ respectively
(c) $\frac{N\sqrt{Q}}{H^{5/4}}$ and $\frac{N\sqrt{P}}{H^{5/4}}$ respectively
(d) $\frac{N\sqrt{Q}}{H^{5/4}}$ and $\frac{N\sqrt{P}}{H^{3/4}}$ respectively

94. Which of the following purposes are served by the volute casing of a centrifugal pump?
1. Increase in the efficiency of the pump.
2. Conversion of part of the pressure head to velocity head.
3. Giving uniform flow of the fluid coming out of the impeller.
Select the correct answer using the codes given below:
(a) 1 and 2
(b) 1 and 3
(c) 2 and 3
(d) 1, 2 and 3

95. The specific speed of a hydraulic pump is the speed of geometrically similar pump working against a unit head and
(a) Delivering unit quantity of water
(b) Consuming unit power
(c) Having unit velocity of flow
(d) Having unit radial velocity

96. Match List-I (Outlet vane angle $\beta_2$) with List-II (Curves labelled 1, 2 and 3 in the given figure) for a pump and select the correct answer using the codes given below the Lists:

**List-I** | **List-II**

A. $\beta_2 < 90^\circ$

B. $\beta_2 = 90^\circ$

C. $\beta_2 > 90^\circ$

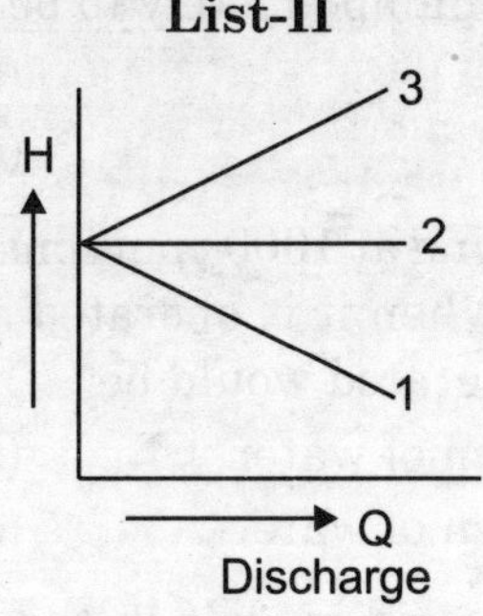


**Codes:**

| | A | B | C |
|---|---|---|---|
| (a) | 1 | 2 | 3 |
| (b) | 1 | 3 | 2 |
| (c) | 2 | 1 | 3 |
| (d) | 3 | 2 | 1 |

97. Consider the following statements regarding the volute casing of a centrifugal pump:
   1. Loss of head due to change in velocity is eliminated.
   2. Efficiency of the pump is increased.
   3. Water from the periphery of the impeller is collected and transmitted to the delivery pipe at constant velocity.

   Which of these statements are correct?
   (a) 1, 2 and 3 (b) 1 and 2
   (b) 2 and 3 (d) 1 and 3

98. Consider the following statements pertaining to a centrifugal pump:
   1. The manometric head is the head developed by the pump.
   2. The suction pipe has, generally, a larger diameter as compared to the discharge pipe.
   3. The suction pipe is provided with a foot valve and a strainer.
   4. The delivery pipe is provided with a foot valve and a strainer.

   Of these statements:
   (a) 1, 2, 3 and 4 are correct (b) 1 and 2 are correct
   (c) 2 and 3 are correct (d) 1 and 3 are correct

99. A centrifugal pump gives maximum efficiency when its blades are
   (a) bent forward (b) bent backward
   (c) straight (d) wave shaped

100. A centrifugal pump was manufactured to couple directly to a 15 H.P. electric motor running at 1950 rpm and delivering 50 liters per minute against a total head of 20 m. It is desired to replace the pump. The head developed by pump is likely to be
(a) 41.4 m (b) 29.6 m
(c) 20 m (d) 9.5 m

101. A pump running at 1000 rpm consumes 1 kW and generates head of 10 m of water. When it is operated at 2000 rpm, its power consumption and head generated would be
(a) 4 kW, 50 m of water (b) 6 kW, 20 m of water
(c) 3 kW, 30 m of water (d) 8 kW, 40 m of water

102. In utilizing scaled models in the designing of turbo-machines, which of the following relationships must be satisfied?
(a) $\frac{H}{ND^3}$ = constant; $\frac{Q}{N^2D^2}$ = constant
(b) $\frac{Q}{D^2\sqrt{H}}$ = constant; $\frac{H}{N^3D}$ = constant
(c) $\frac{\rho}{QH}$ = constant; $\frac{H}{N^2D^2}$ = constant
(d) $\frac{NQ^{\frac{1}{2}}}{H^{\frac{3}{2}}}$ = constant; $\frac{NP^{\frac{1}{2}}}{H^{\frac{3}{4}}}$ = constant

103. The correct sequence of the centrifugal pump components through which the fluid flows is
(a) Impeller, Suction pipe, Foot valve and strainer, Delivery pipe
(b) Foot valve and strainer, Suction pipe, Impeller, Delivery pipe
(c) Impeller, Suction pipe, Delivery pipe, Foot valve and strainer
(d) Suction pipe, Delivery pipe, Impeller, foot valve and strainer

104. A centrifugal pump driven by a directly coupled 3 kW motor of 1450 rpm speed, is proposed to be connected to another motor of 2900 rpm speed. The power of the motor should be
(a) 6 kW (b) 12 kW
(c) 18 kW (d) 24 kW

105. Which one of the following is the correct statement? For a given centrifugal pump,
(a) the discharge varies directly as the speed
(b) the head varies inversely as the speed
(c) the power varies as the square of the speed
(d) the discharge varies as the square of the speed

106. If $N_{S1}$ is the specific speed of a pump handling larger discharges at low heads and $N_{S2}$ is that of a pump handling low discharge at high heads, then which one of the following is correct?
   (a) $N_{S1} > N_{S2}$
   (b) $N_{S1} < N_{S2}$
   (c) $N_{S1} = N_{S2}$
   (d) All the above three are possible under different situations

107. Match the following:

| List-I | List-II |
|---|---|
| A. Wicket gates | 1. Gradually converts the velocity head to pressure head |
| B. Volute tube | 2. Pressure head recovery |
| C. Draft tube | 3. High value of $N$ and $N_S$ |
| D. Axial flow pumps | 4. Adjustable by governor according to load |

**Codes:**

| | A | B | C | D |
|---|---|---|---|---|
| (a) | 4 | 1 | 2 | 3 |
| (b) | 1 | 3 | 2 | 4 |
| (c) | 2 | 3 | 4 | 1 |
| (d) | 2 | 3 | 1 | 4 |

### *(d) Fluid Systems*

108. An accumulator is a device to store
   (a) Sufficient quantity of liquid to compensate the change in discharge
   (b) Sufficient energy to drive the machine when the normal energy source does not function
   (c) Sufficient energy in case of machine which work intermittently to supplement the discharge from the normal source
   (d) Liquid which otherwise would have gone to waste

109. The function of which of the following hydraulic devices is analogous to that of the flywheel of a reciprocating engine and an electric storage battery?
   (a) Hydraulic ram (b) Hydraulic accumulator
   (c) Hydraulic intensifier (d) Hydraulic jack

110. A hydraulic press has a ram of 20 cm diameter and plunger of 5 cm diameter. The force required at the plunger to lift a weight of $16 \times 10^4$ N shall be:
   (a) $256 \times 10^4$ N (b) $256 \times 10^4$ N
   (c) $4 \times 10^4$ N (d) $1 \times 10^4$ N

111. If $\omega_s$ and $\omega_p$ represent the angular velocities of drive and driving members of a fluid coupling respectively, then the slip is equal to

(a) $1-\frac{\omega_s}{\omega_p}$ (b) $\frac{\omega_s}{\omega_p}$

(c) $\frac{\omega_p}{\omega_s}$ (d) $1-\frac{\omega_p}{\omega_s}$

112. Consider the following statement regarding torque converter:
1. It has a stationary set of blades in addition to the primary and secondary rotor.
2. It can be used for multiplication of torques.
3. The maximum efficiency of a inverter is less than that of a fluid coupling
4. In a converter designed to give a large increase of torque, the efficiency .....................speed ratio approaches unity.

Of these statements
(a) 1, 2, 3 and 4 are correct
(b) 1, 2 and 3 are correct
(c) 1, 2 and 4 are correct
(d) 3 and 4 are correct

113. In contrast to fluid couplings, torque converters are operated:
(a) while completely filled with liquid
(b) while partially filled with liquid
(c) without liquid
(d) while completely filled with air

114. Hydraulic ram is a pump which works on the principle of
(a) Water hammer
(b) Centrifugal action
(c) Reciprocation action
(d) Hydraulic press

115. Consider the following statements regarding a torque converter:
1. Its maximum efficiency is less than that of the fluid coupling.
2. It has two runners and a set of stationary vanes interposed between them.
3. It has two runners.
4. The ratio of secondary to primary torque is zero for the zero value of angular velocity of secondary.

Which of these statements are correct?
(a) 1 and 2 (b) 3 and 4
(c) 1 and 4 (d) 2 and 4

116. In a fluid coupling, the torque transmitted is 50 kNm, when the speed of the driving and driven shaft is 900 rpm and 720 rpm respectively. The efficiency of the fluid coupling will be

(a) 20% (b) 25%
(c) 80% (d) 90%

117. Consider the following statements regarding the fluid coupling:

1. Efficiency increase with increase in speed ratio.
2. Neglecting friction the output torque is equal to input torque.
3. At the same input speed, higher slip requires higher input torque.

Which of these statements are correct?

(a) 1, 2 and 3 (b) 1 and 2
(c) 2 and 3 (d) 1 and 3

118. Hydraulic transmission through fluid coupling is suitable for

(a) Unsteady operation and increasing torque
(b) Unsteady operation and increasing speed
(c) Unsteady operation and low starting torque
(d) Increasing torque and low starting load

119. Fluid flow machines are using the principle of either (i) supplying energy to the fluid, or (ii) extracting energy from the fluid. Some fluid flow machines are a combination of both (i) and (ii). They are classified as

(a) compressors (b) hydraulic turbines
(c) torque converters (d) wind mills

120. A hydraulic coupling transmits 1 kW of power at an input speed of 200 rpm, with a slip of 2%. If the input speed is changed to 400 rpm, the power transmitted with the same slip is

(a) 2 kW (b) $\frac{1}{2}$ $kW$
(c) 4 kW (d) 8 kW

121. In a jet pump

(a) Kinetic energy of fluid is converted into potential energy
(b) Energy of high velocity stream is converted into pressure energy
(c) Energy of high pressure fluid is converted into energy of flow pressure fluid
(d) Potential energy of fluid is converted into kinetic energy

***(e) Miscellaneous Questions***

122. A hydraulic reaction turbine working under a head of 16 m develops 640 kW of power. What is the unit power of the turbine?

(a) 10 kW (b) 40 kW
(c) 60 kW (d) 160 kW

123. A Francis turbine working at 400 rpm has a unit speed of 50 rpm and develops 500 kW of power. What is the effective head under which this turbine operates?
(a) 62.5 m
(b) 64.0 m
(c) 40.0 m
(d) 100 m

124. A centrifugal pump with radial vane tips at the impeller of 100 mm outer diameter. If the rotative speed is 3000 rpm and manometric efficiency 0.8 than what is the net head developed?
(a) 10 m (b) 20 m
(c) 30 m (d) 40 m

125. The speed ratio of a Pelton wheel operating under a head of 900 m is 0.45. What is the peripheral velocity of the turbine wheel?
(a) 28 m/s (b) 96 m/s
(c) 42 m/s (d) 60 m/s

126. A mixed flow pump is driven by a 8 kW motor running at1000 rpm. It delivers water at the rate of 1000 litres/min against a total head of 25 m. What is the specific speed of the pump in meter-minutes?
(a) 90 (b) 50
(c) 45 (d) 75

127. Match List I with List II and select the correct answer using the code given below the Lists:

| List I | List II |
|---|---|
| A. Pelton turbine | 1. Specific speed from 300 to 1000 + axial flow with fixed runner vanes |
| B. Francis turbine | 2. Specific speed from 10 to 50+ Tangential flow |
| C. Propeller turbine | 3. Specific speed from 60 to 300+ mixed flow |
| D. Kaplan turbine | 4. Specific speed from 300 to 1000+ axial flow with adjustable runner vanes |

**Code:**

| | A | B | C | D |
|---|---|---|---|---|
| (a) | 2 | 1 | 3 | 4 |
| (b) | 4 | 1 | 3 | 2 |
| (c) | 2 | 3 | 1 | 4 |
| (d) | 4 | 3 | 1 | 2 |

128. Which one of the following graphs represents the characteristics of a torque converter? Where suffix *r* stands for turbine runner and *p* stands for pump impeller.

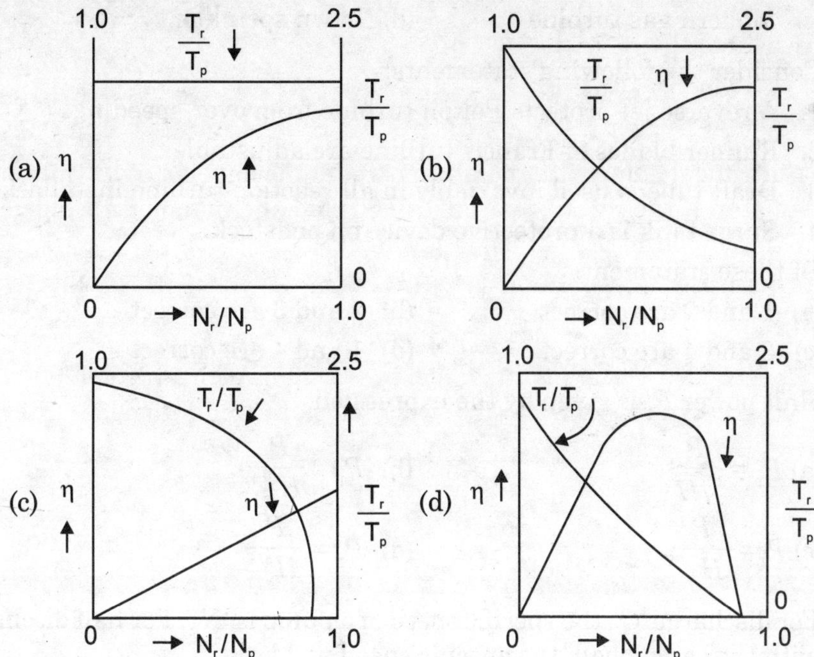


129. Which one of the following combination represents the power transmission systems?
   (a) Pump, hydraulic accumulator, hydraulic intensifier and hydraulic coupling
   (b) Pump, turbine, hydraulic accumulator and hydraulic coupling
   (c) Turbine, accumulator, intensifier and hydraulic coupling
   (d) Accumulator, intensifier, hydraulic coupling and torque converter

130. Which one of the following is not correct regarding both Kaplan and propeller turbines?
   (a) The runner is axial
   (b) The blades are wing type
   (c) There are four to eight blades
   (d) The blades can be adjusted

131. Based on the direction of flow, which one of the following turbines is different from the other three?
   (a) Pelton turbine
   (b) Kaplan turbine
   (c) De-Laval turbine
   (d) Parson's turbine

132. Which one of the following is an example of a pure (100%) reaction machine?
(a) Pelton wheel
(b) Francis turbine
(c) Modern gas turbine
(d) Lawn sprinkler

133. Consider the following statements:
1. A reverse jet protects Pelton turbine from over-speeding.
2. Runner blades of Francis turbine are adjustable.
3. Draft tube is used invariably in all reaction turbine installations.
4. Surge tank is a protective device on penstock.

Of these statements
(a) 1 and 2 are correct
(b) 2 and 3 are correct
(c) 3 and 4 are correct
(d) 1 and 4 are correct

134. Unit power $P_1$ is given by the expression:
(a) $P_1 = \dfrac{P}{\sqrt{H}}$
(b) $P_1 = \dfrac{P}{H^{3/2}}$
(c) $P_1 = \dfrac{P}{H^{3/2}}$
(d) $P_1 = \dfrac{P}{H^{5/4}}$

135. For discharge '$Q$', the specific speed of a pump is '$N_S$'. For half discharge with the same head, the specific speed will be
(a) $N_S$
(b) $\dfrac{N_S}{\sqrt{2}}$
(c) $\sqrt{2}N_S$
(d) $2N_S$

136. Which one of the following figures represents theoretical head versus discharge curves for a centrifugal pump with forward, radial and backward curved vanes?

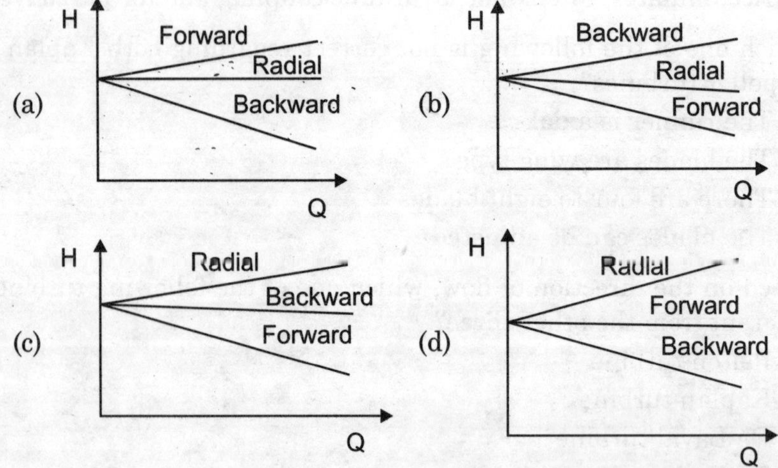

137. For an impulse turbine with exit angle '$\phi$', the maximum hydraulic efficiency is

(a) $\left(1-\dfrac{\cos\phi}{2}\right)$ (b) $\left(\dfrac{1}{2}+\cos\phi\right)$

(c) $\left(\dfrac{1+\cos\phi}{2}\right)$ (d) $\left(\dfrac{1-\cos\phi}{2}\right)$

138. The location of a surge tank in a high head hydraulic power plant would be

(a) nearer to the dam (b) at the powerhouse
(c) nearest to the powerhouse (d) immaterial

139. Match List-I with List-II and select the correct answer using the codes given below the lists:

| List-I (Water turbines) | List-II (Application) |
|---|---|
| A. Pelton | 1. High head and low discharge |
| B. Francis | 2. High head and high discharge |
| C. Kaplan | 3. Medium head and medium discharge |
| | 4. Low head and high discharge |

**Codes:**

| | A | B | C |
|---|---|---|---|
| (a) | 1 | 3 | 2 |
| (b) | 1 | 3 | 4 |
| (c) | 2 | 4 | 3 |
| (d) | 3 | 2 | 4 |

140. Cavitation damage in the turbine runner occurs near the

(a) inlet on the concave side of the blades
(b) outlet on the concave side of the blades
(c) outlet on the convex side of the blades
(d) inlet on the convex side of the blades

141. In hydraulic power-generation systems, surge tanks are provided to prevent immediate damage to

(a) draft tube (b) turbine
(c) tail race (d) penstocks

142. The function of the draft tube in a reaction turbine is

(a) to enable the shaft of the turbine to be vertical
(b) to transform a large part of pressure energy at turbine outlet into kinetic energy
(c) to avoid whirl losses at the exit of the turbine
(d) to transform a large part of kinetic energy at turbine outlet into pressure energy

143. Consider the following characteristics:
  1. The fluid enters the pump axially and is discharged radially.
  2. Maximum efficiency may be of the order of 90%.
  3. Development of a low head.
  4. A limited suction capacity.

  Which of the above characteristics are possessed by axial flow pumps?

  (a) 1 and 2 (b) 2 and 3
  (c) 2 and 4 (d) 3 and 4

144. A centrifugal pump delivers water at the rate of 50 liter against a total head of 40 meters. Then the power required to drive the pump is

  (a) 3 kW (b) 15.2 kW
  (c) 19.6 kW (d) 25.8 kW

145. Match List-I with List-II and select the correct answer using the codes given below the lists:

| List-I | List-II |
|---|---|
| A. Head race | 1. Channel, tunnel or pipes through which water is carried from reservoir to the turbine |
| B. Tail race | 2. Reservoir water level |
| C. Penstock | 3. Diverging tube discharging water from the turbine to the atmosphere |
| D. Draft tube | 4. The level at which water is discharged at atmospheric pressure |

**Codes:**

| | A | B | C | D |
|---|---|---|---|---|
| (a) | 1 | 3 | 2 | 4 |
| (b) | 2 | 4 | 1 | 3 |
| (c) | 1 | 4 | 2 | 3 |
| (d) | 2 | 3 | 1 | 4 |

146. Consider the following statements in respect of centrifugal pumps:
  1. Head developed is proportional to the square of the speed of rotation.
  2. Backward curved bladed impellers are generally used in centrifugal pumps.
  3. These pumps generally do not require priming.
  4. Multistage pumps would give higher discharge proportional o the number of stages.

  Which of these statements are correct?

  (a) 1 and 2 (b) 2 and 3
  (c) 3 and 4 (d) 1 and 4

147. The relationship between overall efficiency ($\eta_o$), mechanical efficiency $\eta_{mech}$, and hydraulic efficiency ($\eta_{hyd}$) of a turbine is:

(a) $\eta_o = \eta_{mech} \times \eta_{hyd}$ (b) $\eta_o = \eta_{mech}/\eta_{hyd}$

(c) $\eta_o = \eta_{hyd}/\eta_{mech}$ (d) $\eta_o = \eta_{mech} - \eta_{hyd}$

148. Which one of the following correctly expresses the specific speed of a turbine and a pump, respectively?

(a) $\dfrac{N\sqrt{Q}}{H^{3/4}}, \dfrac{N\sqrt{P}}{H^{5/4}}$ (b) $\dfrac{N\sqrt{P}}{H^{3/4}}, \dfrac{N\sqrt{Q}}{H^{5/4}}$

(c) $\dfrac{N\sqrt{P}}{H^{5/4}}, \dfrac{N\sqrt{Q}}{H^{3/4}}$ (d) $\dfrac{N\sqrt{P}}{H^{7/4}}, \dfrac{N\sqrt{Q}}{H^{3/4}}$

149. The above figure shows characteristic of three centrifugal pumps *A*, *B* and *C*. If *E* = Efficiency, *H* = Head and *P* = power, than *A*, *B* and *C*, respectively represent which one of the following?

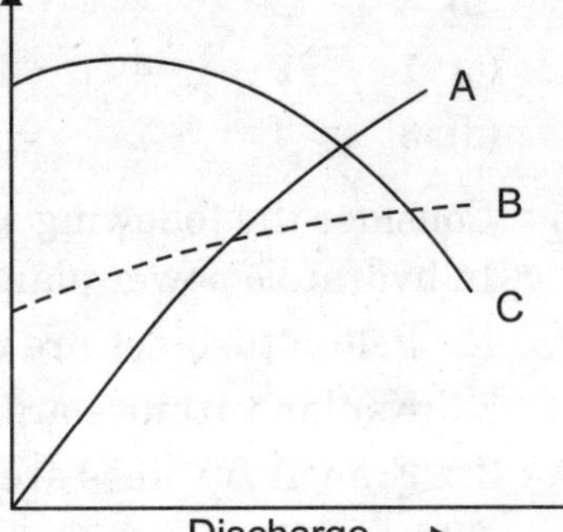


(a) *E*, *P*, *H*

(b) *P*, *E*, *H*

(c) *P*, *H*, *E*

(d) *H*, *P*, *E*

150. Which one of the following helps in avoiding cavitation in centrifugal pumps?

(a) Low suction pressure (b) High delivery pressure

(c) Low delivery pressure (d) High suction pressure

151. Consider the following statements with regard to the specific speeds of different type of turbine:

1. High specific speed implies that it is a Pelton wheel
2. Medium specific speed implies that it is an axial flow turbine
3. Low specific speed implies that it is a Francis turbine

Which of these statements given above is/are correct?

(a) 1 only (b) 2 only

(c) 3 only (d) None

152. Which one of the following is the purpose of a surge tank in a Pelton turbine station?

(a) It acts as a temporary storage during load changes

(b) It prevents hydraulic jump

(c) It prevents surges at the transformer

(d) It prevents water hammer due to sudden reduction in load

153. Match List-I (Type of Turbine) with List-II (Specific Speed) and select the correct answer using the code given below the lists:

| List-I (Type of Turbine) | List-II (Specific Speed) |
|---|---|
| A. Pelton | 1. 0.1 |
| B. Francis | 2. 3.0 |
| C. Propeller | 3. 4.0 |
| D. Kaplan | 4. 1.0 |

**Code:**

| | A | B | C | D |
|---|---|---|---|---|
| (a) | 1 | 4 | 2 | 3 |
| (b) | 3 | 2 | 4 | 1 |
| (c) | 1 | 2 | 4 | 3 |
| (d) | 3 | 4 | 2 | 1 |

154. Consider the following statements:
In hydraulic power plants
1. Pelton turbines are used for high heads.
2. Kaplan turbines are used for medium heads.
3. Francis turbines are used for low head.

Which of the above statements is/are correct?
(a) 1 only (b) 2 only
(c) 3 only (d) 1, 2 and 3

155. The speed ratio for Pelton wheel varies from
(a) 0.45 - 0.50 (b) 0.6 - 0.7
(c) 0.30 - 0.40 (d) 0.8 - 0.9

156. Match List-I with List-II and select the correct answer using the code given below the lists:

| List-I | List-II |
|---|---|
| A. Draft tube | 1. Impulse turbine |
| B. Surging | 2. Reciprocating pump |
| C. Air Vessel | 3. Reaction turbine |
| D. Nozzle | 4. Centrifugal pump |

**Code:**

| | A | B | C | D |
|---|---|---|---|---|
| (a) | 1 | 2 | 4 | 3 |
| (b) | 3 | 4 | 2 | 1 |
| (c) | 1 | 4 | 2 | 3 |
| (d) | 3 | 2 | 4 | 1 |

157. Match List-I with List-II and select the correct answer using the code given below the lists:

| List-I | List-I |
|---|---|
| A. Specific Speed of pump | 1. $\left(\frac{N\sqrt{Q}}{H^{3/4}}\right)$ |
| B. Low specific speed of turbine | 2. Screw pump |
| C. Pumping viscous oil | 3. Reaction turbine |
| D. Draft tube | 4. Turbine Pump |
| | 5. Very high specific speed |
| | 6. $\left(\frac{N\sqrt{P}}{H^{5/4}}\right)$ |
| | 7. Impulse turbine |

**Codes:**

| | A | B | C | D |
|---|---|---|---|---|
| (a) | 1 | 3 | 2 | 5 |
| (b) | 6 | 7 | 4 | 3 |
| (c) | 1 | 7 | 2 | 3 |
| (d) | 6 | 3 | 4 | 5 |

**Answers**

| | | | | | | | | | |
|---|---|---|---|---|---|---|---|---|---|
| 1.(b) | 2.(c) | 3.(a) | 4.(b) | 5.(a) | 6.(a) | 7.(b) | 8.(b) | 9.(d) | 10.(a) |
| 11.(b) | 12.(b) | 13.(b) | 14.(a) | 15.(c) | 16.(a) | 17.(c) | 18.(a) | 19.(b) | 20.(b) |
| 21.(d) | 22.(a) | 23.(d) | 24.(b) | 25.(b) | 26.(b) | 27.(d) | 28.(d) | 29.(c) | 30.(b) |
| 31.(a) | 32.(a) | 33.(b) | 34.(a) | 35.(c) | 36.(a) | 37.(a) | 38.(b) | 39.(b) | 40.(d) |
| 41.(b) | 42.(c) | 43.(c) | 44.(b) | 45.(d) | 46.(b) | 47.(b) | 48.(c) | 49.(d) | 50.(b) |
| 51.(b) | 52.(c) | 53.(c) | 54.(d) | 55.(b) | 56.(d) | 57.(b) | 58.(c) | 59.(c) | 60.(d) |
| 61.(b) | 62.(c) | 63.(a) | 64.(c) | 65.(d) | 66.(b) | 67.(a) | 68.(a) | 69.(d) | 70.(c) |
| 71.(b) | 72.(c) | 73.(b) | 74.(b) | 75.(b) | 76.(d) | 77.(d) | 78.(c) | 79.(b) | 80.(b) |
| 81.(a) | 82.(a) | 83.(b) | 84.(c) | 85.(c) | 86.(a) | 87.(c) | 88.(a) | 89.(b) | 90.(d) |
| 91.(c) | 92.(a) | 93.(a) | 94.(d) | 95.(a) | 96.(a) | 97.(a) | 98.(c) | 99.(b) | 100.(d) |
| 101.(d) | 102.(d) | 103.(b) | 104.(d) | 105.(a) | 106.(c) | 107.(a) | 108.(c) | 109.(b) | 110.(d) |
| 111.(d) | 112.(b) | 113.(a) | 114.(a) | 115.(d) | 116.(a) | 117.(c) | 118.(c) | 119.(c) | 120.(b) |
| 121.(c) | 122.(a) | 123.(b) | 124.(a) | 125.(d) | 126.(a) | 127.(c) | 128.(c) | 129.(c) | 130.(d) |
| 131.(b) | 132.(b) | 133.(c) | 134.(b) | 135.(b) | 136.(a) | 137.(c) | 138.(c) | 139.(b) | 140.(b) |
| 141.(d) | 142.(d) | 143.(d) | 144.(c) | 145.(b) | 146.(a) | 147.(a) | 148.(c) | 149.(a) | 150.(d) |
| 151.(c) | 152.(a) | 153.(a) | 154.(a) | 155.(a) | 156.(b) | 157.(c) | | | |

# Subject Index

## A



## B



## C

## D